T0309193

Small Bite, Big Threat

Small Bite, Big Threat

Deadly Infections Transmitted by *Aedes* Mosquitoes

edited by

Jagriti Narang
Manika Khanuja

Published by

Jenny Stanford Publishing Pte. Ltd.
Level 34, Centennial Tower
3 Temasek Avenue
Singapore 039190

Email: editorial@jennystanford.com
Web: www.jennystanford.com

British Library Cataloguing-in-Publication Data
A catalogue record for this book is available from the British Library.

Small Bite, Big Threat: Deadly Infections Transmitted by *Aedes* Mosquitoes

ISBN 978-981-4800-86-0 (Hardcover)
ISBN 978-1-003-00329-8 (eBook)

Contents

Preface xi
About Lead Author, Prof. Vinod Joshi xiii

1. *Aedes* Mosquitoes: The Universal Vector **1**
Annette Angel, Bennet Angel, Neelam Yadav, Jagriti Narang, Surender Singh Yadav, and Vinod Joshi
1.1 Background 2
1.2 Life Cycle 3
1.2.1 The Egg 4
1.2.2 The Larvae 5
1.2.3 The Pupae 7
1.2.4 The Adult 8
1.3 Anatomy 10
1.4 Gonotrophy 15
1.5 Host Preference 16
1.6 Conclusion 17

2. The *Aedes* Fauna: Different *Aedes* Species Inhabiting the Earth **21**
Annette Angel, Bennet Angel, Neelam Yadav, Jagriti Narang, Surender Singh Yadav, and Vinod Joshi
2.1 *Aedes aegypti* 22
2.1.1 Morphology 22
2.1.2 Origin and Distribution 23
2.1.3 Breeding Habitats 24
2.1.4 Role in Disease Transmission 27
2.1.5 Modern Biology 27
2.2 *Aedes albopictus* 29
2.2.1 Morphology 29
2.2.2 Origin and Distribution 29
2.2.3 Breeding Habitats 31
2.2.4 Role in Disease Transmission 31
2.2.5 Modern Biology 32
2.3 *Aedes vittatus* 33
2.3.1 Morphology 33

	2.3.2	Origin and Distribution	34
	2.3.3	Breeding Habitats	34
	2.3.4	Role in Disease Transmission	36
	2.3.5	Modern Biology	37
2.4	*Aedes atropalus*		38
	2.4.1	Morphology	38
	2.4.2	Origin and Distribution	39
	2.4.3	Breeding Habitats	39
	2.4.4	Role in Disease Transmission	40
	2.4.5	Modern Biology	40
2.5	*Aedes japonicus*		41
	2.5.1	Morphology	41
	2.5.2	Origin and Distribution	41
	2.5.3	Breeding Habitats	42
	2.5.4	Role in Disease Transmission	42
	2.5.5	Modern Biology	43
2.6	*Aedes koreicus*		43
	2.6.1	Morphology	43
	2.6.2	Origin and Distribution	45
	2.6.3	Breeding Habitats	45
	2.6.4	Role in Disease Transmission	46
	2.6.5	Modern Biology	46
2.7	*Aedes triseriatus*		47
	2.7.1	Morphology	47
	2.7.2	Origin and Distribution	48
	2.7.3	Breeding Habitats	48
	2.7.4	Role in Disease Transmission	48
	2.7.5	Modern Biology	48
2.8	*Aedes polynesiensis*		49
	2.8.1	Morphology	49
	2.8.2	Origin and Distribution	50
	2.8.3	Breeding Habitats	50
	2.8.4	Role in Disease Transmission	50
2.9	*Aedes vexans*		51
	2.9.1	Morphology	51
	2.9.2	Origin and Distribution	51
	2.9.3	Breeding Habitats	52
	2.9.4	Role in Disease Transmission	52
2.10	Other Species		54

3. Viral Pathogens: A General Account **81**

Vinod Joshi, Bennet Angel, Annette Angel, Neelam Yadav, and Jagriti Narang

3.1 Introduction 82
3.2 Viruses Transmitted by Arthropods 84
3.2.1 Togaviruses 85
3.2.1.1 Chikungunya virus 86
3.2.1.2 Eastern equine encephalitis virus 87
3.2.2 Bunyaviruses 87
3.2.2.1 Crimean–Congo hemorrhagic fever virus 88
3.2.2.2 Rift Valley fever virus 89
3.2.2.3 La Crosse virus 89
3.2.3 Flaviviruses 90
3.2.3.1 Dengue virus 91
3.2.3.2 Japanese encephalitis virus 92
3.2.3.3 Yellow fever virus 93
3.2.3.4 West Nile virus 94
3.3 Conclusion 94

4. Dengue Fever: A Viral Hemorrhagic Fever of Global Concern **99**

Bennet Angel, Neelam Yadav, Jagriti Narang, Annette Angel, and Vinod Joshi

4.1 Epidemiology 100
4.2 Virus Morphology 103
4.2.1 Structural Proteins 106
4.2.1.1 The nucleocapsid protein 106
4.2.1.2 The membrane protein 107
4.2.1.3 The envelope protein 108
4.2.2 Nonstructural Proteins 109
4.2.2.1 NS1 protein 109
4.2.2.2 NS2 protein 109
4.2.2.3 NS3 protein 110
4.2.2.4 NS4A protein 110
4.2.2.5 NS5 protein 111
4.3 Transmission Route 116

4.3.1 Replication of Dengue Virus within Systems: The Extrinsic System 118
4.3.2 Replication of Dengue Virus within Systems: The Intrinsic System 119
4.3.3 Intracellular Replication of Dengue Virus 121
4.4 Pathogenicity 122
4.5 Treatment and Diagnostics 126
4.5.1 Dengue Vaccine Update 126
4.5.2 Disease Management 127
4.5.3 Laboratory Diagnosis of Dengue Virus 127
4.5.3.1 Virus isolations 127
4.5.3.2 Serological diagnostic tests 130
4.5.3.3 Molecular diagnostic assays 131
4.5.3.4 Nanotechnology-based detection methods 132
4.6 Risk Factors 137
4.7 Conclusion 140

5. Chikungunya Fever: Emergence and Reality **165**

Neelam Yadav, Bennet Angel, Jagriti Narang, Surender Singh Yadav, and Vinod Joshi

5.1 Epidemiology 166
5.2 Virus Morphology 168
5.2.1 Structural Proteins 169
5.2.2 Nonstructural Proteins 171
5.2.2.1 NS1 protein 171
5.2.2.2 NS2 protein 172
5.2.2.3 NS3 protein 172
5.2.2.4 NS4A protein 172
5.2.2.5 NS5 protein 173
5.3 Transmission Route 173
5.3.1 Replication of Chikungunya Virus 174
5.4 Pathogenicity 175
5.5 Treatment and Diagnostics 177
5.5.1 Molecular Diagnostics 178
5.5.2 Laboratory Diagnosis of Chikungunya Virus 179
5.5.3 Advanced Approaches 179
5.6 Risk Factors 180
5.7 Conclusion 180

6. Zika: An Ancient Virus Incipient into New Spaces **191**

Bennet Angel, Neelam Yadav, Jagriti Narang, Surender Singh Yadav, Annette Angel, and Vinod Joshi

6.1 Epidemiology 191
6.2 Virus Morphology 194
6.2.1 Structural Proteins 195
6.2.1.1 Membrane protein 195
6.2.2 Nonstructural Proteins 195
6.3 Transmission Route 196
6.3.1 Replication of Zika Virus within Systems: The Extrinsic System 197
6.3.2 Replication of Zika Virus within Systems: The Intrinsic System 198
6.3.3 Intracellular Replication of Zika Virus 199
6.4 Pathogenicity 199
6.5 Treatment and Diagnostics 201
6.5.1 Zika Vaccine Update 201
6.5.2 Disease Management 202
6.5.3 Laboratory Diagnosis of Zika Virus 203
6.5.3.1 Virus isolation studies 203
6.5.3.2 Immunological diagnostic assays 203
6.5.3.3 Molecular diagnostic assays 204
6.5.4.4 Nanotechnology-based detection methods: Biosensors 205
6.6 Risk Factors 206
6.7 Conclusion 208

7. Yellow Fever: Emergence and Reality **215**

Neelam Yadav, Bennet Angel, Jagriti Narang, Surender Singh Yadav, Vinod Joshi, and Annette Angel

7.1 Epidemiology 216
7.2 Virus Morphology 220
7.3 Historical Overview 222
7.4 Pathogenicity 222
7.5 Transmission/Reservoirs 223
7.5.1 Incubation/Colonization 224
7.6 Diagnosis and Treatment of YFV Infection 225
7.6.1 Clinical Diagnosis of YF 225

7.6.2 Laboratory Diagnosis of YF 225
7.6.3 Treatment of YFV Infection 226
7.6.4 Vaccination for YFV 226
7.6.4.1 Development of live attenuated YF vaccines 226
7.6.4.2 Immune response generated during YFV vaccination 227
7.6.4.3 Adverse effects of YFV vaccines 227
7.7 Modern Biology 227
7.8 Conventional Methods for Detection of YFV 228
7.9 Risk Factors 230
7.10 Conclusion 230

8. West Nile Virus: The Silent Neuro-Invasive Terror 237

Vinod Joshi, Annette Angel, Bennet Angel, Neelam Yadav, Jagriti Narang, and Surender Yadav

8.1 Epidemiology 238
8.2 Virus Morphology 240
8.3 Transmission Route 242
8.4 Pathogenicity 243
8.5 Detection Methods for WNV 245
8.5.1 Conventional Methods 245
8.5.2 Nanotechnology-Based Methods for WNV Detection 246
8.6 Treatment 247
8.7 Modern Biology 248
8.8 Conclusion 249

Index 267

Preface

Aedes mosquitoes play a vital role in transmitting various pathogens to humans and other mammals. They are omnipresent and easily breed when the climate seems favorable. As can be inferred from the title of the book, a small bite of a mosquito can cause a serious disease, which can often be life threatening. Currently, vector-borne diseases, such as dengue, show an increasing trend in developed and developing countries. Dengue fever, associated with its more severe forms dengue hemorrhagic fever (DHF) and dengue shock syndrome (DSS), is posing tremendous challenges to clinicians and public health stakeholders due to the unavailability of a proper vaccine or a therapeutic molecule. Controlling disease transmission in endemic communities is the only effective tool to prevent outbreaks. A crucial reason for disease transmission becoming rampant is the fact that very little and scattered knowledge is available on different species of *Aedes* mosquitoes, many of which play vectorial role in disease transmission. In this book, we have attempted to consolidate the knowledge available on different species of *Aedes* mosquitoes, covering their taxonomic points, vector bionomics, and virological aspects. The chapter authors are experts, scientists, and teachers of the subject and have presented their latest findings generously on vector-borne diseases.

The book contains comprehensive accounts of the entomological, virological, epidemiological, and other aspects of diseases such as dengue, chikungunya, Zika, West Nile, and yellow fevers. It is structured in such a manner that it journeys the readers firstly through the most important and crucial vector inhabiting the earth, the *Aedes* mosquito. After they have understood about this mosquito species, it introduces them to the different viruses that the vector can transmit and imparts a preliminary knowledge of the different arthropod-borne diseases. It treads with the readers further to show how the *Aedes* species is linked with some important diseases that it transmits and at the same time shares detailed knowledge about each one of them.

The book will serve as a helpful guide for public health experts, teachers, and students dealing with *Aedes*-transmitted diseases. It can be a great resource for and budding virology researchers as well as undergraduate and postgraduate students who intend to pursue a career in science and research. Since it presents the most explored areas of research in this field, it will be useful for setting the direction for the current diagnostic boom.

We would like to thank all authors of this book for their valuable contributions, especially Prof. Vinod Joshi, who has spent his entire life discovering and understanding viral diseases caused by mosquitoes.

Jagriti Narang

Manika Khanuja

Spring 2020

About Lead Author, Prof. Vinod Joshi

Vinod Joshi is professor and deputy director of the Amity Institute of Virology and Immunology, Amity University, Noida, India. After earning his PhD, Prof. Joshi worked as a prominent scientist at the Desert Medicine Research Centre, Rajasthan, India, of the Indian Council of Medical Research (ICMR), Ministry of Health and Family Welfare, Government of India. He joined ICMR's nodal virology institute, National Institute of Virology, Pune, in 1984, and dedicated 33 years doing significant research in the area of medical virology, especially on arboviruses. In 1996, he became the first researcher in the Indian subcontinent to establish that dengue viruses undergo transovarial transmission (TOT) from parent to progeny in the vector *Aedes aegypti*. Through his subsequent experimental studies on interaction of *Aedes aegypti* and dengue viruses, he further reported in 2002 that the phenomenon of TOT of dengue viruses can continue up to 7 generations of the mosquito, which could be the mechanism of virus retention in nature during inter-epidemic periods. Prof. Joshi along with his PhD students and coauthors of this book Dr. Bennet Angel and Dr. Annette Angel continued this study further and discovered how a protein of 200 kDa with acidic pH can block TOT of dengue virus. His research group reported the first whole genome study of dengue type-3 virus from Rajasthan, elaborating the significant mutations observed in the genomic constitution. Through his translational research, he and his group demonstrated in 34 districts of Rajasthan as to how by eliminating the vertically infected *Aedes aegypti* mosquitoes during summer season, prospective transmission of dengue virus in the community could be checked. Prof. Joshi has published several research papers in journals of international repute, which have received numerous citations. He represents a peer group of scientists and is currently continuing his study on *Aedes aegypti* and dengue viruses. His other areas of research include malaria, cutaneous leishmaniasis, dracunculiasis, Japanese encephalitis, and influenza A (H1N1) viruses. He is reviewer of several international journals, associate editor of *Transactions of Royal Society of Tropical Medicine and Hygiene*, and council member of the World Society for Virology.

Chapter 1

Aedes Mosquitoes: The Universal Vector

Annette Angel,[a] Bennet Angel,[b] Neelam Yadav,[c,d] Jagriti Narang,[e] Surender Singh Yadav,[f] and Vinod Joshi[b]

[a]*Division of Zoonosis, National Center for Disease Control, 22 Sham Nath Marg, Civil Lines, Delhi, India*
[b]*Amity Institute of Virology and Immunology, Amity University, Sector 125, Noida, India*
[c]*Centre for Biotechnology, Maharshi Dayanand University, Rohtak, India*
[d]*Department of Biotechnology, Deenbandhu Chhotu Ram University of Science and Technology, Murthal, Sonipat, India*
[e]*Department of Biotechnology, Jamia Hamdard University, New Delhi, India*
[f]*Department of Botany, Maharshi Dayanand University, Rohtak, India*
annetteangel_15@yahoo.co.in, vinodjoshidmrc@gmail.com, bennetangel@gmail.com

Mosquitoes play a vital role in transmitting various pathogens to humans and other mammals. They are considered to be one of the successful creatures of this planet. Their different forms occupy every corner of the earth, be it land or water. In this chapter, a general account of the mosquito, its morphology and anatomy as well as different stages of its development are discussed. Though the book

Small Bite, Big Threat: Deadly Infections Transmitted by Aedes *Mosquitoes*
Edited by Jagriti Narang and Manika Khanuja

ISBN 978-981-4800-86-0 (Hardcover), 978-1-003-00329-8 (eBook)
www.jennystanford.com

is specifically on *Aedes* mosquitoes, yet it is very important to have a general understanding about mosquitoes before moving forward.

1.1 Background

Mosquitoes act as an important vectoral component for disease transmission of many of the classified pathogens and parasites, including viruses, bacteria, fungi, protozoa, and nematodes. This is because when they lay eggs, they require blood, for which they feed on vertebrate hosts. During this event, the pathogens residing in them find a route to travel from the infected host to the mosquito and vice versa, though in some cases humans may serve as a dead end. Hence the term vector has been functionally assigned to them (Arbovirus Summary Archives, 2008). Of the vast population of mosquitoes inhabiting the planet, the ones belonging to the family Culicidae (order: Diptera) are known to play a crucial role as vectors of arbovirus transmission (Arbovirus Summary Archives, 2008). They are found throughout the world except places that are permanently frozen, and they occupy the tropics and sub-tropics where the climate seems favorable and efficient for their development (Clements, 2000). Within this family Cuclicidae, the genus *Aedes* is involved in the vicious cycle of transmission of many diseases of infectious and communicable nature, such as dengue fever, Chikungunya fever, West Nile fever, Zika, yellow fever, etc. (CDC, 2016). There are many species of *Aedes* depending on ecological distribution and spatiation and spread across the globe (Arbovirus Summary Archives, 2008). Although there is not much dissimilarity or variation among them, the differences may have been incorporated due to the genetic and environmental changes that might have occurred due to adaptation or continental drift since time immemorial. The genus *Aedes* originated mainly from Africa and might have found ways to spread throughout the world via seaports, which were the main places for export of goods during the 18th century. With increasing globalization, the species might have evolved from dwelling in forests to colonizing urban areas (CDC, 2016).

1.2 Life Cycle

The life cycle of *Aedes* species is similar to that of any other mosquitoes of its genera. These mosquitoes undergo an indirect development with well-defined stages: eggs → larvae → pupae → adult (Figs. 1.1 and 1.2). The larval stage can be further divided into four stages or instars reflected by its size and development (Gubler, 1989).

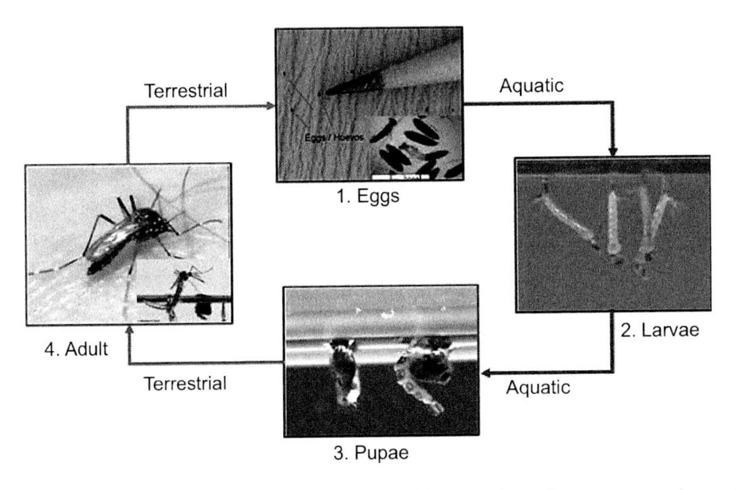

Figure 1.1 Stages of *Aedes* mosquito. Figure taken from www.cdc.gov/dengue/ entomologyecology/m_lifecycle.html#stages.

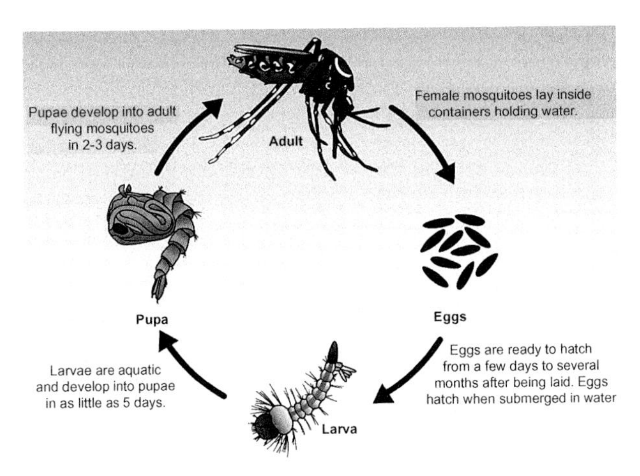

Figure 1.2 Time duration of development of *Aedes* mosquito under favorable environmental conditions. Figure taken from https://www.cdc.gov/dengue/resources/factSheets/MosquitoLifecycleFINAL.pdf

1.2.1 The Egg

Aedes are known as indoor breeding mosquitoes, that is, they breed in freshwater containers usually found inside or around human dwellings (domestic and peri-domestic containers, respectively). For laying eggs in foci favorable within these dwellings, the females search for damp or moist places that form little water bodies (Gubler, 1989). The preferable sites where eggs are seen are the water collected in tree holes, on the edges of artificial containers such as cement, clay, metallic, and plastic tanks. (Fig. 1.3). They lay around 50–100 eggs at one time. The eggs when laid appear white but change to black as they meet the atmosphere after 2 h (Nelson, 1986). They are oval shaped, covered with shell, and 1 mm in length (Fig. 1.4). The shell is partly secreted by the mother and partly by the embryo. They are soft and flexible when laid but later become hard and waterproof. The embryo develops inside the eggs within 2–3 days depending on atmospheric conditions. These eggs can remain viable up to a year (Foster and Walker, 2002). If an infected female has laid the eggs, then there are chances of the progeny getting infected; in other words, the female is capable of vertical transmission (egg to progeny). The process of laying eggs at a suitable place is termed oviposition, and this is done with the help of fine sensory hairs present on the lower parts of the abdomen (Clements, 2000).

Figure 1.3 Some breeding containers preferred by *Aedes* mosquitoes for laying eggs.

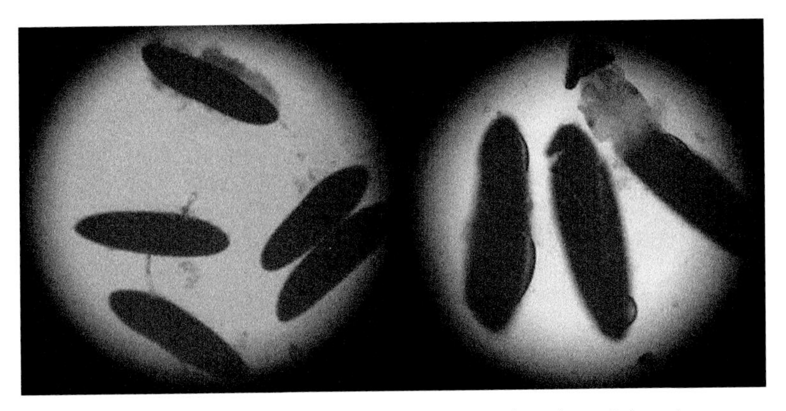

Figure 1.4 Eggs of *Aedes* mosquitoes as observed under a light microscope (10× magnification).

1.2.2 The Larvae

White minute larvae, approximately 2.5 mm in length, emerge from the eggs and start wriggling around in the water. This is known as the first instar (Fig. 1.4, right). The larvae soon start feeding and use atmospheric air for breathing (Gubler, 1989). Thus, it ensures its survival by two important features: use of water-borne particles for food and use of atmospheric oxygen for respiration. But they move away from light and prefer dark places. They feed on particulate matter and aquatic microorganisms such as bacteria, diatoms, algae, and their detritus (Nelson, 1986). The mouth brushes provided in front of their mouth help them to locate food particles. Continuous beating of these brushes creates a water current, which helps them to sense food particles ahead. The atmospheric oxygen is taken with the help of spiracles situated in the rear end of the abdomen (Fig. 1.5). The larvae of *Aedes* species can be distinguished from the other members of the family through their respiratory pattern. Every time they need air, the larvae travel to the water surface and align themselves 90° such that the spiracles are held at the upper end (Figs. 1.6 and 1.7) (Nelson, 1986). Inside the larval body, the spiracles reach through long tracheal tracts that extend throughout the body part. For ion uptake and regulation, the larvae have anal papillae. Undifferentiated cells present within the larva later form organs when they develop into adults. A well-fed larva grows in size four times, also referred to as four molts, and every time it molts, its sheds its skin, which is visible as loose skin in the breeding water bodies. During the fourth molt, the imaginal disks destined to form adult appendages develop rapidly (Fig. 1.8). The larvae are collected from the field sites (domestic and peri-domestic breeding sites) and brought to laboratory plastic jars. Then they are transferred to beakers filled with water (Foster and Walker, 2002).

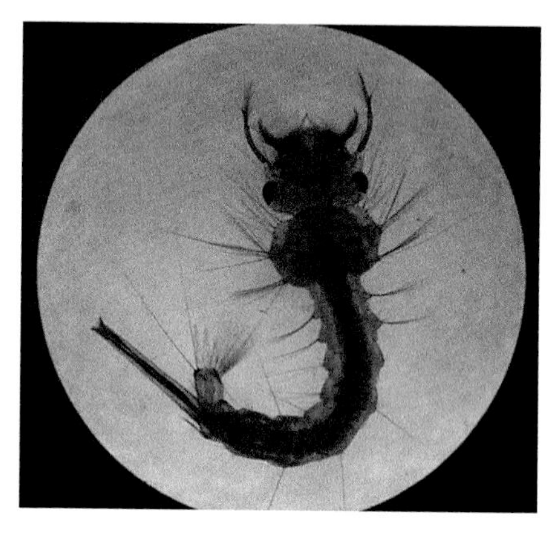

Figure 1.5 A second instar *Aedes* larva as observed under a light microscope (10× magnification).

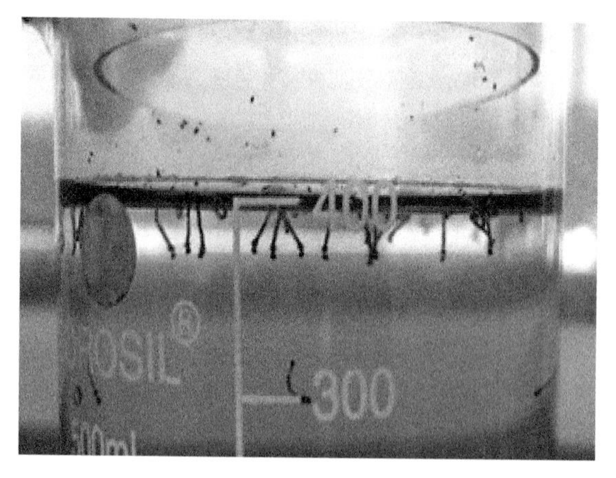

Figure 1.6 *Aedes* larvae positioning themselves at an angle of 90° for respiration.

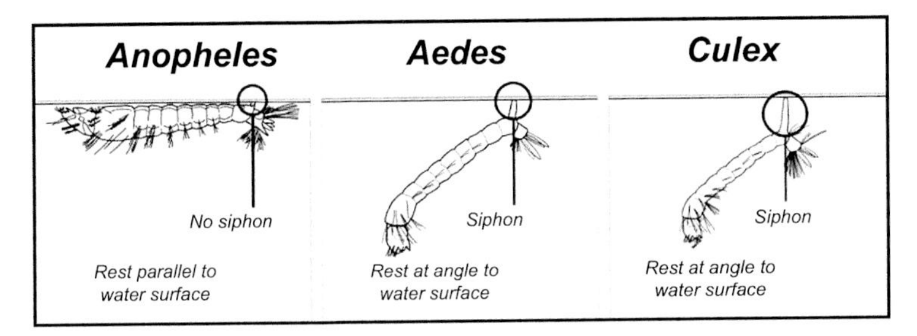

Figure 1.7 Difference between the breathing positions of the three mosquito larvae. Figure taken from www.astc.org/wp-content/uploads/2017/05/GlobalExperiment-science Behind Mosquitoes And Disease.pdf.

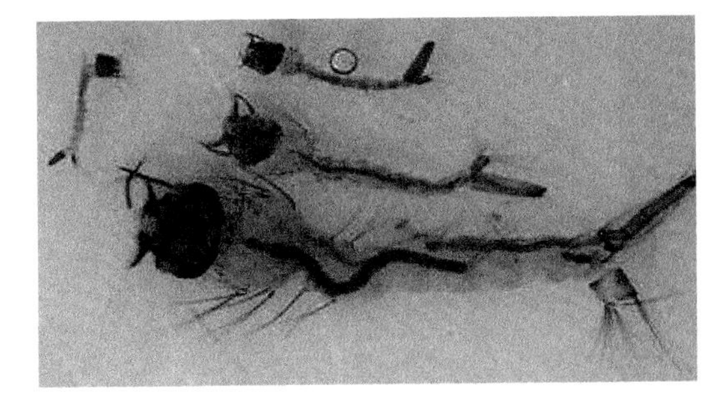

Figure 1.8 The four instars of *Aedes* larvae (first larval instar measures 2.5 mm, second larval instar measures 3.90 mm, third larval instar measures 5.05 mm, and fourth larval instar measures about 7.30 mm). Figure taken from https://www.smsl.co.nz/site/southernmonitoring/images/NZB/MossieAwareness/The%20Biology%20of%20Mosquitoes.pdf.

1.2.3 The Pupae

Sometimes after 48–72 h, the fourth instar larvae stop feeding. The head, thorax, and the elongated appendages now fuse to form a structure termed "cephalothorax." The abdomen part terminates into two paddles, which help in propulsion. The pupae appear to be "comma shaped," lying over the water surface (Fig. 1.9) (Gubler, 1989). Initially they appear brown but change to black in color due to continuous exposure to air. The cephalothorax is formed when an air bubble gets enclosed between the appendages. This helps the pupae to float on the water surface. The pupa does not feed but only breathes through respiratory trumpets (Nelson, 1986). As far as the organs are concerned, some of the larval organs such as alimentary canal get destroyed during the development stage, while others such as heart and fat body remain as such in the pupal stage. Under laboratory conditions, when the process of development of pupae to adult form is near, a "glass gola" is placed over the beaker. The glass gola is an elliptical structure (resembles a traditional candle/lamp shade) open on both sides such that its one end fits on the beaker and the other open end is covered by a muslin cloth (for proper aeration). The gola provides a resting place for the adults and helps in collecting the adults emerged (CDC, 2016) (Fig. 1.10).

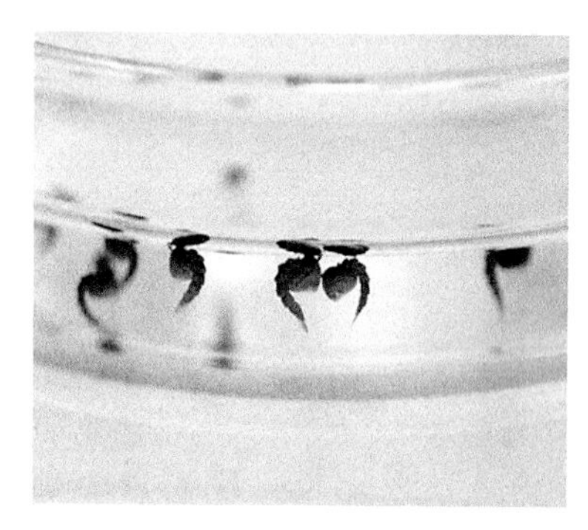

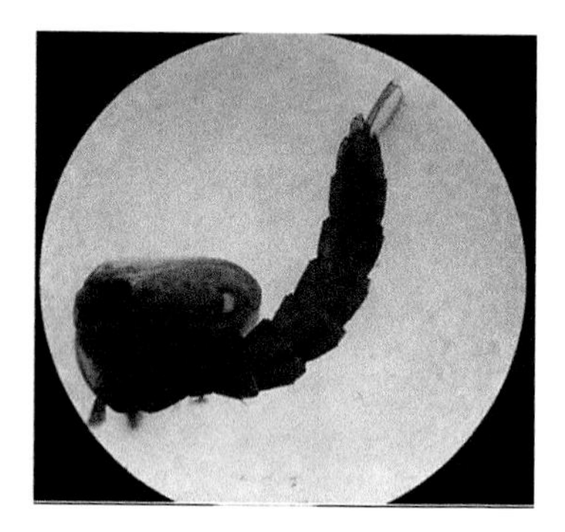

Figure 1.9 (Left) Pupae of *Aedes* species resting on the water surface and (right) a pupa as observed under a light microscope at 10× magnification.

1.2.4 The Adult

An interesting phenomenon occurs during the molting of pupa to adult form; the pupa positions itself on the water surface and starts to swallow in air. This leads to an increase in the pressure within the pupal body, and the cuticle splits open through the midline. A naive form of adult carefully emerges from the pupal cuticle and rests a while on the water surface (Gutsevich, 1970). After gaining enough strength, the adult now takes small flights and resides on places and niches suitable for its further development and nourishment. Under laboratory conditions, the adults are made to survive on 4% glucose solution mixed with dog biscuit powder (Fig. 1.11). Barraud cages are used, which are made of muslin cloth fitted on a square iron frame (Fig. 1.12). An important point to learn here is that while the larval stages live in an aquatic environment, the adult form is terrestrial in nature (Gutsevich, 1970).

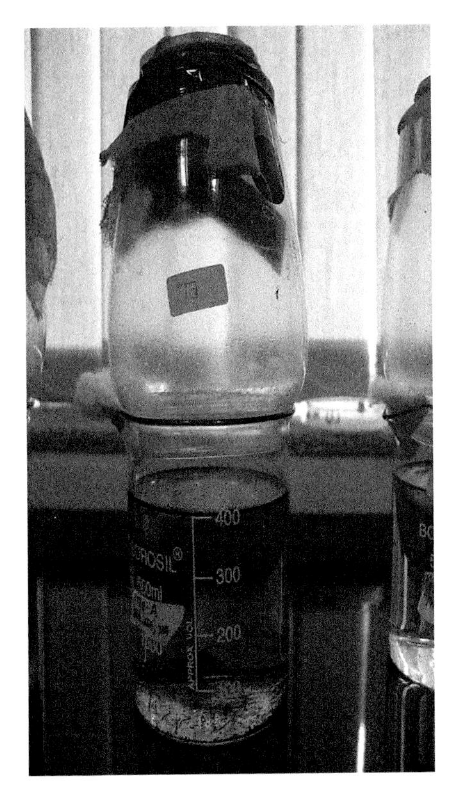

Figure 1.10 Beaker containing larval, pupal, and adult forms of *Aedes* mosquitoes.

Figure 1.11 Adults emerged and resting on twigs inside a beaker (laboratory rearing).

Figure 1.12 Barraud cage prepared for maintaining adults in laboratory.

1.3 Anatomy

The adults (males and females) have a well-developed body consisting of head, thorax, and abdomen (Figs. 1.13 and 1.14). The *Aedes* mosquitoes have a similar body structure as that of other insect species. The head region has a pair of compound eyes, antennae, and skilled mouth parts (Gutsevich, 1970). The thorax region is divided into prothorax, mesothorax, and metathorax and bears wings and appendages. The abdomen is segmented and bears modified last segment containing ovipositor in the case of females and aedeagus in the case of males (Snodgrass, 1959).

The adult is aerodynamically stable and has halteres, which are modified hind legs for sensory control and for assisting flight movements. The adults are sexually differentiated into males and females, both of which feed on nectar. Later when they attain maturity (i.e., after mating), the females engorge on vertebrate blood for blood proteins that help them develop good batches of eggs (Darsie and Samanidou-Voyadjoglou, 1997).

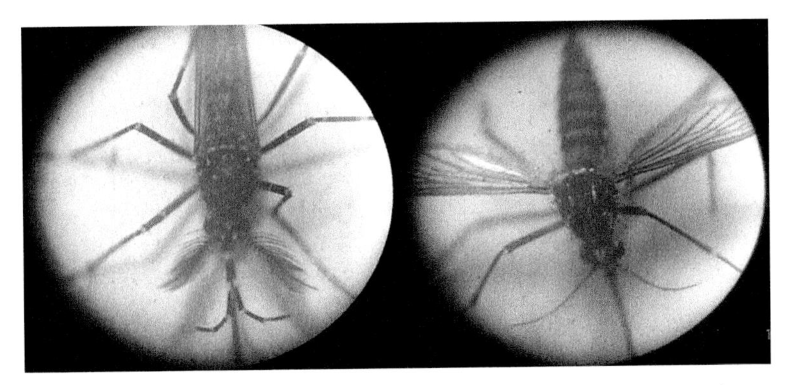

Figure 1.13 Adult form of *Aedes* mosquito; male and female, respectively.

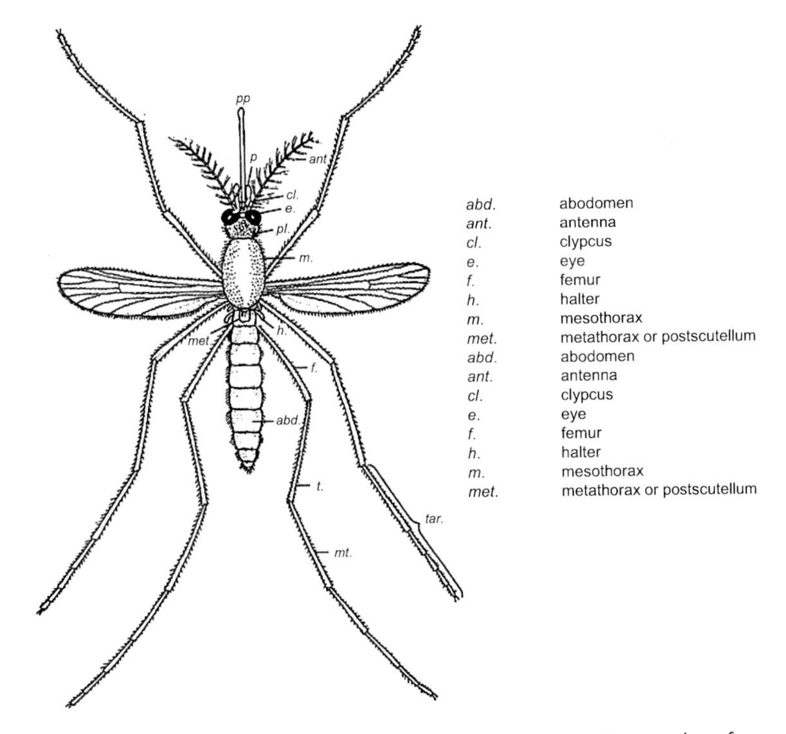

Figure 1.14 General body parts of an adult mosquito. Figure taken from www.biologydiscussion.com/invertebrate-zoology/arthropods/transmission-of-diseases-by-vectors-parasitology/62376.

The male and the female adults can be easily distinguished from one another. Since only females require blood in their later stages, their proboscis is piercing and sucking type, which can penetrate into the skin to reach the blood vessels. The males feed on nectar throughout their life, so the proboscis brushes help them to brush off the pollen grains that stick to their mouth while feeding (Gutsevich, 1970). The female proboscis lies between the eyes, and below the antennae is a pair of maxillary palps, which are short in females and elongated in males. The proboscis is composed of upper labium and lower labella. The interior of the labium has six structures: a pair of maxilla, a pair of mandibles, the labrum, and the hypopharynx (Ribeiro and Ramos, 1999).

Inside the hypopharynx, there is a salivary canal through which saliva flows. The saliva consists of many important proteins that include anticoagulant, vasodilating properties (Snodgrass, 1959). The digestive or alimentary canal consists of pharynx, esophagus, stomach divided into foregut, midgut, and hindgut, followed by intestine, rectum, and anus (Fig. 1.15). The nervous system consists of central, peripheral, and sympathetic systems with brain and ganglions connecting to all parts of the body (Fig. 1.16). On each side of the head is a compound eye consisting of hundreds of small units called ommatidia (Jirakanjanakit et al., 2008). The respiratory system includes spiracles on either side of the body, which open into tracheal trunks toward the inside of the body cavity (Mill, 1985, 1998; Sláma, 1999).

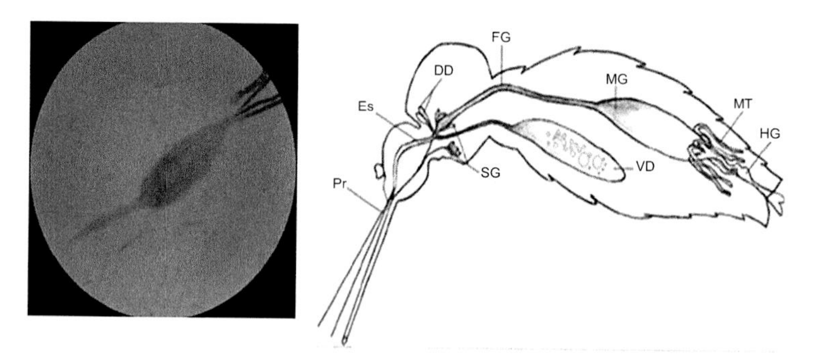

Figure 1.15 Alimentary canal of *Aedes aegypti* as observed under a light microscope (10× magnification). Different parts—DD: dorsal diverticulum, FG: foregut, HG: hindgut, MD: midgut, MT: Malpighian tubules, ES: esophagus, Pr: proboscis, SG: salivary glands, VD: ventral diverticulum. Reprinted from Gusmão et al., 2010, with permission from Elsevier.

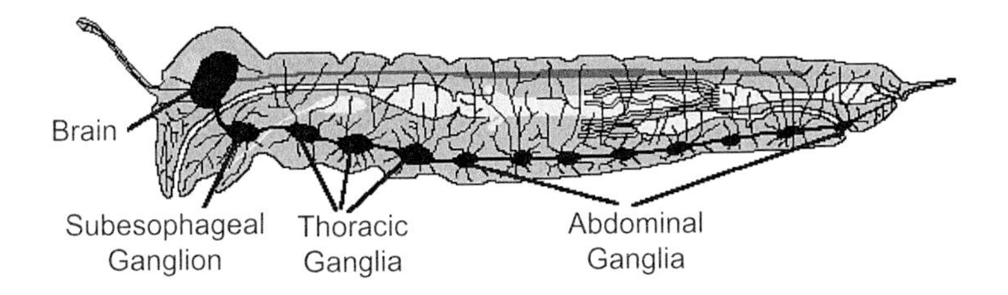

Figure 1.16 Nervous system of an adult mosquito. Figure taken https://www.earthlife.net/insects/anatomy.html.

The trunks then bifurcate into tracheoles, which blindly terminate into body tissue. The exchanges of gases occur at this point through the process of diffusion (Fig. 1.17) (Wigglesworth, 1983).

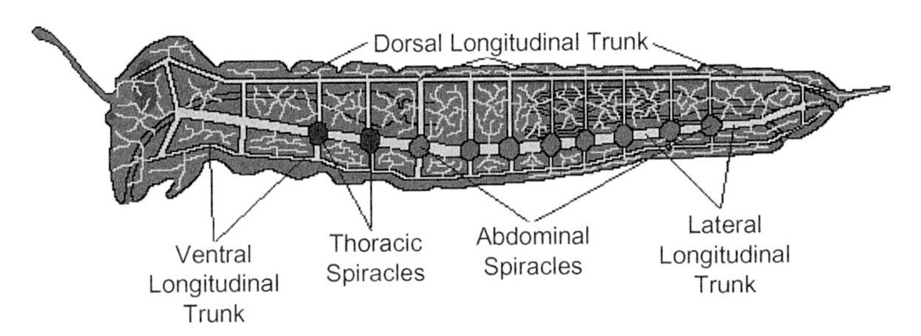

Figure 1.17 Respiratory system of an adult mosquito. Figure taken from https://www.earthlife.net/insects/anatomy.html.

The circulatory system of *Aedes* mosquitoes is similar to that of other insects, that is, open type comprising a heart enclosed in the pericardial sinus and vessels (Klowden, 2007). The heart is tubular in shape and has pores called ostia, which communicate with the pericardial sinus (Fig. 1.18). The blood is also called hemolymph and is composed of organic phosphates, uric acid, and trehalose. The hemolymph does not contain any pigment (Andereck et. al., 2010; Glenn et. al., 2010).

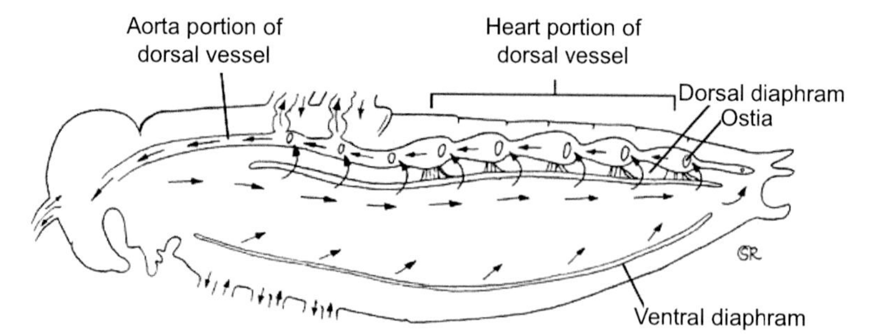

Figure 1.18 Circulatory system of an adult mosquito. Figure taken from www.bioexplorer.net/do-insects-have-hearts.html.

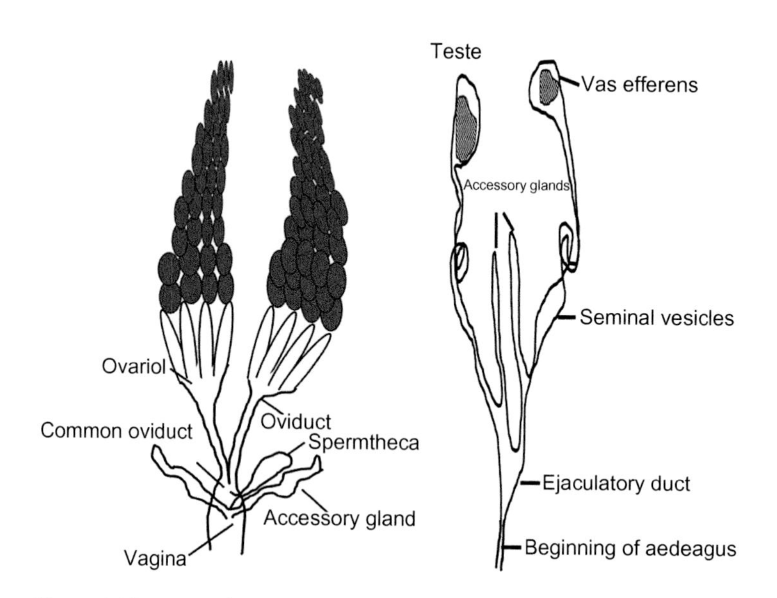

Figure 1.19 Reproductive morphology structure of male and female mosquito.

The reproductive tract of females consists of a pair of ovaries (which are bifurcated into many ovarioles connected to the main oviduct by lateral ducts), vagina, accessory glands, and spermatheca (stores sperms released by males during mating) (Figs. 1.19 and 1.20). Males, on the other hand, are known to be involved in insemination only. They possess a pair of testes, a pair of vas deferens leading into the ejaculatory duct, and a pair of accessory glands (Figs. 1.19 and 1.20). They exhibit a swarming pattern like any other insect and attract females to enter their swarm (Christophers, 1961). The process of mating by the selected male with the female occurs outside the swarm. During mating, the male deposits its sperms into the female's spermatheca.

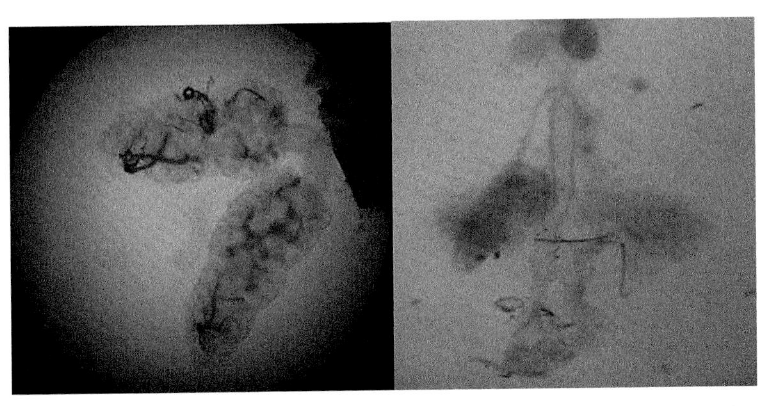

Figure 1.20 Pair of ovaries and pair of testes, respectively, as observed under a light microscope (10× magnification).

1.4 Gonotrophy

The gonotrophic cycle of mosquitoes, as defined by Beklemishev (1940), is the search for a host, ingestion of a blood meal, digestion of the meal, maturation of ovaries, and laying of mature eggs after searching an oviposition site (Beklemishev, 1940). A fully matured female mosquito prepares itself for the next stage by finding a host, preferably a vertebrate host. It feeds on the blood until its gut is filled, and sometimes it can consume blood even more than its body size (Fig. 1.21). This can be seen sometimes when blood starts oozing out of its body. After blood sucking, the female then rests on some comfortable surface for some time (Birley and Rajagopalan, 1981). Blood is necessary for the development of oocytes. For egg laying, the female prefers containers or small niches filled with water and lays eggs on the surface (ovipositioning), a bit above the water line, ensuring delivery of the first larval instars born in the water. The egg released by the female is a fertilized one, formed when the oocyte meets the sperm while traveling down the common oviduct before releasing out from the vagina (Briegel, 1990). The sperm enters the oocyte through the micropyle. The female then looks for some site for ovipositioning and starts laying eggs. A single mosquito can lay hundreds of eggs in batches. Usually it undergoes three gonotrophic cycles, but in each cycle, it can consume more than one blood meal from a host as per its body needs (Beklemishev, 1940). After releasing the eggs, a permanent structural change can be seen in the female's abdomen, which can be observed if one wants to keep a record of the female's reproductive history and, thereafter, her age (Birley and Rajagopalan, 1981).

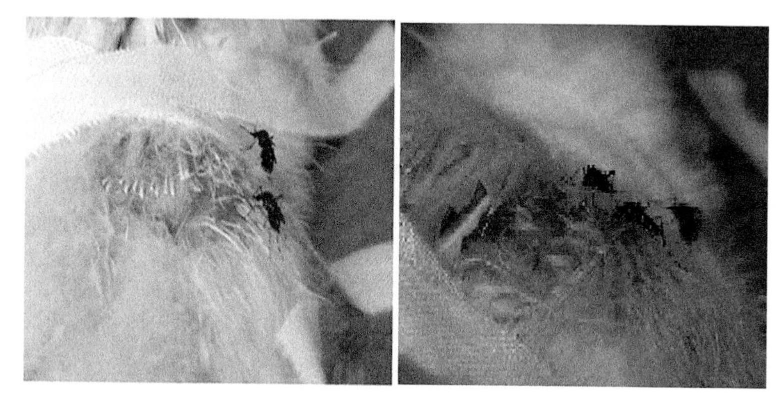

Figure 1.21 Blood feeding by female *Aedes* under laboratory condition, an earlier practice.

The length of this gonotrophic cycle is very important for determining the frequency at which the mosquito interacts with the host and is an indirect measure for estimating the transmission and disease acquisition (Klowden and Briegel, 1994).

1.5 Host Preference

After mating when the females have the urge to suck blood, they sense the host by body odor and CO_2 dispersed by a host closest to its vicinity. Studies have shown that in 99% of the cases, *Aedes aegypti* prefer human blood, while in less than 1% cases, they may feed on other vertebrate hosts, mainly bovine, swine, cat, rat, and chicken (Ponlawat and Harrington, 2005). Some studies also hypothesize the role of ADP and ATP identification by mosquitoes in identifying their preferred host (Clements, 2000). During the ingestion of blood meal, a mosquito also injects its saliva into the host. It is during this time when any virus present inside a mosquito system is released into the host's blood. On the other hand, if the mosquito is drawing a blood meal from a human host infected with any viral disease, the virus reaches the mosquito's midgut along with the blood. In the midgut, the blood components get digested. On the other hand, if the mosquito is infected, the saliva that enters the human blood from the salivary glands carries the virus and infects the host.

1.6 Conclusion

This chapter summarizes the various developmental stages of *Aedes* mosquitoes, such as eggs, larvae, pupae, and adults. The difference between the breathing positions of the three mosquito larvae has also been discussed. Anatomical features revealed that adult mosquitoes (males and females) have a well-developed body consisting of head, thorax, and abdomen. The adults are sexually differentiated into males and females, both of which feed on nectar. Later when they attain maturity (i.e., after mating), the females engorge on vertebrate blood for blood proteins that help them develop good batches of eggs. All the features of developmental stages and anatomical features have been concluded.

References

Andereck, J. W.; King, J. G.; and Hillyer, J. F. Contraction of the ventral abdomen potentiates extracardiac retrograde hemolymph propulsion in the mosquito hemocoel. *PLoS One* 2010, **5**: e12943. [PMC free article] [PubMed] [Google Scholar].

Arbovirus Summary Archives. Repellents, Traps, Virus Information, Maps, etc. Pest Alert. http://entomology.ifas.ufl.edu/pestalert/arbovirus/arbovirus.htm (14 May 2008).

Beklemishev, W. N. Gonotrophic rhythm as a basic principle of the biology of (Anopheles). *Vopr. Fiziol. Ekol. Malar. Komara* 1940, **1**: 3–22 (cited in Detinova, 1962).

Birley, M. H. and Rajagopalan, P. K. Estimation of the survival and biting rates of *Culex quinquefasciatus* (Diptera: Culicidae). *J. Med. Entomol.* 1981, **18**: 181–186.

Briegel, H. Metabolic relationship between female body size, reserves, and fecundity of *Aedes aegypti*. *J. Insect Physiol.* 1990a, **36**: 165–172.

Centers for Disease Control. Mosquito life cycle. Centers for Disease Control. www.cdc.gov/dengue. 2016.

Centers for Disease Control. Surveillance and Control of *Aedes aegypti* and *Aedes albopictus* in the United States. Division of Vector-Borne Diseases: Centers for Disease Control. http://www.cdc.gov/chikungunya/resources/vector-control.html. 2016.

Christophers, S. R. *Aedes aegypti (L), the Yellow Fever Mosquito: Its Life History, Bionomics and Structure*. Cambridge University Press, London, 1961, 676.

Clements, A. N. *The Biology of Mosquitoes: Development, Nutrition and Reproduction*, Vol. 1, CABI, Cambridge, MA, 2000, 509.

Darsie, R. E. Jr. and Samanidou-Voyadjoglou, A. Keys for the identification of the mosquitoes of Greece. *J. Am. Mosq. Control Assoc.* 1997, **13**(3): 247–254.

Foster, W. A. and Walker, E. D. Mosquitoes (Culicidae), In *Medical and Veterinary Entomology* (Mullen, G. and Durden, L., eds.), Academic Press, San Diego, CA, 2002, 203–262.

Glenn, J. D.; King, J. G.; and Hillyer, J. F. Structural mechanics of the mosquito heart and its function in bidirectional hemolymph transport. *J. Exp. Biol.* 2010, **213**: 541–550.

Gubler, D. J. *Aedes aegypti* and *Aedes aegypti*-borne disease control in the 1990's: Top down or bottom up. Centers for Disease Control, 1989, http://wonder.cdc.gov/wonder/prevguid/p0000434/p0000434.asp (11 April 2008).

Gusmão D. S.; Santos A.V.; Marini D. C.; Bacci Jr. M.; Berbert-Molina M.A.; Lemos F.J.A.; Culture-dependent and culture-independent characterization of microorganisms associated with *Aedes aegypti* (Diptera: Culicidae) (L.) and dynamics of bacterial colonization in the midgut. *Acta. Tropica*. 2010, **115**(3): 275–281.

Gutsevich, A. V. The determination of mosquito females by microscopic preparations of head. *Mosq. Syst.* 1974, **6**(4): 243–250.

Jirakanjanakit, N.; Leemingsawat, S.; and Dujardin, J. P. The geometry of the wing of *Aedes (Stegomyia) aegypti* in isofemale lines through successive generations. *Infect. Genet. Evol.*, 2008, **8**: 414–421.

Klowden, M. J. and Briegel, H. Mosquito gonotrophic cycle and multiple feeding potential: Contrasts between *Anopheles* and *Aedes* (Diptera: Culicidae). *J. Med. Entomol.* 1994, **31**(4): 618–622.

Klowden, M. J. *Physiological Systems in Insects.* Elsevier Academic Press, 2007, 688.

Mill, P. J. Structure and physiology of the respiratory system, In *Comprehensive Insect Physiology, Biochemistry and Pharmacology* (Kerkut, G. A. and Gilbert, L. I., eds.), Pergamon Press, Oxford, UK, 1985, 517–593.

Mill, P. J. Trachea and tracheoles, In *Microscopic Anatomy of Invertebrates, Insecta* (Harrison, F. W. and Locke, M., volume eds.; Harrison, F. W. and Ruppert, E. E., treatise eds.), Wiley-Liss, Inc, New York , USA, 1998, Vol. 11: 303–336.

Nelson, M. J. *Aedes aegypti: Biology and Ecology.* Pan American Health Organization, Washington, D. C., 1986.

Ponlawat, A. and Harrington, L. C. Blood feeding patterns of *Aedes aegypti* and *Aedes albopictaus* in Thailand. *J. Med. Entomol.* 2005, **42** (5): 844–849.

Ribeiro, H. and Ramos, H. C. Identification keys of the mosquitoes (Diptera: Culicidae) of Continental Portugal, Acores and Madeira. *Eur. Mosq. Bull.* 1999, **3**: 1–11.

Sláma, K. Active regulation of insect respiration. *Ann. Entomol. Soc. Am.* 1999, **92**: 1–2.

Snodgrass, R. E. *The Anatomical Life of the Mosquito.* The Smithsonian Institute, Washington, D. C., 1959.

Wigglesworth, V. B. The physiology of insect tracheoles. *Adv. Insect Physiol.* 1983, **17**: 85–148.

Chapter 2

The *Aedes* Fauna: Different *Aedes* Species Inhabiting the Earth

Annette Angel,[a] Bennet Angel,[b] Neelam Yadav,[c,d] Jagriti Narang,[e] Surender Singh Yadav,[f] and Vinod Joshi[b]
[a]*Division of Zoonosis, National Center for Disease Control, 22 Shamnath Marg, Civil lines, Delhi, India*
[b]*Amity Institute of Virology and Immunology, Amity University, Sector-125, Noida, India*
[c]*Centre for Biotechnology, Maharshi Dayanand University, Rohtak, India*
[d]*Department of Biotechnology, Deenbandhu Chhotu Ram University of Science and Technology, Murthal, Sonepat, India*
[e]*Department of Biotechnology, Jamia Hamdard University, New Delhi, India*
[f]*Department of Botany, Maharshi Dayanand University, Rohtak, India*
annetteangel_15@yahoo.co.in

The global spread of *Aedes* mosquitoes has created serious issues related to public health. These mosquitoes are distributed in tropical and subtropical areas. The different species of *Aedes* mosquitoes act as vectors for transmission of diseases and, therefore, affect the health of people worldwide. Hence, solutions to the challenges of controlling these mosquito vectors can be improved by increasing the knowledge of their biology, ecology, and vector competence.

Small Bite, Big Threat: Deadly Infections Transmitted by Aedes *Mosquitoes*
Edited by Jagriti Narang and Manika Khanuja

ISBN 978-981-4800-86-0 (Hardcover), 978-1-003-00329-8 (eBook)
www.jennystanford.com

This chapter provides a detailed account of the distribution of different *Aedes* mosquitoes, their role as a vector, and methods to control them. Researchers have been taking interest in mitigating the threat caused by *Aedes* mosquitoes by investigating their history, characterizing the present circumstances, and collaborating on future efforts.

The name of genus *Aedes* was given by Johann Wilhelm Meigen in 1818. *Aedes* is a genus of mosquitoes that was mainly found in tropical and subtropical regions. Nowadays they have been reported in all continents, except the coldest end, Antarctica. The genus *Aedes* includes species like *Aedes aegypti*, *Aedes atropalpus*, *Aedes albopictus*, *Aedes vittatus*, *Aedes japonicus*, *Aedes koreicus*, and *Aedes trsieriatus*. Some of these species are thought to spread to extreme localities due to human activity. For example, *Aedes albopictus* has been considered as the most invasive species transmitted to the United States via tire trade business. *Aedes* mosquitoes are easily identifiable by the presence of peculiar marks on their body. They are characteristically different from other mosquitoes as they have black-and-white stripes on their bodies and legs. These mosquitoes bite mostly during daytime, and peak biting timings are early morning and before dusk (http://www.who.int/denguecontrol/faq/en/index5. html). Let's study the different species of *Aedes* mosquitoes that inhabit the world.

2.1 *Aedes aegypti*

2.1.1 Morphology

Aedes aegypti is the most common circulating species of the genus *Aedes*. Since its identification, it has been commonly called as the yellow fever mosquito. The name was coined by Linnaeus in 1762. The adult mosquito is black in color and has white and black stripes on its legs. In the mesonotum, which is the thorax region, there is presence of semilunar markings (Fig. 2.1). The species has now been renamed as *Stegomyia* (sensu Reinert et al., 2004; Wilkerson et al., 2015).

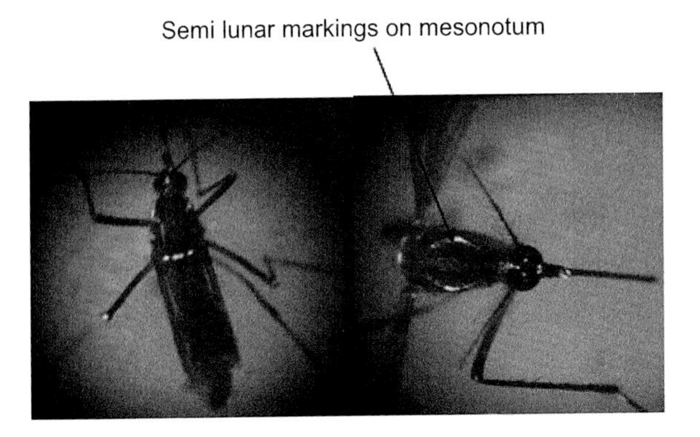

Figure 2.1 *Aedes (Stegomyia) aegypti.*

2.1.2 Origin and Distribution

Although the origin of *Aedes aegypti* dates back to 1400, it was not officially known then. It was only in the year 1495 that reports of its existence were recorded (Cloudsley-Thompson, 1976). Few records also mention that the *Aedes aegypti* (*Aae*) that is found distributed as urban and domestic forms emerged from the sylvatic (zoonotic) ancestor, *Aedes aegypti formosus* (*Aaf*), of the sub-Saharan Africa (Bennett et al., 2016; Brown et al., 2011, 2014; Gloria-Soria et al., 2016; Moore et al., 2013; Powell and Tabachnick, 2013; Salgueiro et al., 2019). The *Aaf* species still exists in the African forest and prefer nonhuman mammals for blood source (Gouck 1972; Lounibos, 1981; McBride et al., 2014; Peterson 1977; Powell et al., 2018). With the expansion of human trade activities, the species slowly colonized and established itself in different parts of the world (Fig. 2.2). It has also been reported that it might have travelled through sea routes from west Africa to Portugal and Spain and to and fro (Murphy 1972, Nelson 1986). But now there is hardly any evidence of its origin from west Africa and has been reported as a re-introductive species in urban areas (Brown et al., 2011).

The Sahara region was the first place of occupancy of this species, but with the expansion of the desert some 4000–6000 years ago, it became insufficient for the species to spread over the increasing desert area and soon they started to thrive in small water bodies of the Mediterranean region (Kropelin et al., 2008). Genetic evidences suggest that this might have been the source of spread to the new world in early 1900s, i.e. to the American continents, and then to Asian and Oceania regions (Brown et al., 2014; Gloria-Soria et al., 2016). This is also evident from the fact that the first report of yellow fever in the new world occurred in Havanna and Yucatan in the year 1648 (McNeill, 1976).

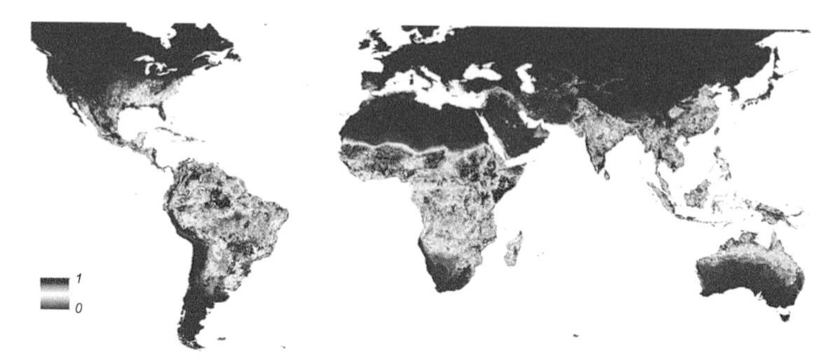

Figure 2.2 Global distribution of *Aedes aegypti* as predicted in 2015 (red = present; blue = absent). Figure taken from https://en.wikipedia.org/wiki/Aedes_aegypti.

In Asia, with the formation of Suez Canal in 1869, the first case of chikungunya was reported in 1870s and that of dengue in 1890s, which shows that *Aedes aegypti* had started urbanizing itself (Carey, 1971; Smith, 1956). As yellow fever is not reported in Asia, hence there are no records of *Aedes* spreading the disease. In 1897, the species was reported from Australia (Hare, 1898; Lee et al., 1987) (Fig. 2.3). With the onset of time, the species has become prevalent in almost all parts of the world.

2.1.3 Breeding Habitats

It is well known that the larval forms of *Aedes* species breed in water collected specifically in domestic areas while the adult forms have an aerial living and rest in various damp and humid places in urban, rural, and peri-urban areas. The gravid females prefer such sites for egg laying possibilities. As they prefer human blood, they are mostly found within a range of 100 meters from human dwellings and can take a flight of approximately 200 meters. Many researchers throughout the world have studied the habitat preferences of larval forms of *Aedes aegypti* and have found that cement tanks,

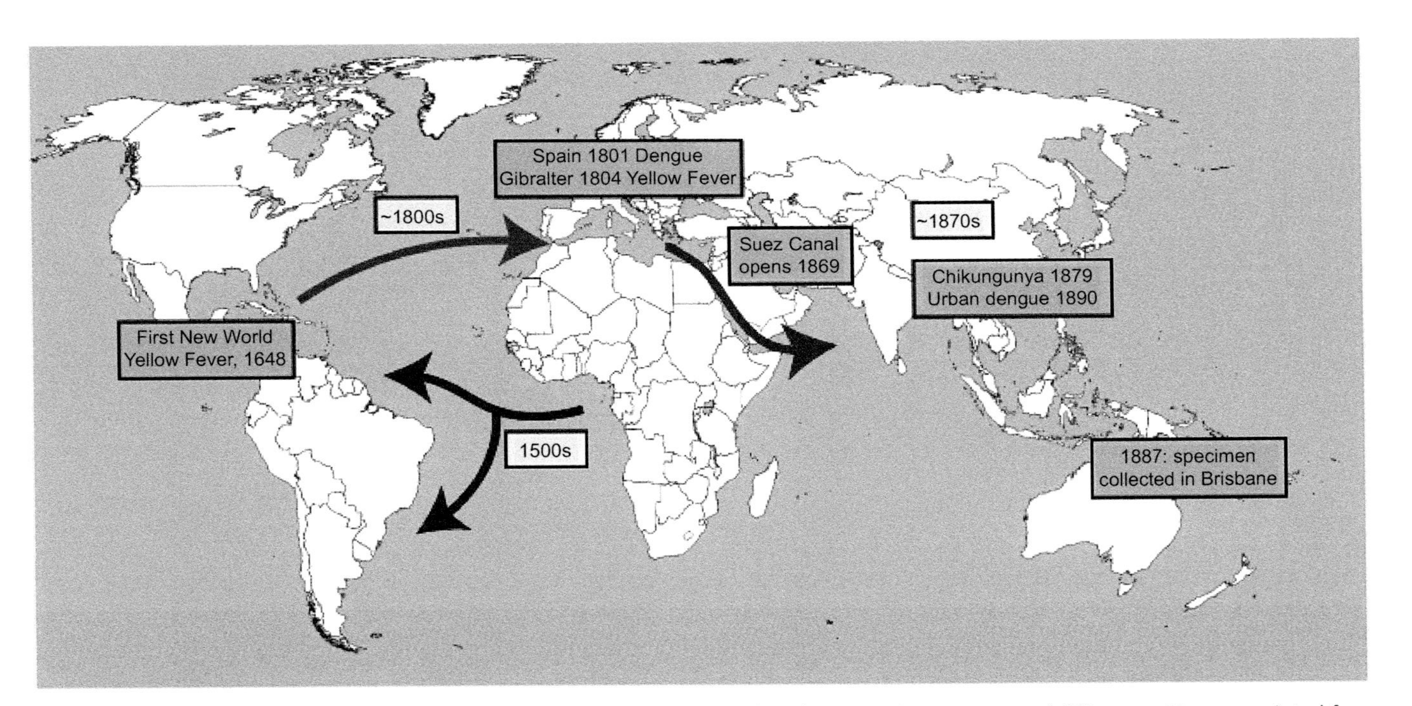

Figure 2.3 Possible routes (shown by arrows) and time of movement of *Aedes aegypti* over a span of 600 years. Figure reprinted from Powell et al., 2018, under Creative Commons license.

earthen vessels, flower vessels, coconut shells, metallic containers, underground tanks, overhead tanks, plastic containers, waste tires, mud dish, ant traps, etc. are the preferred ones (Angel et al., 2016; Boornema et al., 2018; Chan et al., 1971; Ferede et al., 2018). Studies on distribution of *Aedes aegypti* under arid conditions of Jodhpur, Rajasthan (India), have shown that peridomestic containers, as a mark of religious habits of people play crucial role in providing outdoor expansion to the domestic mosquito species *Aedes aegypti* (Joshi et al., 2006). Further in another study conducted in the state of Rajasthan, a total number of 1,30,525 domestic containers in 32 districts of Rajasthan were examined out of which 2288 (1.75 %) of containers were positive for breeding of dengue vectors and only 1.0 % of the 1,30,525 containers examined showed virus presence. This was then highlighted as an important observation that vector control strategy can be an option and and larvicide can be applied only to those 1.0% containers and the dengue infection can be controlled (Angel et al., 2016). During more epidemiological investigations, it was observed that artificial collection of domestic water, such as in peridomestic water sources for cattle drinking (Fig. 2.4), hanging water pots for birds, tree holes in zoos and parks/gardens, and small earthen vessels (commonly called "kunda" in Rajasthan) that are kept beneath rounded earthen water containers for support (Fig. 2.5), can also form crucial habitats for larval breeding and egg laying (Joshi et al., 2013) (Fig. 4.31 of Chapter 4).

Figure 2.4 (a) Peridomestic water pots kept for birds and (b) peridomestic water container kept for cattle.

Figure 2.5 Picture showing small earthen vessel or "kunda" kept beneath drinking water containers to collect dripping water from the main reservoir.

2.1.4 Role in Disease Transmission

Aedes aegypti is the main vector behind transmission of diseases like yellow fever, dengue fever, chikungunya, and Zika fever. Researchers who have studied the evolutionary biology and genomics of these mosquitoes indicate it to be the most suitable vector for disease transmission. They have concluded that since the vector lives abundantly in habitats close to human dwellings, feeds on human blood, and easily accumulates pathogen from the human host, and allows it to replicate and multiply in its salivary glands, it is considered to be a good vector (Powell, 2018; Ritchie, 2014).

2.1.5 Modern Biology

The genome of *Aedes aegypti* has been sequenced and it appeared to be ~1.38 GBp. This was done by a large group of researchers Nene et al. in the year 2007. For sequencing the whole genome, newly hatched larvae of an inbred sub-strain (LVP^{ib12}) of the Liverpool strain was used. The whole genome of *Aedes* is summed up into 3 chromosomes (Nene et al., 2007). The gene data generated (AaegL3, though was only a draft) in terms of the various proteins coded, the miRNA profile, the transposable elements present, and most importantly the receptors present will help us understand a lot about the microorganisms *Aedes* species can facilitate in its system and explore possibilities of disease regime (Bosio et al., 2000). For example, R67 and R80 are receptors that have been found in the midgut of the adult mosquitoes are considered to be receptors of dengue viruses (all four serotypes) (Curiel et al., 2006).

Another whole genome sequencing (AaegL4) was attempted using the Hi-C sequencing technology that generated chromosome length scaffolds (Dudchenko et al., 2017). Mathews and coworkers attempted for a more refined genome sequencing (AaegL5) as was done by Nene et al. but could not provide the wholesome data. They used the Pacific Biosciences sequencing and Hi-C sequencing method, which presented the genome with a decrease in the percentage of contigs (more contigs were found in Dudchenko et al.'s work). Their current data speaks that the genome is actually 1.25 Bp in size.

As the larval forms survive in aquatic habitats and feed on the microorganisms within the habitats, research keeps pace to find out the role of the microbiota in influencing the disease transmission capabilities of the adult forms (Coon et al., 2016; Duguma et al., 2013). It has also come to notice that these micro-organism can at length interfere or alter the immune response; modify the mi-RNA profile; play a significant role in the process of nutrient acquisition, reproduction, and other life processes of the mosquito; and may also lead to variation in vectoral capacity (Caragata et al., 2014; Charan et al., 2016; Coon et al., 2014, 2017; De Gaio et al., 2011; Dong et al., 2009; Gendrin et al., 2015; Hussai et al., 2011; Mayoral et al., 2014; Ramirez et al., 2014; Xi et al., 2008; Oliver et al., 2003). *Wolbachia* is a bacterium which lives in the gut of *Aedes* mosquitoes and studies have found out that when present naturally it does not allow transmission of diseases like yellow fever, dengue, chikungunya, and Zika (Aliota et al., 2016; Balgrove et al., 2013; Moreira et al., 2009; Walker et al., 2011). Hence *Wolbachia* is now being looked upon for vector control strategy. Headed by Prof. Scott O' Neill, research is being done in 12 countries since 2011 to assess the risk of introducing *Aedes* infected with *Wolbachia*. If the reserch succeeds, it will pave the way to vector control and disease reduction (worldmosquitoprogram.org). The role of miRNAs are also being looked upon because while some of them are known to upregulate viral replication process (Slonchak et al., 2014; Zhang et al., 2014) others are known to downregulate the replication (Hussain et al., 2013; Liu et al., 2015).

2.2 *Aedes albopictus*

2.2.1 Morphology

Aedes albopictus is yet another circulating species of the genus *Aedes* considered second in disease transmission. It was first described by a British–Australian entomologist, Frederick A. Askew Skuse in 1894 who named it as *Culex albopictus* initially but it was later given the name "*Aedes albopictus*" (Skuse, 1894). Now the exact name for *Aedes albopictus* is *Stegomyia albopicta* (sensu Reinert et al., 2014). It is commonly called as "Asian tiger mosquito" or "forest's mosquito." The adult mosquito is black in color and has well-defined white and black stripes on its legs. In the mesonotum, which is the thorax region, there is presence of straight white streak as shown in Fig. 2.6.

Figure 2.6 Details of mesonotum scaling of *Aedes albopictus.*

2.2.2 Origin and Distribution

Aedes albopictus is thought to have originated in the dense forests of south Asia and *Aedes aegypti* is now distributed in all the continents except Antarctica (Fig. 2.7). The spread of this species is also a contribution of international trade especially of used tires and bamboo (Adhami and Reiter, 1998; Benedict et al., 2007; Bonizzoni et al., 2013; Hofhuis et al., 2009; Kamgang et al., 2011; Madon et al., 2002; Paupy et al., 2009; Reiter, 1998; Tedjou et al., 2019; Watson, 1967). Due to this, it quickly spread to North and South America (Forattini, 1986; Nawrocki and Hawley, 1987; Reiter, 1984), Africa (Cornel and Hunt, 1991), and Europe (Medlock et al., 2015). In Europe it was Albania

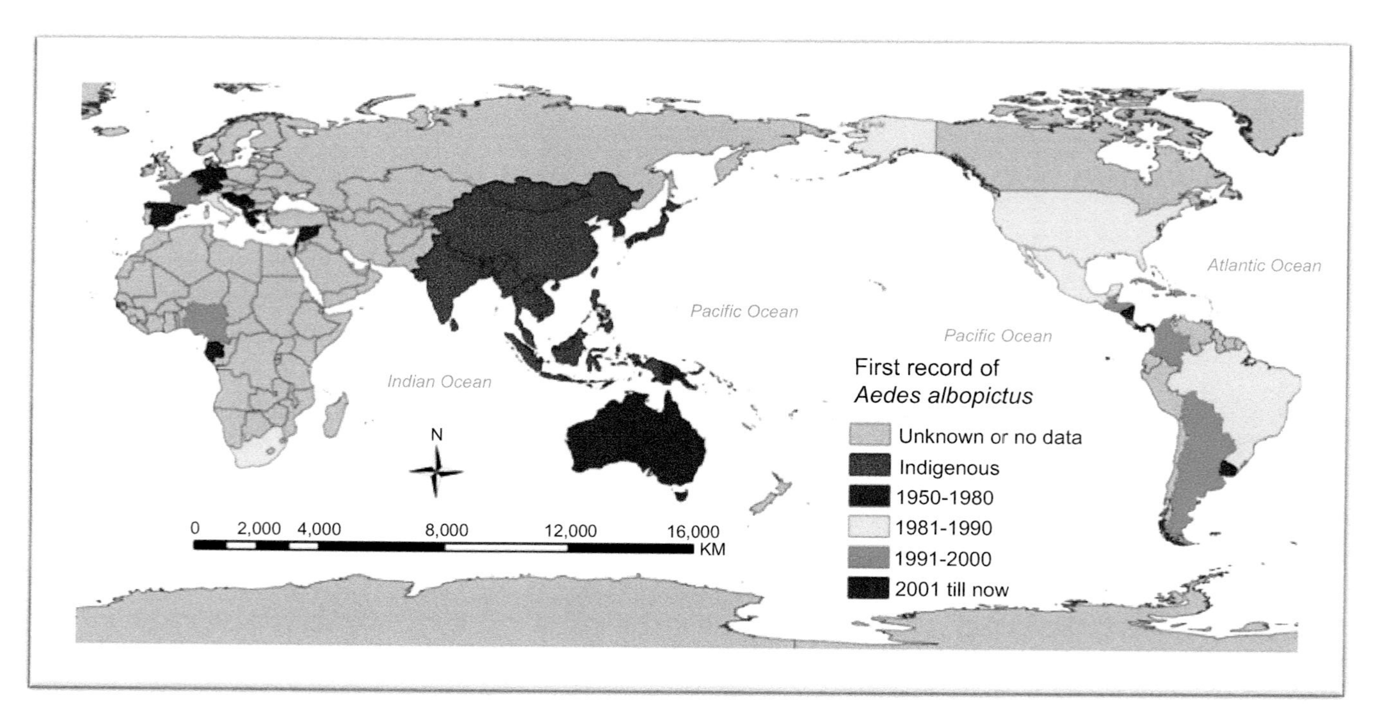

Figure 2.7 Distribution of *Aedes albopictus* across the globe. Figure reprinted from Bonizzoni et al., 2013, with permission from Elsevier.

where the species was first reported through a shipment that came from China. In United states, it was first reported in Memphis, Tennessee, in the year 1983 (Reiter and Darsie, 1984). In middle east, in Lebanon in 2003 and Syria in the year 2005 (Haddad et al., 2007) but till 2005 it was not native to Australia due to good surveillance and quarantine measures at the international seaports by the government (Russel et al., 2005). Till 2006, there are no reports of the species' presence in New Zealand (Derraik, 2006) and also in Antarctica (Kraemer et al., 2015). The global invasive database has recorded *Aedes albopictus* as one of the 100 worst invasive species in the world (http://www.issg.org/database/).

Studies have indicated that *Aedes albopictus* can survive varied ecological conditions by producing dormant eggs in cold temperatures, especially in temperate areas. In tropical areas it is known to reproduce continuously (Hansen and Criag, 1995; Hawley et al., 1987; Knudsen et al., 1996; Lounibos et al., 2014; Mori, 1981; Paupy et al., 2009; Tippelt et al., 2019).

2.2.3 Breeding Habitats

The species *Aedes albopictus* is generally observed to be a tree hole-breeding mosquito (peri-urban) thus playing the role of a maintenance vector for dengue than a frequent transmitter (Joshi et al., 2006). Though generally found in dense forests, in some countries it is also found to inhabit urban and rural areas in containers like cement tanks, flower pots, bird baths, abandoned containers and tires (Eritja et al., 2005; Invasive species compendium from CABI.org; Joshi et al., 2006). Researchers have also found co-existence of *Aedes aegypti* and *Aedes albopictus* in many parts of the world (Gilotra et al., 1967; Juliano et al., 2004; Kamgang et al., 2017; Lounibos, 2002; O'Meara et al., 1995; Rudnick and Chan, 1965; Tedjou et al., 2019; Weetman et al., 2018).

2.2.4 Role in Disease Transmission

Aedes albopictus is known to transmit a total of 26 viruses from five different families that includes dengue virus, chikungunya virus, West Nile virus, yellow fever virus, Rift Valley fever, Sindbis virus, Zika virus, Eastern equine encephalitis virus, La Crosse virus, Japanese encephalitis virus, and Venezuelan equine encephalitis virus (de Lamballerie et al., 2008; Paupy et al., 2009). This is strengthened by the fact that these viruses have been isolated from *Aedes albopictus* (Paupy et al., 2009). On the other hand some viruses like Potosi virus, Cache Valley virus, Mayaro virus, Ross River virus, Western equine encephalitis virus, Oropouche virus, Jamestown Canyon virus, San Angelo virus, and Trivittatus virus have been experimentally seen to be transmitted (Medlock et al., 2015; Schaffner et al., 2013). Although dengue is mainly transmitted by *Aedes aegypti* but in many outbreaks *Aedes albopictus* have been observed to be the lone vector present (Gasperi et al., 2012; Lambrechts et al., 2010; Rezza 2012; Wu et al., 2010). Thus, this species proves to play a crucial role in the etiology of dengue as its presence demonstrates persistence of virus within a setting. In desert areas of Rajasthan where rains are erratic, *Aedes albopictus* has been reported to occur post monsoon (September). During this period, the mosquito has been reported to breed in zoos/parks or in the areas situated in close vicinity of trees (Joshi et al., 2006). *Aedes albopictus* have been seen in outbreaks of chikungunya virus and in some parts of the world have become a predominant vector of chikungunya virus transmission (de Lamballerie et al., 2008).

Another interesting and recently reported observation is the transmission of a parasitic nematode *Dirofilaria* by *Aedes albopictus* (Bonizzoni et al., 2013; Grard et al., 2014; Medlock et al., 2010; Paupy et al., 2009). This is a parasite that is transmitted between dogs and mosquitoes but is also known to infect humans (Cancrini et al., 2003; Genchi et al., 2009; Giangaspero et al., 2013; Pampiglione et al., 2001; Poppert et al., 2009) and has been so far reported in Asia, North America, and Europe (Paupy et al., 2009).

2.2.5 Modern Biology

The genome of *Aedes albopictus* has been sequenced in the year 2015 (Chen et al., 2015). Sequencing results showed that its size is the largest in whole of the mosquito family, that is, 689.59 Gbp. The Foshan strain of *Aedes albopictus* was used for the sequencing (procured from CDC, Atlanta). Slight variations in the genome size has been observed from samples sequenced from different geographical locations. The genome has 68% of repetitive sequences. Interestingly, the sequence comparison results showed similarities between genome of flaviviruses and this species (Crochu et al., 2004; Rizzo et al., 2014; Roiz et al., 2009). The genome has 86 types of odorant-binding proteins (OBPs) and 158 odorant-receptor (OR) genes. These proteins help the mosquito to move to different environments in search of food and blood meal and for mating and ovipositioning activities. Like *Aedes aegypti, Aedes albopictus* also has *Wolbachia* species residing in its gut; the wA1bA and the wA1bB strains (Sinkins et al., 1995). Some *Wolbachia* strains, such as *w*Ri, *w*MelPop, *w*Pip, and *w*Mel, were introduced artificially (Blagrove et al., 2012, 2013; Calvitti et al., 2010; Fy et al., 2010; Suh et al., 2009; Xi et al., 2006). This bacterium is capable of not only inducing cytoplasmic instability but is also resistant to viral infections (Kambhampati et al., 1993; Mousson et al., 2012). The mosquito cell line that has been successfully used by many researchers worldwide is derived from the larvae of *Aedes albopictus* and are now being commercially supplied everywhere for research purposes. This was first developed by Dr. K. R. P. Singh in the year 1967 (Singh, 1967).

2.3 *Aedes vittatus*

2.3.1 Morphology

Aedes vittatus is a very interesting species of the genus *Aedes* and was earlier referred to as the non-refractory species. It was first discovered in 1861 by a French entomologist Jacques Marie Frangile Bigot from Corsica, Europe. Initially this was also placed under *Stegomyia* but in the year 2000, Reinert placed it under a new subgenus *Fredwardsius* (Reinert, 2000). The adult mosquito is black in color and is very robust in appearance compared to the rest of the *Aedes* species. Its peculiar characteristic is the presence of three pairs of white dots on its thorax region as shown in Fig. 2.8.

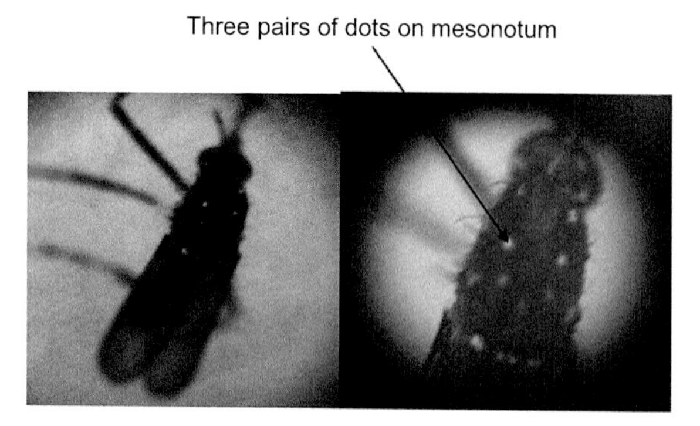

Figure 2.8 Details of mesonotum scaling of *Aedes vittatus*.

2.3.2 Origin and Distribution

The species is found in the Mediterranean region of Europe, Asia, and Africa (Sudeep and Shil, 2017) as shown in Fig. 2.9.

In Europe, the species have been reported from France, Italy, Portugal, and Spain while in Asia it has been seen in Bangladesh, China, India, Malaysia, Nepal, Pakistan, Sri Lanka, Thailand, and many other countries. In African region, a total of 37 countries including Algeria, Cameroon, Central African Republic, Ethiopia, Gambia, Ghana, Guinea, Kenya, Nigeria, Somalia, South Africa, Sudan, Tanzania, and Zimbabwe have reported the species. Though the species is available in these countries, but not all have reported outbreaks or epidemics due to this species and not much have been documented about this. In India, in many areas, a co-existence of *Aedes aegypti* and *Aedes vittatus* as urban and rural species and *Aedes albopictus* as the peri-urban/sylvatic species has been observed (Angel et al., 2008). Seasonal appearance of the species has also been reported from Rajasthan, India, comparatively more during the rainy season (Angel, 2008).

2.3.3 Breeding Habitats

Aedes vittatus is found in the urban, peri-urban, and rural areas (as mentioned in the earlier section) but is more specifically observed during the rainy seasons in Rajasthan, India (Angel, 2008). Tewari et al. (2004) observed larval forms throughout the year in Vellore

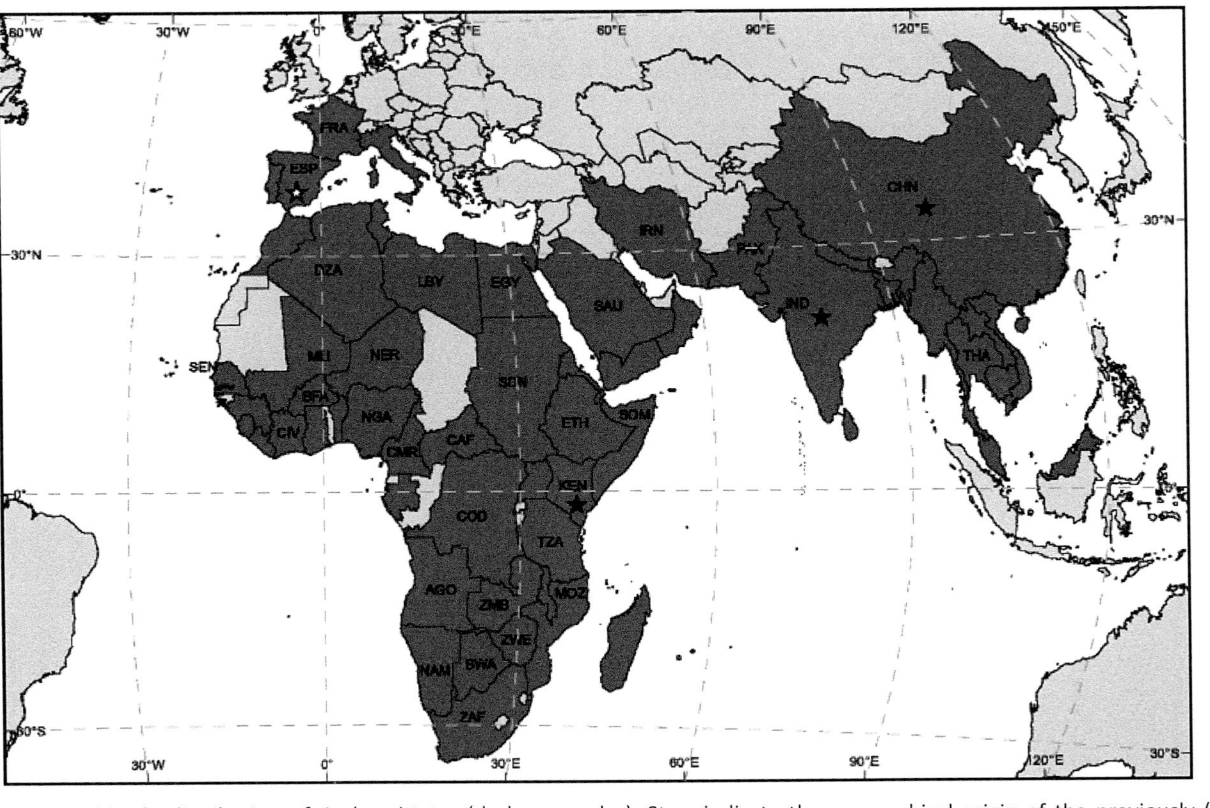

Figure 2.9 Worldwide distribution of *Aedes vittatus* (dark grey color). Stars indicate the geographical origin of the previously (black) and new (white) described genetic sequences of the barcoding region. Figure reprinted from Díez-Fernández et al., 2018, under Creative Commons license.

district of Tamil Nadu. The species has also been found to occupy Mangrove areas of Karnataka and Kerala in India (Rajavel et al., 2006). Like the other two species studied above, the *vittatus* larvae prefers hoofprints, boats, wells, tree holes, waste or abandoned cups and pots, occasional utensils, and granite rock pools (Irving et al., 1991; Huang, 1977). It is also reported to be seen in open concrete floodwater drains (Huang, 1977). The larval forms are majorly seen in Africa in puddles, rock holes, discarded containers, fresh fruit husks, septic tanks, etc. (Adebote et al., 2008; Adeleke et al., 2013; Service, 1970, 1974). Seasonal variations leading to change in breeding sites was also observed here. A study done in the African region concluded that during the months of June to October, they occupy the forested land cover (Diallo et al., 2012).

2.3.4 Role in Disease Transmission

Although the species prefers human blood, not much is known about the possibilities of it being a vector for disease transmission. In laboratory conditions, it has been known to transmit yellow fever in monkeys while another report mentions its role in yellow fever in Africa (Service, 1974). Yet another report of an epidemic in Sudan (Nuba mountain) in 1940, where an estimated of 15,000 cases and 1500 deaths were recorded, it was suspected that *Aedes vittatus* might have played the role of the transmitting vector (Huang, 1977). Lee et al. (1972) and Germain et al. (1978) have also mentioned *vittatus* during epidemics in African regions of Nigeria and Gambia, respectively.

In the 1990s, there was an epidemic of dengue fever in Jaipur, Rajasthan, India, where upon investigating for the presence of main vector *Aedes aegypti*, *Aedes vittatus* was observed instead (unpublished reports). The virus was also isolated from *Aedes vittatus* of the sylvatic populations in Senegal (1999–2000) and in Cote d'Ivoire (2016) but did not seem to cause human infections then (Cordellier et al., 1983; Diallo et al., 2005; Zahouli et al., 2016). In vitro studies have shown higher rate of dissemination of dengue virus in *Aedes vittatus* species but have also recorded a low infection rate that further concludes that they may have little or no role in transmission (Diallo et al., 2005; Mavale et al., 1992; Tewari et al., 2004).

Besides yellow fever and dengue fever, other viruses isolated from *Aedes vittatus* are chikungunya and Zika viruses. The first reports of chikungunya virus detection from *vittatus* was from Kedougou in Senegal (Diallo et al., 1999). Experimental studies done on mice models also proved transmission of the virus from these mosquitoes (Mourya and Banerjee, 1987). However, whether *vittatus* can transmit the virus through vertical route could not be proved. Chikungunya virus has also been isolated from *Aedes vittatus* from west African region (Diagne et al., 2014).

Zika virus has also been isolated from *Aedes vittatus* mosquitoes from Kedougou in Senegal in 2011 and from Cote d'Ivoire in 1999 during the outbreak of yellow fever virus (Hayes, 2009; Diallo, et al., 2011). In vitro studies done showed high dissemination rate but was not competent enough to transmit Zika virus (Akoua-Koffi et al., 2001; Diagne et al., 2015). Thus, looking at the above scenario of virus transmission, it appears that further investigations are required to figure out the possibilities of the species as a vector.

2.3.5 Modern Biology

Genome sequencing of *Aedes vittatus* has not been done as of date. But identification of the species based on molecular techniques have been attempted in Spain and data has been submitted to the NCBI database (Díez-Fernández et al., 2018) for the first time ever. Cytochrome c oxidase subunit 1 (COX 1) gene is a mitochondrial gene which is generally used for characterization of mosquito species (Hebert et al., 2003; Ondrejicka et al., 2014). This sequence MF429950.1 is available in the NCBI site. Molecular identification of larval forms of *Aedes aegypti, albopictus*, and *vittatus* have also been done using the 18s rDNA. The 18sDNA is a highly conserved gene is many species and therefore has been used by many researchers (Das et al., 2012; De Jong et al., 2009; Gale and Crampton, 1989; Hill et al., 2008). Proteomic studies done by our group for determining the vector competence of the adult mosquitoes in transmitting dengue virus have shown presence of a 200 kDa protein band in *Aedes vittatus* when compared to that present in *Aedes aegypti* (Angel et al., 2008, 2014). The protein was highly observable in the midgut or the ovaries of *Aedes vittatus* (individual mosquitoes were assayed) but was in very low amounts in *Aedes aegypti*. Viral detection studies when done simultaneously (on individual mosquitoes) concluded that those mosquitoes that had this protein did not allow the dengue virus to be transmitted through their system thus making them individual refractories and those that did not have it became competitive for viral transmission.

2.4 *Aedes atropalus*

2.4.1 Morphology

Aedes atropalus is another species of the genus *Aedes.* It was first described by Coquillett in 1902 from Virginia, Maryland, Pennsylvania, and New Hampshire (Hedeen, 1953). Initially this was placed under *Culex* but then the larval instars showed close resemblance to *Aedes* and so was assigned to *Aedes* in 1906 by Dyar. It is commonly called as the American rock pool mosquito. Other names in use are *Ochlerotatus atropalpus* (Reinert, 2000) and *Georgecraigius atropalpus* (Reinert et al., 2006). The adult mosquito is black in color and has similarity to the marks present on thorax of *Aedes aegypti*, the difference being that in case of *Aedes aegypti* the marks are semilunar or lyre or bracket shaped while in *Aedes atropalus* they are two lateral lines on the thorax region (Fig. 2.10) (https://www.ecdc.europa.eu/en/disease-vectors/facts/mosquito-factsheets/aedes-atropalpus).

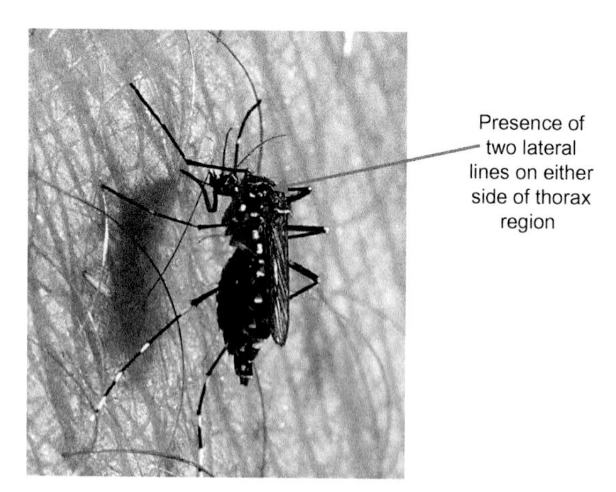

Figure 2.10 Identification of *Aedes atropalus* by marks on thorax region. Figure taken from ecdc.europa.eu.

2.4.2 Origin and Distribution

The species is believed to have been originated in North America in 1980s (Carpenter and LaCasse, 1955). Initially it was restricted to eastern Canada and eastern part of Mississippi river but soon spread to 54 U. S. states that included Indiana, Kentucky, New York, and Ohio (Andreadis, 1988; Berry and Craig, 1984; Berry et al., 1988; Nawrocki and Craig, 1989). When larval forms were investigated at various tire disposal areas in St. Joseph Co., Indiana, USA, it was found that larvae of *Culex* and *Aedes* breeded in shaded and sun-exposed tires. *Aedes atropalus* was found breeding in the exposed tires (Beier et al., 1983). The species then spread to Europe through scrap tire business. In 1996, larval forms of *Aedes atropalus* was interestingly discovered from Veneto region, Italy, during an investigation actually conducted for *Aedes albopictus* larvae in different tire storage areas where tires were imported from North America (Romi et al., 1997). By the year 2003 till 2005, it spread to France (Adege-EID Mediterranee, 2003, 2006; Chouin and Schaffner, unpublished data; Koban et al., 2019) but investigations suggest that they were eliminated from Italy and France by adopting control measures (Schaffner, personal communication; Scholte et al., 2009). Investigations similar to those done in Italy were conducted in 2009 in two tire companies of the Netherlands in which larval forms of *Aedes albopictus* in used tires were to be observed but the observation revealed the presence of larval forms of *Aedes atropalus* (Scholte et al., 2009). There is no further evidence whether the species has been observed in any other parts of the world. Although studies have reported tire industries and companies as sites from where larval forms have been collected, none have reported the urban and rural colonization of these species.

2.4.3 Breeding Habitats

As the name suggests, American rock pool mosquito is a species abundantly found in water-filled depressions of rocks like granite. Section 2.4.2 also indicates yet another habitat, tires, which become their habitats due to international trade. Many researchers have reportedoccurrence of *Aedes atropalus* in non-rock pool habitats such as canisters, plastic pans, broken light fixtures, auto wheel wells, bird baths, cups and jars, metal buckets, septic tanks (Berry et al., 1988; Covell and Brownell, 1979; Restifo and Lanzaro, 1980; Shields, 1938; White and White, 1980). An interesting feature of this species is that they can produce non-dessicating eggs autogenously, that is, without a blood meal and these eggs can survive outside water for a long time (Juliano and Lounibos, 2005; Yee, 2008). This does not infer that they do not feed on mammalian host; they do feed on humans and reportedly on deer and canaries also (Frier and Beier, 1984; Medlock et al., 2005; Turell et al., 2005) but as they have a limited flight range (Turell et al., 2005), it is unknown whether they have the ability to cause epidemics or outbreaks.

2.4.4 Role in Disease Transmission

Not much is available in records as to whether any outbreaks have been caused by this species or whether they are able to sustain any virus within them, but yes, experimentally they have been observed to be a good vector for viruses such as La Crosse, West Nile virus, Japanese encephalitis virus, St. Louis encephalitis virus, Murray valley encephalitis virus, Western and Eastern equine encephalitis virus (Frier and Beier, 1984; King et al., 1960; Turell et al., 2001). Also, in vitro studies have concluded occurrence of vertical transmission of La Crosse and St. Louis encephalitis virus by the species (Frier and Beier, 1984; Plez and Freier, 1990). As it has also been seen to co-exist with *Aedes triseriatus* and *Aedes albopictus*, thus futuristic role of this species in becoming an efficient vector cannot be neglected.

2.4.5 Modern Biology

Not much data is available about the genome sequence of *Aedes atropalus* except for some proteins that have been submitted to the NCBI databank. Larval forms of the *Aedes* species have been subjected to molecular identification using the cytochrome oxidase 1 (COX1) gene to distinguish between presence of *Aedes albopictus* and *Aedes atropalus* in the Netherlands (Scholte et al., 2009). Genetic differentiation studies have also been done using the chromosomes of *Aedes atropalus* and *Aedes epactius* (Scymczak and Rai, 1987). The cytochrome oxidase subunit 1 protein details are available on the NCBI database. Besides this protein, data for catepsin, vitellogenin, putative transposase, alpha amylase, hexamerin, enolase, arginine kinase-like protein, etc. are also enlisted (ncbi.mlm.nih.gov/protein).

2.5 *Aedes japonicus*

2.5.1 Morphology

Aedes japonicus belongs to subgenus *Finlaya* of the genus *Aedes.* It was first described by an English entomologist Frederick Vincent Theobald in 1901 from Tokyo, Japan (Theobald, 1901; Wikipedia.org). It is commonly known as the Asian bush mosquito or the Asian rock pool mosquito while other names in use are *Ochlerotatus japonicus japonicus* (Reinert, 2000) and *Hulecoeteomyia japonica japonica* (Reinert et al., 2006; www. Ecdc.europa.eu). The adult mosquito is brownish-black in color and comparatively large in size and has bronze-yellowish lyre-shaped markings on the scutum (Fig. 2.11). *Aedes japonicus* is basically a group of four subspecies namely *Aedes j. japonicus, Aedes j. shintienensis, Aedes j. yaeyamensis,* and *Aedes j. amamiensus* (Kaufman and Fonseca, 2014).

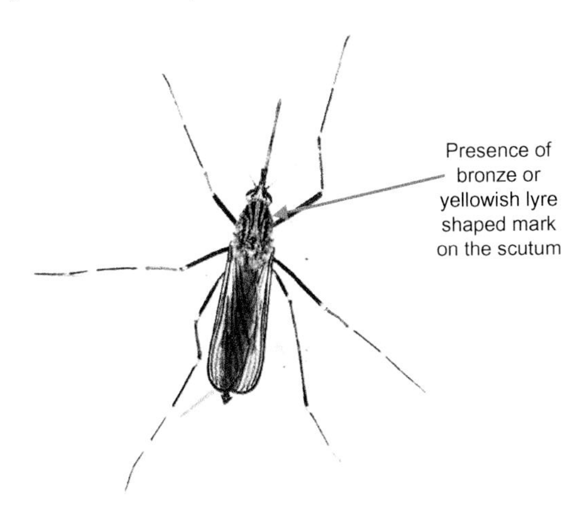

Figure 2.11 Identification of *Aedes japonicus* by marks on thorax region.

2.5.2 Origin and Distribution

The species first originated in eastern Asia specifically in Japan and the Korean penninsula. Slowly it spread to areas within north-eastern Russia to southern China and Taiwan and further occupying nearby islands (Tanaka et al., 1979). Though the species then appeared in New Zealand in the year 1993, but whether they established themselves there or not is not clear (Laird et al., 1994; Derraik, 2004). It was in August 1998, in Southold, Suffolk County, New York, that the first adult *Aedes japonicus* was captured followed by another in September 1998. More adults were captured from New Jersey and then search began to identify the larval habitats nearby, which led to first report of *Aedes japonicus* from the United States (Andreadis and Wolfe, 2010; Fonseca et al., 2001, 2010; Peyton et al., 1999). Here also the main route of entry was the international tire trade. Now the species is found in all U. S. states except Florida and Louisiana. It was in the autumn of 1999, that during the routine entomological investigations for *Aedes albopictus* species, *Aedes japonicus* was observed in the tire storage sites in France by the team of Atlantic coast mosquito-control agency (Schaffner et al., 2009; Scott et al. 1999). This was in a small village of Montsecret. Thus, the species has successfully sustained and distributed itself as eggs and larval forms from its native area in Asia to Central Europe and as the tire trade continues chances of further distribution is possible.

2.5.3 Breeding Habitats

As its common name suggests, the larval forms of the "Asian rock pool" mosquitoes have often been seen inhabiting the rocky pools and shallow vents. Besides this they have also been collected from tires, buckets, tree holes, kinked bamboo trunks, plant dishes, rainwater catchments, trash cans, discarded snack bags, fountains, etc. (Barlett-Healy et al., 2012; Kampen et al., 2012; Kauffman et al., 2012; Miyagi, 1971; Zielke et al., 2015). They feed on humans and other mammalian hosts like white-tailed deer, fallow deer, horses, and birds (Molaei et al., 2009; Wiligies et al., 2008). They lay eggs that are non-dessicating and enter pre-diapause stage in unfavorable conditions (Bova et al., 2019).

2.5.4 Role in Disease Transmission

As of date there are not many reports pertaining to disease transmission through *Aedes japonicus*, but the risk is there as their competence to transmit viruses have been observed through in vitro studies by few authors (Kampen and Werner, 2014). The species have been shown to carry West Nile virus (Riccardo et al., 2018; Veronesi et al., 2018; Wagner et al., 2018), Eastern equine encephalitis (Sardelis et al., 2002), La Crosse virus (Sardelis et al., 2002), St. Louis encephalitis virus (Sardelis et al., 2003), dengue and chikungunya viruses (Schaffner et al., 2011), Japanese encephalitis (Takashima et al., 1989), Rift Valley fever virus (Turell et al., 2013), etc. Field caught adults when detected for virus presence have shown to carry La Crosse, Cache Valley, and West Nile viruses (Harris et al., 2015; Novello et al., 2000; Sardelis and Turell, 2001; Yang et al., 2018). Takashima and Rosen in 1989 have reported transmission of Japanese encephalitis virus under laboratory conditions both horizontally and vertically.

Besides transmission of viruses, the species have also been reported to transmit nematode, *Dirofilaria immitis and Dirofilaria repens* (Silaghi et al., 2017).

2.5.5 Modern Biology

Detection of *Aedes japonicus* has also been done on similar lines as done for other species studied above. For reporting the first-ever occurrence of the species in Spain, Belgium, western Germany, and Italy, the researchers carried out molecular detection assays using COX 1 gene in the samples collected (Eritja et al., 2009; Versteirt et al., 2009; Kampen et al., 2012). Montarsi and coworkers (2009) attempted two more genes besides the COX 1 gene for diagnosis of the species; a nuclear one, that is, *β tubulin* gene (BTUB) and a mitochondrial one, that is, nicotinamide adenine dinucleotide dehydrogenase subunit 4 gene (*nad*4). The NCBI database enlists a total of 467 partial protein sequences from different geographical area. The protein data submitted are mostly of the COX 1 gene, enolase, arginine kinase–like protein, NAD protein, prohibitin-1, and β tubulin.

2.6 *Aedes koreicus*

2.6.1 Morphology

Aedes koreicus is named so because it is a native of Korea. Like *Aedes japonicus*, it belongs to the subgenus *Finlaya*. It was first described

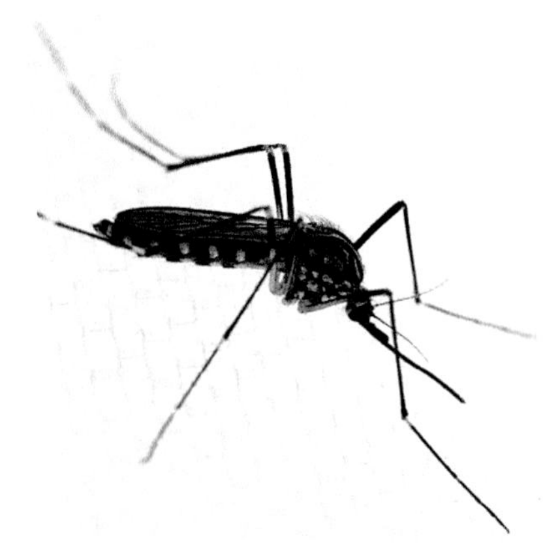

Figure 2.12 *Aedes koreicus.*

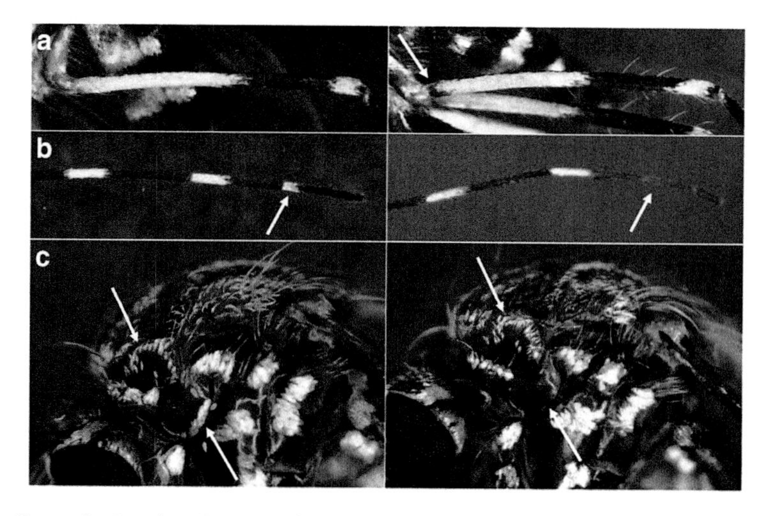

Figure 2.13 Identification of *Aedes koreicus* females (left) in comparison to *Aedes j. japonicus* (right). (a) Hind femur. The arrow shows the dark subbasal band in *Aedes j. japonicus*, which is missing in *Aedes koreicus*. (b) Hind tarsus with tarsomeres II–V. Arrows show hindtarsomere IV. (c) Lateral view of thorax. Arrows show the postpronotum and the subspiracular patch (missing in *Aedes j. japonicus*). The costa can also be seen. Figure reprinted from Pfitzner et al., 2018, under Creative Commons license.

by a British entomologist Frederick Wallace Edward in 1912. Other names in use are *Ochlerotatus koreicus* (Reinert, 2000) or *Hulecoeteomyia koreica* (Reinert et al., 2006). The adult mosquito appears brownish-black in color and is similar to *Aedes japonicus* with clear longitudinal lines on the thorax region but has a complete basal band on hind tarsomere IV which differentiates it from *Aedes japonicus* (www.ecdc.europa.eu) (Figs. 2.12 and 2.13).

2.6.2 Origin and Distribution

Aedes koreicus, as mentioned earlier, is native to Korea and also north-east China, Japan, and eastern Russia (Tanaka et al., 1979; ecdc.europa.eu). Until 2008, there were no reports of its existence outside its native place, but during a National Mosquito Survey, MODIRISK, in Belgium, the species made its first invasive appearance. Since the species closely resembles *Aedes japonicus*, hence minute morphological differences helped to conclude its presence in Belgium (Versteirt et al., 2012). Soon it spread to other European countries via international trade namely in Belluno Province, Italy, in the year 2011 (Capelli et al., 2011), Sochi city in Russia in 2013 (Bezzhonova et al., 2014), in a village Lovrenc na Dravskem Polju in Slovenia in 2013 (Kalan et al., 2017), in Germany in the 2015 (Werner et al., 2016) and in Pécs, south-west Hungary, in 2016 (Kurucz et al., 2016). Also in 2016, a single larva was recorded in Wiesbaden, central Germany, during a KABS survey (Kommunale Aktionsgemeinschaft zur Bekämpfung der Schnakenplage) program (Pfitzner et al., 2018). Since Wiesbaden is situated west of Frankfurt/ river Main, is also next to river Rhine, is bound by High Taunus range in the north-west, and has mouth of river Main in the south-east, surveys were conducted near the river and airport areas for possibility of its further distribution in and outside the city. As the Rhine plant centres were nearby and the Frankfurt airport is also within a 20 km distance, it was assumed that the introduction of the species here might have been possible either through transport of plants or through transport of goods via airport. The ovitraps laid in these areas confirmed presence of eggs of this species. The species is seen to co-exist in certain areas with *Aedes japonicus*, hence possibilities of more distribution may be possible in near future.

2.6.3 Breeding Habitats

Though not much information is available about the species, yet entomological studies conducted suggest that the larvae breeds in garden ponds, water drums and other water-containing vessels (Ho, 1931), tree holes and stone cavities (Feng, 1938), discarded tires, as well as in more natural sites such as road tracks and ditches (Versteirt et al., 2012), manholes, plastic buckets, flower pots, etc. (Capelli et al., 2011). They are found in all the areas including urban, rural, and peri-urban sites and prefer human, mammals, and birds as host for their blood meal. They are known to be more tolerate to cold climatic conditions than other *Aedes* species and lay eggs that hatch in spring when the snow melts (Knight, 1947; Myiagi, 1971).

2.6.4 Role in Disease Transmission

The vector competence of *Aedes koreicus* has not been explored much, but as they seem to exist in areas where disease transmission is being done by other members of the *Aedes* genus, hence it is likely that this may come out to be a potential vector in the time to come or even in absence of the primary vector. For example, in Italy and its surrounding areas, *Aedes albopictus* is a very strong and prominent vector causing epidemics of chikungunya fever. With the arrival of *Aedes koreicus* in the area, possibilities of transmission of chikungunya virus though them is quite feasible. Studies conducted proved that the CHIKV *'La Reunion'* strain was transmitted by *Aedes koreicus* but of a low level (Ciocchetta et al., 2018). They are also thought to transmit Japanese encephalitis (Takashima and Rosen, 1989) and nematodes *Dirofilaria immitis* (Myiagi, 1971) and *Brugia malayi* (Capelli et al., 2011).

2.6.5 Modern Biology

Nad4 gene was used for identification of the species in Germany and Liguria (Pfitzner et al., 2018; Ballardini et al., 2019). A total of 62 proteins are enlisted in the NCBI database that includes COX 1 subunit, NAD 4, and β tubulin. Microbiota present in *Aedes koreicus* has been demonstrated by Alfano and his team (2019) using 16srRNA for characterizing purposes. V3 and V4 regions of the 16s rDNA was used for sequencing of microbiota. The results showed presence of *Proteobacteria* species in the observed adults (84%) and larvae (66%) , specifically *Gammaproteobacteria* followed by *Bacteroidetes* and *Actinobacteria* in the water, larvae, and pupae collected and *Alphaproteobacteria* in the adult forms collected. Firmicutes were also observed in all, except water samples. *Pseudomonas, Gilliamella, Dyella* and *Pantoea*, and *Enterobacteriacea* family were also seen in adults. When the trend of microbial fauna in the water samples and the larval and adult forms were compared, it was seen that only 10% of those found in water reached the larval gut and established themselves there, and as the life cycle continued, only few microbial fauna were able to invade the pupal and then the adult system (Alfani et al., 2019).

2.7 *Aedes triseriatus*

2.7.1 Morphology

Aedes triseriatus is yet another species of the genus *Aedes.* It was first described by an English entomologist Thomas Say in 1823. It is commonly called as the eastern tree-hole mosquito or the American tree-hole mosquito. Other names in use are *Ochlerotatus triseriatus* (Reinert, 2000). The adult mosquito has large, dark-scaled pattern on its thorax and another important characteristic is the absence of white bands or pale rings on the legs as evident in the Fig. 2.14 (Grimstad et al., 1974; www.ecdc.europa.eu).

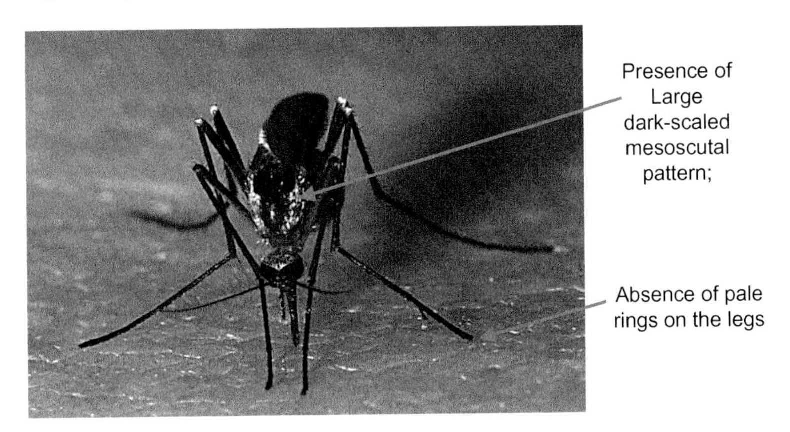

Figure 2.14 Identification of *Aedes triseriatus* by marks on thorax region. Figure taken from BioLb.cz.

2.7.2 Origin and Distribution

As the common name of the species suggests, American tree-hole mosquito is native to many parts of America such as Florida, Canada, Ontario, Utah, Texas, New Brunswick, and Quebec (Borucki et al., 2002; ecdc.europa.ec; wikipedia.org; Williams et al., 2007). It has not known to travel outside America but in 2001, a specimen was recovered from a shipment of used tires that was transported from the United States to France. Immediate control measures were applied for its elimination and now there are no traces of the species in Europe (https://animaldiversity.org/accounts/Aedes_triseriatus/).

2.7.3 Breeding Habitats

The larvae of the species prefer tree holes for breeding specifically in the forest areas. Besides it has also been reported to breed in tires, urban woodlogs, artificial containers near urban areas, etc. The eggs are laid in tree holes (Walker, 1952), which then hatch into larvae during spring. The species generally prefer human blood but besides this they have also been reported to feed on deer, cats, opossums, rats, squirrels, birds, reptile, and amphibians (Frier and Beier 1984; Molaei et al., 2008; Moore et al., 1993; Platt et al., 2007; Turell et al., 2005).

2.7.4 Role in Disease Transmission

Aedes triseriatus is known to be the primary vector for La Crosse virus and West Nile virus transmission (Barker et al., 2003; Berry et al., 1974; Borucki et al., 2002; Pantuwatana et al., 1974; Watts et al., 1974; Williams et al., 2007). Vertical transmission of the La Crosse virus has also been established (Miller et al., 1977; Tesh and Gubler, 1975; Watts et al., 1973). It has also been known to transmit *Dirofilaria immitis* (Liu et al., 2011). In vitro studies indicate its vectoral competence for viruses like yellow fever, Eastern encephalitis, Venezuelan encephalitis, and Western encephalitis.

2.7.5 Modern Biology

Not much data is available on the gene sequencing of the species *Aedes triseriatus*, however, salivary gland proteins have been sequenced, which is important from virus transmission point of view. Like other

Aedes mosquitoes, the saliva of *Aedes triseriatus* is highly allergic and also has vasodilating properties (Edwards et al., 1998; Peng et al., 1998; Reno and Novak, 2005; Ribeiro et al., 1994). Studies were undertaken by Calvo and team to develop a salivary repertoire or sialome that would be useful for understanding the causation of allergy and thus indirectly provide new knowledge for development of vaccines and chemotherapeutics (Calva et al., 2010). Other than this, Beck and coworkers have studied a gene known as the *AtIAP1 (Aedes triseriatus* inhibitor of apoptosis 1) which is responsible for the ability to vertically transmit the La Crosse virus (Beck et al., 2009). Searching the NCBI database for protein submission of the species indicates partial sequence available for proteins like COX 1, putative disulfide isomerase, inhibitor of apotosis protein 1-like protein, putative chaperonin containing t-complex polypeptide 1 CCT delta subunit, putative large subunit ribosomal protein rpL44, enolase, arginine kinase-like protein, actin, argonaute 2, and vitellogenin C.

2.8 *Aedes polynesiensis*

2.8.1 Morphology

Aedes polynesiensis is yet another species of the genus *Aedes.* It was first described by Marks in 1951. It is commonly called as the Polynesian tiger mosquito. Other names in use are *Stegomyia polynesiensis* (Reinert et al., 2004). The adult mosquito is similar to *Aedes albopictus* in appearance as shown in Fig. 2.15. There is a supra-alar white line but no lyre-shaped silvery markings on the mesonotum and is a bit brownish in color.

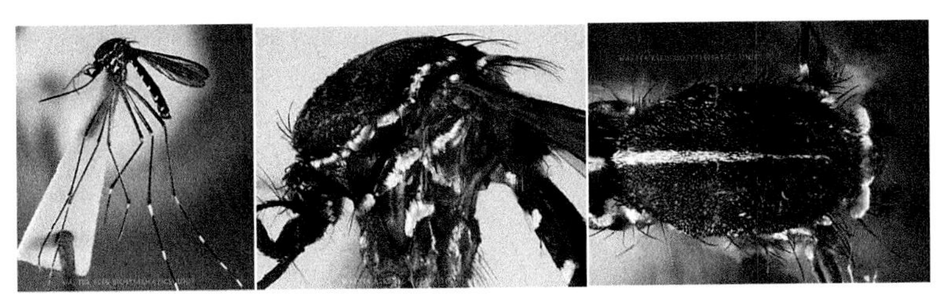

Figure 2.15 *Aedes polynesiensis*. Figure taken from Walter Reed Biosystematics Unit (WRBU).

2.8.2 Origin and Distribution

The species is found in small islands including Pacific islands, Fiji islands, Austral Islands, Cook Islands, Ellice Islands, Hoorn Islands, Marquesas Islands, Pitcairn Islands, Samoa Islands, Society Islands, Tokelau Islands, Tuamotu Archipelago, Alofi island, Magarewa isalnds, and Wallice and Futuna islands (Lee et al., 1987; Wikipedia.org; www.smsl.co.nz)

2.8.3 Breeding Habitats

The larval forms *of Aedes polynesiensis* prefers to breed in tree holes, rock pools and urban containers, waste coconut shells, banana stumps, used tires, drums, used tins and cans, crab holes, cement tanks, machine parts, fallen leaves, cisterns, drinking pans of animals, ant guards, etc. (Belkin, 1962; Bonnet and Chapman, 1956, 1958; Laird, 1956; Lee et al., 1987; Samarawickrema et al., 1993). The eggs laid are non-dessicant and can survive for long time. The adults have a short flight range (Jachowski and Otto, 1953) and therefore remain restricted to small places until transported via some means. They feed on humans but alternatively can feed on domestic mammals such as pigs, dogs, cats, sheep, and goats. (Ramalingan, 1968; Symes, 1961; Symes and Matika, 1959).

2.8.4 Role in Disease Transmission

The species is known to be the vector for *Wucheria bancrofti* and thus responsible for transmitting lymphatic filariasis (Belkin, 1962; Samarawickrema et al., 1993). Dengue virus has also been transmitted by *Aedes polynesiensis* (Maguire et al., 1971; Rosen et al., 1985; Russell et al., 2005). In vitro studies performed have also shown the species to transmit Murray Valley encephalitis, chikungunya virus, and Ross river virus (Gilotra and Shah, 1967; Gubler, 1981; Kessel, 1971; Maguire et al., 1971; Richard et al., 2016; Rosen, 1955).

2.9 *Aedes vexans*

2.9.1 Morphology

Aedes vexans also belongs to the genus *Aedes.* It was first described by a German entomologist Johann Wilhelm Meigen in 1830. It is commonly called as the inland floodwater mosquito or tomguito (Wikipedia.org). Initially it was named *Culex vexans* but in 1904, J. B. Smith gave it the name *Culex sylvestris* or the "swamp mosquito" (Carpenter and LaCasse, 1955; Headlee, 1945). Other names with which this species is referred is *Aedimorphus vexans or Aedes sylvestris.* The adult mosquito is brownish-black in color, the thorax is also uniformly brown in color, and the abdomen has bands that are constricted in the center and in the edges. This gives a "B"-shaped appearance to the bands that are visible at the base of each segment. This band is clearer in the terminal segments of the abdomen as is evident from Fig. 2.14 (Komp, 1923). It is considered to be a nuisance mosquito.

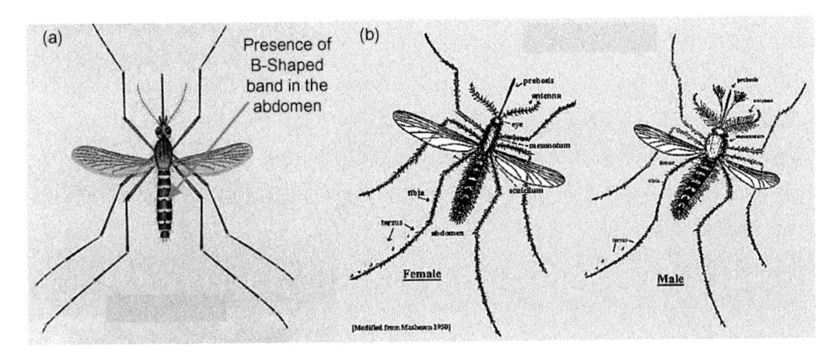

Figure 2.16 Identification of *Aedes vexans* by marks on thorax region (a and b). Figure (a) taken from http://phil.cdc.gov/ and (b) taken from https://faculty.ucr.edu/.

2.9.2 Origin and Distribution

This *Aedes* species is distributed widely throughout the Nearctic and Palearctic regions, the African west coast, and Oriental regions south and east to Samoa (Horsfall, 1972). It is also distributed mostly throughout the United States covering more than 45 countries and territories (Darsie and Ward, 2005; O'Malley, 1990; Pratt and Moore, 1993; Thompson and Dicke, 1965).

2.9.3 Breeding Habitats

The larval forms of *Aedes vexans* is known to breed in floodplains of rivers or lakes where water level is shallow, in sheet water or open rain pools, used tires, dredge spoil sites, salt marsh reservoirs, ditches, around the edges of swamps and bogs, temporary rain pool, etc. The adults are voracious feeders. They prefer human as well as domestic animals' blood. The eggs laid can remain viable up to 3–5 years (Bradford, 2005; James and Harwood, 1969). A peculiar feature of the eggs they lay is that not all of the eggs hatch at a time. The eggs first dry for a brief period. During the first rains or the first flooding time, most of the eggs get hatched while the remaining ones hatch during the next watery time. It is also studied that dissolved oxygen and presence of micro-organisms accelerate the process of hatching (Bates, 1970). The larval forms of *Aedes vexans* are known to co-exist with a variety of other mosquito species like *Aedes cantator, Aedes sollicitans, Culex pipiens,* and *Culex restuans*. The adult forms have a good flight range and are known to travel quite some distance after they emerge out from the pupa. Some researchers found that the species can fly 5 to 8 miles, some say they can fly up to 10 miles, and some say they can fly upto 15 miles, which speaks of their wide dispersal tendency and so about their disease transmission capabilities (Carpenter and LaCasse, 1955; Headlee, 1945).

2.9.4 Role in Disease Transmission

Aedes vexans are known to be vectors of parasite *Dirofiliaria immitis* since the 1980s (Lewandowski et al., 1980). Besides this they are also known to transmit diseases such as St. Louis encephalitis, Eastern equine encephalitis, Western equine encephalitis, and La Crosse encephalitis (Bates, 1970; Pratt and Moore, 1993), Rift Valley fever (Ndiaye et al., 2016). Recent studies have reported that it can be a possible vector for Zika virus transmission probably because of its high dispersal and abundancy and productivity compared to *Aedes aegypti* (Gendernalik et al., 2017).

Table 2.1 List of various *Aedes* species

S. No.	Name	Another name used	Identified by	Native place	Diseases transmitted
1	*Aedes atlanticus*	–	Dyar and Knab, 1906	USA	Yellow fever
2	*Aedes cantator (Ochlerotatus cantator)*	Brown salt marsh mosquito		Canada, New England, American mid-atlantic states	Eastern equine encephalitis, Jamestown Canyon virus, West Nile virus
3	*Aedes cinereus*	–	Meigen, 1818	North America and Europe	Bunyamwera virus, California encephalitis virus, Eastern equine encephalitis, Jamestown Canyon virus, Sindbis virus, West Nile virus
4	*Aedes taeniorhynchus*	Black salt marsh mosquito	Weidemann, 1821	North America	West Nile virus, Venezuelan equine encephalitis
5	*Aedes vigilax*	*Ochlerotatus vigilax*	Skuse	Oriental and Australasian regions of Australia, Fiji, New Guinea, Taiwan, Indonesia, etc.	Ross river virus, Murray Valley encephalitis

Source: https://en.wikipedia.org/wiki/List_of_Aedes_species

2.10 Other Species

A total of nine important *Aedes* species have been discussed in detail in this chapter, but besides these, there are more members in the *Aedes* genus. Only those which are known to carry certain diseases are enlisted in Table 2.1.

References

100 of the World's Worst Invasive Alien Species. Global Invasive Species Database. Retrieved 21 August 2008.

Adebote, D. A.; Oniye, S. J.; and Muhammed, Y. A. (2008). Studies on mosquitoes breeding in rock pools on inselbergs around Zaria, northern Nigeria. *Journal of Vector Borne Diseases,* **45**(1), 21–28.

Adege-EID Méditerranée. Éléments entomologiques relatifs au risque d'apparition du virus Chikungunya en métropole. [Entomological facts related to the risk of appearance of chikungunya virus in Metropolitan France]. Study report. Montpellier: Entente interdépartementale pour la démoustication du littoral (EID) Méditerranée; March 2006. In French.

Adege-EID Méditerranée. Surveillance committee of *Aedes albopictus* - Meeting report at DGS, Paris, 17 Dec. 2003. Montpellier: Entente interdépartementale pour la démoustication du littoral (EID) Méditerranée. In French.

Adeleke, M. A.; Adebimpe, W. O.; Hassan, A. O.; Oladejo, S. O.; Olaoye, I.; Olatunde, O. G.; et al. (2013). Larval habitats of mosquito fauna in Osogbo metropolis, southwestern Nigeria. *Asian Pacific Journal of Tropical Biomedicine,* **3**(9), 673–677.

Adhami, J. and Reiter, P. (1998). Introduction and establishment of *Aedes (Stegomyia) albopictus* Skuse (Diptera: Culicidae) in Albania. *Journal of the American Mosquito Control Association,* **14**(3), 340–343. pmid:9813831.

Akoua-Koffi, C.; Diarrassouba, S.; Benie, V. B.; Ngbichi, J. M.; Bozoua, T.; Bosson, A.; et al. (2001). Investigation surrounding a fatal case of yellow fever in Cote d'Ivoire in 1999. *Le Bulletin de la Société de Pathologie Exotique,* **94**(3), 227–230.

Alfano, N.; Tagliapietra, V.; Rosso, F.; Manica, M.; Arnoldi, D.; Pindo, M.; and Rizzoli, A. (2019). Changes in microbiota across developmental stages of *Aedes koreicus*, an invasive mosquito vector in Europe: Indications for microbiota-based control strategies. *Frontiers in Microbiology*, **10**(2832), 1–15. doi: 10.3389/fmicb.2019.02832

Aliota, M. T.; et al. (2016). The wMel strain of Wolbachia reduces transmission of chikungunya virus in *Aedes aegypti*. *PLOS Neglected Tropical Diseases*, **10**, 28792.

Andreadis, T. G. (1988). A survey ofmosquitoes breeding in stockpiles in Connecticut. *Journal of the American Mosquito Control Association*, **4**, 256–260.

Andreadis, T. G. and Wolfe, R. J. (2010). Evidence for reduction of native mosquitoes with increased expansion of invasive *Ochlerotatus japonicus japonicus* (Diptera: Culicidae) in the northeastern United States. *Journal of Medical Entomology*, **47**(1), 43–52.

Angel, A.; Angel, B.; and Joshi, V. (2016). Rare occurrence of natural transovarial transmission of dengue virus and elimination of infected foci as a possible intervention method. *Acta Tropica*, **155**, 1–5.

Angel, A.; Angel, B.; Bohra, N.; and Joshi, V. (2014). Structural study of mosquito ovarian proteins participating in transovarial transmission of dengue viruses. *International Journal of Current Microbiology and Applied Sciences*, **3**(4), 565–572.

Angel, B. (2008). Proteomics of *Aedes* Mosquitoes of Rajasthan for development of molecular markers of vector competence for dengue viruses. Ph.D. Thesis.

Angel, B. and Joshi, V. (2008). Distribution and seasonality of vertically transmitted dengue viruses in *Aedes* mosquitoes in arid and semi-arid areas of Rajasthan, India. *Journal of Vector Borne Diseases*, **45**, 56–59

Angel, B.; Sharma, K.; and Joshi, V. (2008). Association of ovarian proteins with transovarial transmission of dengue viruses by Aedes mosquitoes in Rajasthan, India. *Indian Journal of Medical Research*, **128**, 181–184.

Ballardini, M.; Ferretti, S.; Chiaranz, G.; et al. (2019). First report of the invasive mosquito *Aedes koreicus* (Diptera: Culicidae) and of its establishment in Liguria, northwest Italy. *Parasites & Vectors*, **12**, 334.

Barker, C. M.; Paulson, S. L.; Cantrell, S.; and Davis, B. S. (2003). Habitat preferences and phenology of *Ochlerotatus triseriatus* and *Aedes albopictus* (Diptera: Culicidae) in Southwestern Virginia. *Journal of Medical Entomology*, **40**(4), 403–410.

Bartlett-Healy, K.; Ünlü, I.; Obenauer, P.; Hughes, T.; Healy, S.; Crepeau, T.; et al. Larval mosquito habitat utilization and community dynamics of *Aedes albopictus* and *Aedes japonicus* (Diptera: Culicidae). *Journal of Medical Entomology*, **49**, 813–824.

Bates, M. (1970). *The Natural History of Mosquitoes.* Gloucester, Mass: Peter Smith.

Beck, E. T.; Lozano Fuentes, S.; Geske, D. A.; Blair, C. D.; Beaty, B. J.; and Black, W. C., 4th (2009). Patterns of variation in the inhibitor of apoptosis 1 gene of *Aedes triseriatus*, a transovarial vector of La Crosse virus. *Journal of Molecular Evolution*, **68**(4), 403–413. doi:10.1007/s00239-009-9216-7

Becker, N.; Petric, D.; Boase, C.; Lane, J.; Zgomba, M.; Dahl, C; and Kaiser, A. (2003). Subfamily Culicinae, in *Mosquitoes and Their Control* (Becker, N., ed.). New York: Kluwer Academic/Plenum Publishers, 193–341.

Beier, J. C.; Travis, M.; Patricoski, C.; and Kranzfelder, J. (1983). Habitat segregation among larval mosquitoes (Diptera: Culicidae) in tire yards in Indiana, USA. *Journal of Medical Entomology*, **20**(1), 76–80.

Belkin, J. N. (1962). Mosquitoes of the South Pacific. Vol. II. Berkeley and Los Angeles: University of California Press, 412.

Benedict, M. Q.; Levine, R. S.; Hawley, W. A.; and Lounibos, L. P. (2007). Spread of the tiger: global risk of invasion by the mosquito *Aedes albopictus. Vector-Borne and Zoonotic Diseases,* **7**, 76–85.

Bennett, K. L.; Gómez-Martínez, C.; Chin, Y.; et al. (2019). Dynamics and diversity of bacteria associated with the disease vectors *Aedes aegypti* and *Aedes albopictus*. *Scientific Reports*, **9**, 12160.

Bennett, K. L.; Shija, F.; Linton, Y.-M.; Misinzo, G.; Kaddumukasa, M.; Djouaka, R; et al. (2016). Historical environmental change in Africa drives divergence and admixture of *Aedes aegypti* mosquitoes: A precursor to successful worldwide colonization? *Molecular Ecology*, **25**, 4337– 4354.

Berry, R. L.; LaLonde, B. J.; Stegmiller, H. W.; Parsons, M. A.; and Bear, G. T. (1974). Isolation of La Crosse virus (California encephalitis group) from field collected *Aedes triseriatus* (Say) larvae in Ohio (Diptera: Culicidae). *Mosquito News*, **34**, 454–457.

Berry, R. L.; Peterson, E. D.; and Restifo, R. A. (1988). Records of imported tire-breeding mosquitoes in Ohio. *Journal of the American Mosquito Control Association,* **4**, 187–189.

Berry, W. J. and Craig Jr., G. B. (1984). Bionomics of *Aedes atropalus* breeding in scrap tires in northern Indiana. *Mosquito News*, **44**, 476–484.

Bezzhonova, O. V.; Patraman, I. V.; Ganushkina, L. A.; Vyshemirskii, O. I.; and Sergiev, V. P. (2014). The first finding of invasive species *Aedes* (*Finlaya*) *koreicus* (Edwards, 1917) in European Russia. *Meditsinskaia parazitologiia i parazitarnye bolezni,* **1**, 16–19.

Bigot, M. J. (1861). Trois Dipteres nouveaux de la Corse. *Annales de la Société Entomologique de France*, **4**(1), 227–229.

Blagrove, M. S. C.; Arias-Goeta, C.; Di Genua, C.; Failloux, A.-B.; and Sinkins, S. P. A. (2013). Wolbachia wMel transinfection in *Aedes albopictusis* not detrimental to host fitness and inhibits chikungunya virus. *PLOS Neglected Tropical Diseases,* **7**, e2152.

Blagrove, M. S.; et al. (2013). A Wolbachia wMel transinfection in *Aedes albopictus* is not detrimental to host fitness and inhibits chikungunya virus. *PLOS Neglected Tropical Diseases,* **7**, e2152.

Blagrove, M. S.; et al. (2012). Wolbachia strain wMel induces cytoplasmic incompatibility and blocks dengue transmission in *Aedes albopictus. Proceedings of the National Academy of Sciences of the United States of America,* **109**, 255–260.

Bonizzoni, M.; Gasperi, G.; Chen, X.; and James, A. A. (2013). The invasive mosquito species *Aedes albopictus*: current knowledge and future perspectives. *Trends in Parasitology,* **29**(9), 460–468.

Bonnet, D. B. and Chapman, H. (1958). The larval habitats of Aedes polynesiensis Marks in Tahiti and methods of control. *The American Journal of Tropical Medicine and Hygiene,* **7**(5), 512–518.

Bonnet, D. D. and Chapman, H. (1956). The importance of mosquito breeding in tree holes, with special reference to the problem in Tahiti. *Mosquito News,* **16**, 301–305.

Boornema, R. and Senthil Murugan, T. K. (2018). Breeding habitats of *Aedes aegypti* mosquitoes and awareness about prevention of dengue in urban Chidambaram: a cross-sectional study. *International Journal of Community Medicine and Public Health,* **5**(10), 4584–4589.

Borucki, M. K.; Kempf, B. J.; Blitvich, B. J.; Blair, C. D.; and Beaty, B. J. (2002). La Crosse virus: replication in vertebrate and invertebrate hosts. *Microbes and Infection,* **4**(3), 341–350.

Bosio, C. F.; Fulton, R. E.; Salasek, M. L.; Beaty, B. J.; Black, W. C. 4th. (2000). Quantitative trait loci that control vector competence for dengue-2 virus in the mosquito *Aedes aegypti. Genetics,* **156**(2), 687–698.

Bova, J.; Soghigian, J.; and Paulson, S. (2019). The prediapause stage of *Aedes japonicus japonicus* and the evolution of embryonic diapause in Aedini. *Insects,* **10**(8), 222–230.

Bradford, C. M. (2005). Effects of weather on mosquito biology, behavior, and potential for West Nile virus transmission on the southern high plains of Texas. PhD. Thesis submitted to Graduate Faculty of Texas Tech University.

Brown, J. E.; Evans, B. R.; Zheng, W.; Obas, V.; Barrera-Martinez, L.; Egizi, A.; et al. (2014). Human impacts have shaped historical and recent evolution in *Aedes Aegypti*, the dengue and yellow fever mosquito. *Evolution*, **68**, 514–525.

Brown, J. E.; McBride, C. S.; Johnson, P.; Ritchie, S.; Paupy, C.; Bossin, H.; et al. (2011). Worldwide patterns of genetic differentiation imply multiple "domestications" of *Aedes aegypti*, a major vector of human diseases. *Proceedings of the Royal Society B: Biological Sciences*, **278**(1717), 2446–2454.

Calvitti, M.; et al. (2010). Characterization of a new *Aedes albopictus* (Diptera: Culicidae)-Wolbachia pipientis (Rickettsiales: Rickettsiaceae) symbiotic association generated by artificial transfer of the wPip strain from Culex pipiens (Diptera: Culicidae). *Journal of Medical Entomology*, **47**, 179–187.

Calvo, E.; Sanchez-Vargas, I.; Kotsyfakis, M.; Favreau, A. J.; Barbian, K. D.; Pham, V. M.; et al. (2010). The salivary gland transcriptome of the eastern tree hole mosquito, *Ochlerotatus triseriatus*. *Journal of Medical Entomology*, **47**(3), 376–386. doi:10.1603/me09226

Cameron, E. C.; Wilkerson, R. C.; Mogi, M.; Miyagi, I.; Toma, T.; Kim, H-C; et al. (2010). Molecular phylogenetics of *Aedes japonicus*, a disease vector that recently invaded western Europe, North America, and the Hawaiian Islands. *Journal of Medical Entomology*, **47**, 527–535.

Cancrini, G.; Frangipane di Regalbono, A.; Ricci, I.; Tessarin, C.; Gabrielli, S.; and Pietrobelli, M. (2003). Aedes albopictus is a natural vector of Dirofilaria immitis in Italy. *Veterinary Parasitology*, **118**(3–4), 195–202.

Cancrini, G.; Romi, R.; Gabrielli, S.; Toma, L.; Dl Paolo, M.; and Scaramozzino P. (2003). First finding of Dirofilaria repens in a natural population of Aedes albopictus. *Medical and Veterinary Entomology*, **17**(4), 448–451.

Capelli, G.; Drago, A.; Martini, S.; Montarsi, F.; Soppelsa, M.; Delai, N.; Ravagnan, S.; Mazzon, L.; Schaffner, F.; Mathis, A.; Di Luca, M.; Romi, R.; and Russo, F. (2011). First report in Italy of the exotic mosquito species Aedes (Finlaya) koreicus, a potential vector of arboviruses and filariae. *Parasites & Vectors*, **4**, 188.

Caragata, E. P.; Rancès, E.; O'Neill, S. L.; and McGraw, E. A. (2014). Competition for amino acids between Wolbachia and the mosquito host, *Aedes aegypti*. *Microbial Ecology*, **67**, 205–218.

Carey, D. E. (1971). Chikungunya and dengue: a case of mistaken identity? *Journal of the History of Medicine and Allied Sciences*, **26**(3), 243–262.

Carpenter, S. J. and LaCasse, W. J. (1955). *Mosquitoes of North America (North of Mexico)*. California: University of California Press.

Carpenter, S. J. and LaCasse, W. J. (1955). *Mosquitoes of North America*. Los Angeles: University of California Press, 360.

Chan, K. L.; Ho, B. C.; and Chan, Y. C. (1971). *Aedes aegypti* (L.) and *Aedes albopictus* (Skuse) in Singapore city. *Bulletin of the World Health Organization*, **4**, 629–633.

Charan, S.; Pawar, K.; Gavhale, S.; Tikhe, C. V.; Charan, N.; Angel, B; Joshi, V.; Patole, M.; and Shouche, Y. (2016). Comparative analysis of midgut bacterial communities of three Stegomyia mosquito species from dengue-endemic and -non-endemic areas of Rajasthan, India. *Medical and Veterinary Entomology*, **30**(3), 264–277.

Chen, X. G.; Jiang, X.; Gu, J.; Xu, M.; Wu, Y.; Deng, Y.; et al. (2015). Genome sequence of the Asian Tiger mosquito, *Aedes albopictus*, reveals insights into its biology, genetics, and evolution. *Proceedings of the National Academy of Sciences of the United States of America*, **112**(44), E5907–E5915. doi:10.1073/pnas.1516410112

Cloudsley-Thompson, J. L. (1976). *Insects and History*, London: Weidenfeld & Nicolson.

Coon, K. L.; Brown, M. R.; and Strand, M. R. (2016). Mosquitoes host communities of bacteria that are essential for development but vary greatly between local habitats. *Microbial Ecology*, **25**, 5806–5826.

Coon, K. L.; Vogel, K. J.; Brown, M. R.; and Strand, M. R. (2014). Mosquitoes rely on their gut microbiota for development. *Molecular Ecology*, **23**, 2727–2739.

Coon, K. L.; et al. (2017). Bacteria-mediated hypoxia functions as a signal for mosquito development. *Proceedings of the National Academy of Sciences*, **114**, E5362–E5369.

Cordellier, R.; Bouchite, B.; Roche, J.-C.; Monteny, N.; Diaco, B.; and Akoliba, P. (1983). The sylvatic distribution of dengue 2 virus in the sub-Sudanese savanna areas of Ivory Coast in 1980: Entomological data and epidemiological study. *Cahiers ORSTOM Serie Entomologie Medicale et Parasitologie*, **21**, 165–179.

Cornel, A. and Hunt, R. (1991). *Aedes albopictus* in Africa? First records of live specimens in imported tires in Cape Town. *Journal of the American Mosquito Control Association*, **7**(1), 107–108.

Covell Jr., C. J. and Brownell, A. J. (1979). Aedes atropalpus in abandoned tires in Jefferson County, Kentucky. *Mosquito News*, **39**, 142.

Crochu, S.; Cook, S.; Attoui, H.; Charrel, R. N.; De Chesse, R.; Belhouchet, M.; Lemasson, J. J.; de Micco, P.; and de Lamballerie, X. (2004). Sequences of flavivirus-related RNA viruses persist in DNA form integrated in the genome of *Aedes* spp. mosquitoes. *Journal of General Virology,* **85**(7), 1971–1980.

Darsie, R. F. and Ward, R. A. (2005). Identification and geographical distribution of the mosquitoes of North America, North of Mexico. Gainesville, Florida: University Press of Florida.

Das, B.; Swain, S.; Patra, A.; Das, M.; Tripathy, H. K.; Mohapatra, N.; Kar, S. K.; and Hazra, R. K. (2012). Development and evaluation of a single-step multiplex PCR to differentiate the aquatic stages of morphologically similar *Aedes* (subgenus: *Stegomyia*) species. *Tropical Medicine and International Health,* **17**(2), 235–243.

De Gaio, A. O.; et al. Contribution of midgut bacteria to blood digestion and egg production in *Aedes aegypti* (Diptera: Culicidae) (L.). *Parasites & Vectors,* **4**, 105.

De Jong, L.; Moreau, X.; Dalia, J.; Coustau, C.; and Thiery, A. (2009). Molecular characterization of the invasive Asian tiger mosquito, *Aedes* (*Stegomyia*) *albopictus* (Diptera: Culicidae) in Corsica. *Acta Tropica,* **112**, 266–269.

de Lamballerie, X.; Leroy, E.; Charrel, R. N.; Ttsetsarkin, K.; Higgs, S.; and Gould, E. A. (2008). Chikungunya virus adapts to tiger mosquito via evolutionary convergence: a sign of things to come? *Virology Journal,* **5**, 33.

Demirci, B.; Lee, Y.; Lanzaro, G. C.; and Alten, B. (2012). Identification and characterization of single nucleotide polymorphisms (SNPs) in Culex theileri (Diptera: Culicidae). *Journal of Medical Entomology,* **49**, 581–588.

Derraik, J. G. B. (2004). Exotic mosquitoes in New Zealand: a review of species intercepted, their pathways and ports of entry. *The Australian and New Zealand Journal of Public Health,* **28**, 433–444.

Derraik, J. G. B. (2006). A scenario for invasion and dispersal of *Aedes albopictus* (Diptera: Culicidae) in New Zealand". *Journal of Medical Entomology,* **43**(1), 1–8.

Diagne, C. T.; Diallo, D.; Faye, O.; Ba, Y.; Faye, O.; Gaye, A.; et al. (2015). Potential of selected Senegalese *Aedes* spp. mosquitoes (Diptera: Culicidae) to transmit Zika virus. *BMC Infectious Diseases,* **15**: 492.

Diagne, C. T.; Faye, O.; Guerbois, M.; Knight, R.; Diallo, D.; Faye, O.; et al. (2014). Vector competence of Aedes aegypti and Aedes vittatus

(Diptera: Culicidae) from Senegal and Cape Verde Archipelago for west African lineages of chikungunya virus. *The American Journal of Tropical Medicine and Hygiene,* **91**(3), 635–641.

Diallo, D.; Diagne, C.; Hanley, K. A.; Sall, A. A.; Buenemann, M.; Ba, Y.; et al. (2012). Larval ecology of mosquitoes in sylvatic arbovirus foci in southeastern Senegal. *Parasites & Vectors,* **5**, 286.

Diallo, D.; Sall, A. A.; Diagne, C. T.; Faye, O.; et al. (2014). Zika virus emergence in mosquitoes in southeastern Senegal, 2011. *PLoS One,* **9**(10), e109442.

Diallo, M.; Sall, A. A.; Moncavo, A. C.; Ba, V.; Fernandez, Z.; Ortiz, D.; et al. (2005). Potential role of sylvatic and domestic African mosquito species in dengue emergence. *The American Journal of Tropical Medicine and Hygiene,* **73**, 445–449.

Diallo, M.; Thonnon, J.; Traore-Lamizana, M.; and Fontenille, D. (1999). Vectors of chikungunya virus in Senegal: Current data and transmission cycles. *The American Journal of Tropical Medicine and Hygiene,* **60**(2), 281–286.

Díez-Fernández, A.; Martínez-de la Puente, J.; Ruiz, S. et al. (2018). *Aedes vittatus* in Spain: current distribution, barcoding characterization and potential role as a vector of human diseases. *Parasites & Vectors,* **11**, 297. doi:10.1186/s13071-018-2879-4.

Díez-Fernández, A.; Puente, J. M.; Ruiz, S.; López, R. G.; Soriguer, R.; and Figuerola, J. (2018). *Aedes vittatus* in Spain: current distribution, barcoding characterization and potential role as a vector of human diseases. *Parasites & Vectors,* **11**, 297–214.

Dong, Y.; Manfredini, F.; and Dimopoulos, G. (2009). Implication of the mosquito midgut microbiota in the defense against malaria parasites. *PLoS* Pathogens, 5, e1000423.

Draft Assessment Report for *Aedes* (*Stegomyia*) *polynesiensis.* (2006). https://www.environment.gov.au › system › files › consultations › files

Dudchenko, O.; et al. (2017). De novo assembly of the *Aedes aegypti* genome using Hi-C yields chromosome-length scafolds. *Science,* **356**, 92–95.

Duguma, D.; et al. (2013). Bacterial communities associated with Culex mosquito larvae and two emergent aquatic plants of bioremediation importance. *PLoS One,* **8**, e72522.

Edwards, J. F.; Higgs, S.; and Beaty, B. J. (1998). Mosquito feeding-induced enhancement of Cache Valley virus (Bunyaviridae) infection in mice. *Journal of Medical Entomology,* **35**(3), 261–265.

Eritja, R.; Escosa, R.; Lucientes, J.; Marque, E.; Molina, R.; Roiz, D.; and Ruiz, S. (2005). Worldwide invasion of vector mosquitoes: Present European distribution and challenges for Spain. *Biological Invasions*, **7**, 87–97.

European Centre for Disease Prevention and Control (ECDC): *Aedes koreicus*.

Feng, L.-C. (1938). The tree hole species of mosquitoes of Peiping, China. *Chinese Medical Journal*, **2**, 503–525.

Ferede, G.; Tiruneh, M.; Abate, E.; Kassa, W. J.; Wondimeneh, Y.; Damtie, D.; and Tessema, B. (2018). Distribution and larval breeding habitats of *Aedes* mosquito species in residential areas of northwest Ethiopia. *Epidemiology and Health*, **40**, e2018015. doi:10.4178/epih.e2018015

Fonseca, D. M.; Campbell, S.; Crans, W. J.; Mogi, M.; Miyagi, I.; Toma, T.; et al. (2001). *Aedes* (*Finlaya*) *japonicus* (Diptera: Culicidae), a newly recognized mosquito in the United States: analyses of genetic variation in the United States and putative source populations. *Journal of Medical Entomology*, **38**, 135–146.

Fonseca, D. M.; Widdel, A. K.; Hutchinson, M.; Spichiger, S. E.; and Kramer, L. D. (2010). Fine-scale spatial and temporal population genetics of *Aedes japonicus*, a new US mosquito, reveal multiple introductions. *Molecular Ecology*, **19**, 1559–1572.

Forattini, O. P. (1986). Identificação de Aedes (Stegomyia) albopictus (Skuse) no Brasil. *Revista de Saúde Pública*, **20**, 244–245.

Freier, J. E. and Beier, J. C. (1984). Oral and transovarial transmission of La Crosse virus by Aedes atropalpus. *The American Journal of Tropical Medicine and Hygiene*, **33**(4), 708–714.

Freier, J. E. and Beier, J. C. (1984). Oral and transovarial transmission of la Crosse Virus by *Aedes atropalpus*. *The American Journal of Tropical Medicine and Hygiene*, **33**(4), 708–714.

Freier, J. E. and Beier, J. C. (1984). Oral and transovarial transmission of La Crosse virus by *Aedes atropalpus*. *The American Journal of Tropical Medicine and Hygiene*, **33**(4), 708–714.

Fu, Y.; et al. (2010). Artificial triple Wolbachia infection in *Aedes albopictus* yields a new pattern of unidirectional cytoplasmic incompatibility. *Applied and Environmental Microbiology*, **76**, 5887–5891.

Gaffigan, T. V.; Wilkerson, R. C.; Pecor, J. E.; Stoffer, J. A.; and Anderson, T. *Aedes Fredwardsius vittatus* (Bigot), *Systematic Catalog of Culicidae, Walter Reed Biosystematics Unit*. http://www.mosquitocatalog.org/taxon_descr.aspx?ID=17622

Gale, K. and Crampton, J. (1989). The ribosomal genes of the mosquito, *Aedes aegypti*. *European Journal of Biochemistry*, **185**, 311–317.

Gasperi, G.; Bellini, R.; Malacrida, A. R.; Crisanti, A.; Dottori, M.; and Aksoy, S. (2012). A new threat looming over the Mediterranean basin: emergence of viral diseases transmitted by *Aedes albopictus* mosquitoes. *PLOS Neglected Tropical Diseases*, **6**(9), e1836.

Genchi, C.; Rinaldi, L.; Mortarino, M.; Genchi, M.; and Cringoli, G. (2009). Climate and Dirofilaria infection in Europe. *Veterinary Parasitology*, **163**(4), 286–292.

Gendernalik, A.; Weger-Lucarelli, J.; Garcia Luna, S. M.; Fauver, J. R.; Rückert, C.; Murrieta, R. A.; et al. (2017). American *Aedes* vexans mosquitoes are competent vectors of Zika virus. *The American Journal of Tropical Medicine and Hygiene*, **96**(6), 1338–1340.

Gendrin, M.; et al. (2015). Antibiotics in ingested human blood affect the mosquito microbiota and capacity to transmit malaria. *Nature Communications*, **6**, 5921.

Germain, M.; Francy, D. B.; Monath, T. P.; Ferrara, L.; Bryan, J.; Salaun, J. J.; et al. Yellow fever in the Gambia, 1978–1979: Entomological aspects and epidemiological correlations. *The American Journal of Tropical Medicine and Hygiene*, **29**(5), 929–940.

Giangaspero, A.; Marangi, M.; Latrofa, M. S.; Martinelli, D.; Traversa, D.; Otranto, D.; et al. (2013). Evidences of increasing risk of dirofilarioses in southern Italy. *Parasitology Research*, **112**(3), 1357–1361.

Gilotra, S. K. and Shah, K. V. (1967). Laboratory studies on transmission of Chikungunya virus by mosquitoes. *American Journal of Epidemiology*, **86**(2), 379–385.

Gilotra, S. K.; Rozeboom, L. E.; and Bhattacharya, N. C. (1967). Observations on possible competitive displacement between populations of *Aedes aegypti* Linnaeus and *Aedes albopictus* Skuse in Calcutta. *Bulletin of the World Health Organization*, **37**, 437–446.

Gloria-Soria, A.; Ayala, D.; Bheecarry, A.; Calderon-Arguedas, O.; Chadee, D. D.; Chiappero, M.; and Powell, J. R. (2016). Global genetic diversity of *Aedes aegypti*. *Molecular Ecology*, **25**, 5377–5395.

Gouck, H. K. (1972). Host preferences of various strains of *Aedes aegypti* and *A. simpsoni* as determined by an olfactometer. *Bulletin of the World Health Organization*, **47**(5), 680–683.

Grimstad, P. R.; Garry, C. E.; and Defoliart, G. R. (1974). *Aedes hendersoni* and *Aedes triseriatus* (Diptera: Culicidae) in Wisconsin: Characterization of larvae, larval hybrids, and comparison of adult and hybrid mesoscutal patterns. *Annals of the Entomological Society of America,* **67**(5), 795–804.

Gubler, D. J. (1981). Transmission of Ross River virus by *Aedes polynesiensis* and *Aedes aegypti. The American Journal of Tropical Medicine and Hygiene,* **30**(6), 1303–1306.

Haddad, N.; Harbach, R. E.; Chamat, S.; and Bouharoun-Tayoun, H. (2007). Presence of *Aedes albopictus* in Lebanon and Syria. *Journal of the American Mosquito Control Association,* **23**(2), 226–228.

Hanson, S. M. and Craig, G. B. (1995). *Aedes albopictus* (Diptera: Culicidae) eggs: field survivorship during northern Indiana winters. *Journal of Medical Entomology,* **32**(5), 599–604.

Hare, R. E. (1898). The 1897 epidemic of dengue in North Queensland. *Australian Medical Gazette,* 98–107.

Harris, M. C.; Dotseth, E. J.; Jackson, B. T.; Zink, S. D.; Marek, P. E.; Kramer, L. D.; et al. (2015). La Crosse virus in *Aedes japonicus japonicus* mosquitoes in the Appalachian Region, United States. *Emerging Infectious Diseases,* **21**, 646–649.

Hawley, W. A.; Reiter, P.; Copeland, R. S.; Pumpuni, C. B.; and Craig, G. B. (1987). *Aedes albopictus* in North America: probable introduction in used tires from northern Asia. *Science,* **236**(4805), 1114–1116.

Hayes, E. B. (2009). Zika virus outside Africa. *Emerging Infectious Diseases,* **15**, 1347–1350.

Headlee, T. J. (1945). *The Mosquitoes of New Jersey and Their Control.* New Brunswick, New Jersey: Rutgers University Press, 316.

Hebert, P. D. N.; Cywinska, A.; Ball, S. L.; and de Waard, J. R. (2003). Biological identifications through DNA barcodes. *Proceedings of the Royal Society B: Biological Sciences,* **270**, 313–321.

Hedeen, R. A. (1953). The Biology of the Mosquito *Aedes atropalpus* Coquillett. *Journal of the Kansas Entomological Society,* **26**(1), 1–10.

Hill, L. A.; Davis, J. B.; Hapgood, G.; Whelan, P. I.; Smith, G. A.; and Ritchie, S. A. (2008) Rapid identification of *Aedes albopictus, Aedes scutellaris,* and *Aedes aegypti* life stages using real-time polymerase chain reaction assays. *American Journal of Tropical Medicine and Hygiene,* **79**, 866–875.

Ho, C. (1931). Study of the adult Culicidae of Peiping. *Bulletin Fan Memorial Institute of Biology,* **11**, 107–175.

Hofhuis, A.; Reimerink, J.; Reusken, C.; Scholte, E.-J.; Boer, Ad.; Takken, W.; et al. (2009). The hidden passenger of lucky bamboo: do imported *Aedes albopictus* mosquitoes cause dengue virus transmission in the Netherlands? *Vector-Borne and Zoonotic Diseases,* **9**(2), 217–220.

Horsfall, W. R. (1972). *Mosquitoes: Their Bionomics and Relation to Disease.* New York: Hafner Pub. Co., 723.

http://www.who.int/denguecontrol/faq/en/index5. html

https://animaldiversity.org/accounts/Aedes_triseriatus/

https://en.wikipedia.org/wiki/Aedes_aegypti

https://en.wikipedia.org/wiki/Aedes_japonicus.

https://en.wikipedia.org/wiki/List_of_Aedes_species

https://www.cabi.org/isc/datasheet/94897#tobiologyAndEcology

https://www.ecdc.europa.eu/en/disease-vectors/facts/mosquito-factsheets/aedes-atropalpus

https://www.ecdc.europa.eu/en/disease-vectors/facts/mosquito-factsheets/aedes-japonicus

https://www.ecdc.europa.eu/en/disease-vectors/facts/mosquito-factsheets/aedes-triseriatus

https://www.smsl.co.nz/site/southernmonitoring/files/NZB/Aedes%20 polynesiensis%20-%20profile%20Apr%2007.doc.pdf

Huang, M. (1977). Medical entomology studies—VIII. Notes on the taxonomic status of *Aedes vittatus.* (Diptera: Culicidae). *Contributions of the American Entomological Institute,* **14**(1), 1–132.

Hussain, M.; Frentiu, F. D.; Moreira, L. A.; O'Neill, S. L.; and Asgari, S. (2011). Wolbachia uses host microRNAs to manipulate host gene expression and facilitate colonization of the dengue vector Aedes aegypti. *Proceedings of the National Academy of Sciences,* **108**, 9250–9255.

Hussain, M.; Walker, T.; O'Neil, S. L.; and Asgari, S. (2013). Blood meal induced microRNA regulates development and immune associated genes in the dengue mosquito vector, *Aedes aegypti. Insect Biochemistry and Molecular Biology,* **43**(2), 146–152.

Integrated Taxonomic Information System (2017). *Aedes aegypti* (Linnaeus). (27 February 2017).

Irving-Bell, R. J.; Inyang, E. N.; and Tamu, G. (1991). Survival of Aedes vittatus (Diptera: Culicidae) eggs in hot, dry rock pools. *Tropical Medicine and Parasitology,* **42**(1), 63–66.

Jachowski, L. A. and Otto, G. F. (1953). Filariasis in American Samoa. V. Bionomics of the principal vector, Aedes polysiensis Marks. *American Journal of Hygiene,* **60**, 186–203.

James, M. T. and R. F. Harwood. (1969). *Hermsimedical Entomology*. New York: Macmillan Co., 484.

Joshi, V.; Angel, A.; Angel, B.; and Kucheria, K. (2013). Egg laying sites of *Aedes aegypti* and their elimination as the crucial etiological intervention to prevent dengue transmission in Western Rajasthan, India. *International Journal of Scientific Research*, **2**(12), 468–469.

Joshi, V.; Sharma, R. C.; Sharma, Y.; Adha, S.; Sharma, K.; Singh, H.; Purohit, A.; and Singhi, M. (2006). Importance of socio-economic status and tree hole distribution in *Aedes* mosquitoes (Diptera: Culicidae) in Jodhpur, Rajasthan, India. *Journal of Medical Entomology,* **43**(2), 330–336.

Juliano, S. A. and Lounibos, L. P. (2005) Ecology of invasive mosquitoes: effects on resident species and on human health. *Ecology Letters*, **8**(5), 558–574.

Juliano, S. A.; Lounibos, L. P.; and O'Meara, G. F. (2004). A field test for competitive effects of *Aedes albopictus* on *A. aegypti* in South Florida: differences between sites of coexistence and exclusion? *Oecologia*, **139**, 583–593.

Kalan, K.; Susnjar, J.; Ivovic, V.; and Buzan, E. (2017). First record of *Aedes koreicus* (Diptera, Culicidae) in Slovenia. *Parasitology Research,* **116**, 2355–2358.

Kambhampati, S.; Rai, K. S.; and Burgun, S. J. (1993). Unidirectional cytoplasmic incompatibility in the mosquito, *Aedes albopictus. Evolution*, **47**, 673–677.

Kamgang, B.; Brengues, C.; Fontenille, D.; Njiokou, F.; Simard, F.; et al. (2011). Genetic structure of the tiger mosquito, *Aedes albopictus*, in Cameroon (Central Africa). *PLoS One*, **6**, e20257. pmid:21629655.

Kampen, H. and Werner, D. (2014). Out of the bush: The Asian bush mosquito *Aedes japonicus japonicus* (Theobald, 1901) (Diptera, Culicidae) becomes invasive. *Parasites & Vectors,* **7**, 59.

Kampen, H.; Zielke, D.; and Werner, D. (2012). A new focus of *Aedes japonicus japonicus* (Theobald, 1901) (Diptera, Culicidae) distribution in western Germany: rapid spread or a further introduction event? *Parasites & Vectors,* **5**, 284.

Kaufman, M. G. and Fonseca, D. M. (2014). Invasion Biology of *Aedes japonicus japonicus* (Diptera: Culicidae). *The Annual Review of Entomology,* **59**, 31–49.

Kaufman, M. G.; Stanuszek, W. W.; Brouhard, E. A.; Knepper, R. G.; and Walker, E. D. (2012). Establishment of *Aedes japonicus japonicus* and its colonization of container habitats in Michigan". *Journal of Medical Entomology,* **49**(6), 1307–1317.

Kessel, J. F. (1971). A review of the filariasis control program in Tahiti from November 1967 to January 1968. *Bulletin of the World Health Organization*, **44**, 783–794.

Kim, H. C.; Chong, S. T.; O'Brien, L. L.; O'Guinn, M. L.; Turell, M. J.; Lee H-C; and Klein T. A. Seasonal prevalence of mosquitoes collected from light traps in the Republic of Korea in 2003. *Entomological Research*, **36**, 139–148.

King, W. L.; Bradley, G. H.; Smith, C. N.; and McDuffie, W. C. (1960). *A Handbook of the Mosquitoes of the Southeastern United States*. Handbook 173. Washington, DC: US Department of Agriculture; 1960.

Knight, K. L. (1947). The *Aedes (Finlaya) chrysolineatus* group of mosquitoes (Diptera: Culicidae). *Annals of the Entomological Society of America*, **40**(4), 624–649.

Knudsen, A.; Romi, R.; and Majori, G. (1996). Occurrence and spread in Italy of *Aedes albopictus*, with implications for its introduction into other parts of Europe. *Journal of the American Mosquito Control Association*, **12**(2), 177–183.

Koban, M. B.; Kampen, H.; Scheuch, D. E. et al. (2019). The Asian bush mosquito *Aedes japonicus japonicus* (Diptera: Culicidae) in Europe, 17 years after its first detection, with a focus on monitoring methods. *Parasites & Vectors*, **12**, 109.

Kraemer, Moritz U. G.; Reiner, Robert C.; Brady, Oliver J.; Messina, Jane P.; Gilbert, Marius; Pigott, David M.; Yi, Dingdong; Johnson, Kimberly; Earl, Lucas (2019). Past and future spread of the arbovirus vectors Aedes aegypti and Aedes albopictus". *Nature Microbiology*, **4** (5), 854–863.

Kröpelin, S.; Verschuren, D.; Lézine, A. M.; Eggermont, H.; Cocquyt, C.; Francus, P.; Cazet, J. P.; Fagot, M.; Rumes, B.; Russell, J. M.; Darius, F.; Conley, D. J.; Schuster, M.; von Suchodoletz, H.; and Engstrom, D. R. (2008). Climate-driven ecosystem succession in the Sahara: the past 6000 years. *Science*, **320**(5877):765–768.

Kurucz, K.; Kiss, V.; Zana, B.; Schmieder, V.; Kepner, A.; Jakab, F.; et al. (2016). Emergence of *Aedes koreicus* (Diptera: Culicidae) in an urban area, Hungary, 2016. *Parasitology Research*, **115**, 4687–4689.

Laird, M. (1956). Studies of mosquitoes and freshwater ecology in the South Pacific. *Bulletin of the Royal Society of New Zealand*, **6**, 1–213.

Laird, M.; Calder, L.; Thornton, R. C.; Syme, R.; Holder, P. W.; and Mogi, M. (1994). Japanese *Aedes albopictus* among four mosquito species reaching New Zealand in used tires. *Journal of the American Mosquito Control Association*, **10**, 14–23.

Lambrechts, L.; Scott, T. W.; and Gubler, D. J. (2010). Consequences of the expanding global distribution of *Aedes albopictus* for dengue virus transmission. *PLOS Neglected Tropical Diseases,* **4**(5), e646.

Lee, D. J.; Hicks, M. M.; Griffiths, M.; Debenham, M. I.; Bryan, J. H.; Russel, R. C.; Geary, M.; and Marks, E. N. (1987). *The Culicidae of the Australian Region*, Vol 4. Canberra: Australian Government Publishing Service.

Lee, V. H. and Moore, D. L. (1972). Vectors of the 1969 yellow fever epidemic on the Jos Plateau, Nigeria. *Bulletin of the World Health Organization,* **46**, 669–673.

Lewandowski Jr., H. B.; Hooper, G. R.; and Newson, H. D. (1980). Determination of some important natural potential vectors of dog heartworm in central Michigan. *Mosquito News*, **40**, 73–79.

Liu, J.; Ma, X.; Li, Z.; Wu, X.; and Sun, N. (2011). Risk analysis of *Aedes triseriatus* in China, in *Computer and Computing Technologies in Agriculture IV*, Springer Berlin Heidelberg, 465–472,

Liu, Y.; Zhou, Y.; Wu, J.; Zheng, P.; Li, Y.; Zheng, X.; et al. (2015). The expression profile of *Aedes albopictus* miRNAs is altered by dengue virus serotype -2 infection. *Cell Bioscience*, **5**, 16.

Lounibos, L. P. (1981). Habitat segregation among African treehole mosquitoes. *Ecological Entomology*, **6**, 129–154.

Lounibos, L. P. (2002). Invasions by insect vectors of human disease. *Annual review of Entomology*, **47**, 233–266.

Lounibos, L. P.; Escher, R. L.; and Lourenço-de-Oliveira, R. (2014). Asymmetric evolution of photoperiodic diapause in temperate and tropical invasive populations of *Aedes albopictus* (Diptera: Culicidae). *Annals of the Entomological Society of America,* **96**(4), 512–518.

Madon, M. B; Mulla, M. S.; Shaw, M. W.; Kluh, S.; and Hazelrigg, J. E. (2002). Introduction of *Aedes albopictus* (Skuse) in southern California and potential for its establishment. *Journal of Vector Ecology,* **27**(1), 149–154.

Maguire, T.; MacNamara, F. N.; Miles, J. A. R.; and Spears, G. F. S. (1971). Mosquito-borne infections in Fiji. II. Arthropod-borne virus infections. *The Journal of Hygiene*, **69**, 287–296.

Marks, E. N. (1951). The vector of filariasis in Polynesia: A change in nomenclature. *Annals of Tropical Medicine and Parasitology*, **45**, 137–140.

Mavale, M. S.; Ilkal, M. A.; and Dhanda, V. (1992). Experimental studies on the susceptibility of *Aedes vittatus* to dengue viruses. *Acta Virologica,* **36**(4), 412–416.

Mayoral, J. G.; Etebari, K.; Hussain, M.; Khromykh, A. A., and Asgari, S. (2014). Wolbachia infection modifies the profile, shuttling and structure of MicroRNAs in a mosquito cell line. *PLoS One*, **9**, e96107.

McBride, C. S.; Baier, F.; Omondi, A. B.; Spitzer, S. A.; Lutomiah, J.; Sang, R.; Ignell, R.; and Vosshall, L. B. (2014). Evolution of mosquito preference for humans linked to an odorant receptor. *Nature*, **515**(7526), 222–227.

McNeill, W. H. (1976). *Plagues and People*, New York: Doubleday.

Medlock, J.; Hansford, K.; Versteirt, V.; Cull, B.; Kampen, H.; Fontenille, D.; et al. (2015). An entomological review of invasive mosquitoes in Europe. *Bulletin of Entomological Research*, **105**(6), 637–663.

Mercado-Curiel, R. F.; Esquinca-Avilés, H. A.; Tovar, R.; Díaz-Badillo, A.; Camacho-Nuez, M.; Muñoz Mde, L. (2006). The four serotypes of dengue recognize the same putative receptors in *Aedes aegypti* midgut and *Ae. albopictus* cells. *BMC Microbiology*, **6**, 85.

Miller, B. R.; DeFoliart, G. R.; and Yuill, T. M. (1977). Vertical transmission of La Crosse virus (California encephalitis group): transovarial and Þlial infection rates in *Aedes triseriatus* (Diptera: Culicidae). *Journal of Medical Entomology*, **14**, 437–440.

Miyagi, I. (1971). Notes on the *Aedes* (*Finlaya*) *chrysolineatus* subgroup in Japan and Korea (Diptera: Culicidae). *Tropical Medicine*, **13**, 141–151.

Molaei, G.; Andreadis, T. G.; Armstrong, P. M.; and Diuk-Wasser, M. (2008). Host-feeding patterns of potential mosquito vectors in Connecticut, U.S.A.: molecular analysis of bloodmeals from 23 species of *Aedes, Anopheles, Culex, Coquillettidia, Psorophora,* and *Uranotaenia*. *Journal of Medical Entomology*, **45**(6), 1143–1151.

Molaei, G.; Farajollahi, A.; Scott, J. J.; Gaugler, R.; and Andreadis, T. G. (2009). Human bloodfeeding by the recently introduced mosquito, *Aedes japonicus japonicus*, and public health implications. *Journal of the American Mosquito Control Association*, **25**, 210–214.

Montarsi, F.; Martini, S.; Michelutti, A.; et al. (2019). The invasive mosquito *Aedes japonicus japonicus* is spreading in northeastern Italy. *Parasites & Vectors*, **12**, 120.

Moore, C. G.; McLean, R. G.; Mitchell, C. J.; Nasci, T. F.; Calisher, C. H.; Marfin, A. A.; et al. (1993). *Guidelines for arbovirus surveillance programs in the United States*. Fort Collins, Colorado: Division of Vector-Borne Infectious Diseases, National Center for Infectious Diseases, Centers for Disease Control and Prevention.

Moore, M.; Sylla, M.; Goss, L.; Burugu, M. W.; Sang, R.; Kamau, L. W.; et al. (2013). Dual African origins of global *Aedes aegypti* s.l. Populations revealed by mitochondrial DNA. *PLoS Neglected Tropical Diseases*, **7**, e2175.

Moreira, L. A.; et al. (2009). A Wolbachia symbiont in Aedes aegypti limits infection with dengue, chikungunya, and plasmodium. *Cell*, **139**, 1268–1278.

Mori, A.; Oda, T.; and Wada, Y. (1981). Studies on the egg diapause and overwintering of *Aedes albopictus* in Nagasaki. *Tropical Medicine*, **23**, 79–90.

Mourya, D. T. and Banerjee, K. (1987). Experimental transmission of chikungunya virus by Aedes vittatus mosquitoes. *Indian Journal of Medical Research*, **86**, 269–271.

Mousson, L.; et al. (2012). The native Wolbachia symbionts limit transmission of dengue virus in *Aedes albopictus*. *PLOS Neglected Tropical Diseases*, **6**.

Myiagi, I. (1971). Notes on the *Aedes (Finlaya) chrysolineatus* Subgroup in Japan and Korea (Diptera: Culicidae). *Tropical Medicine*, **13**(3), 141–151.

Nawrocki, S. and Hawley, W. (1987). Estimation of the northern limits of distribution of *Aedes albopictus* in North America. *Journal of the American Mosquito Control Association*, **3**(2), 314–317.

Nawrocki, S. J. and Craig Jr., G. B. (1989). Further extension of the range of the rock pool mosquito, *Aedes atropalpus*, via tire breeding. *Journal of the American Mosquito Control Association*, **5**(1), 110–114.

Ndiaye, E. H.; Fall, G.; Gaye, A.; et al. (2016). Vector competence of Aedes vexans (Meigen), Culex poicilipes (Theobald) and Cx. quinquefasciatus Say from Senegal for West and East African lineages of Rift Valley fever virus. *Parasites & Vectors*, **9**, 94.

Nelson, M. J. (1986). *Aedes aegypti*: Biology and Ecology. *Pan American Health Organization*. http://iris.paho.org/xmlui/handle/123456789/28514

Nene, V.; Wortman, J. R.; Lawson, D.; Haas, B.; Kodira, C.; Tu, Z. J.; et al. (2007). Genome sequence of *Aedes aegypti*, a major arbovirus vector. *Science*, **316**(5832), 1718–1723.

Novello, A.; White, D.; Kramer, L; and Trimarchi, C. (2000). West Nile virus activity: New York and New Jersey, 2000. *The Morbidity and Mortality Weekly Report*, **49**, 640–642.

Oliver, K. M.; Russell, J. A.; Moran, N. A.; and Hunter, M. S. (2003). Facultative bacterial symbionts in aphids confer resistance to parasitic wasps. *Proceedings of the National Academy of Sciences of the United States of America,* **100**, 1803–1807.

O'Malley, C. M. (1990). *Aedes vexans* (Meigen): An old foe. *Proceedings - New Jersey Mosquito Control Association*, 90–95.

O'Meara, G. F.; Evans Jr., L. F.; Gettman, A. D.; and Cuda, J. P. (1995). Spread of *Aedes albopictus* and decline of *Aedes aegypti* (Diptera: Culicidae) in Florida. *Journal of Medical Entomology,* **32**, 554–562.

Ondrejicka, D. A.; Locke, S. A.; Morey, K.; Borisenko, A. V.; and Hanner, R. H. (2014). Status and prospects of DNA barcoding in medically important parasites and vectors. *Trends in Parasitology,* **30**, 582–591.

Pampiglione, S.; Rivasi, F.; Angeli, G.; Boldorini, R.; Incensati, R. M.; Pastormerlo, M.; et al. (2001). Dirofilariasis due to Dirofilaria repens in Italy, an emergent zoonosis: report of 60 new cases. *Histopathology,* **38**(4), 344–354.

Pantuwatana, S.; Thompson, W. H.; Watts, D. M.; Yuill, T. M.; and Hanson, R. P. (1974). Isolation of La Crosse virus from field collected Aedes triseriatus larvae. *The American Journal of Tropical Medicine and Hygiene,* **23**, 246–250.

Paupy, C.; Delatte, H.; Bagny, L.; Corbel, V.; and Fontenille, D. (2009). *Aedes albopictus,* an arbovirus vector: from the darkness to the light. *Microbes and Infection,* **11**, 1177–1185.

Pelz, E. G. and Freier, J. E. (1990). Vertical transmission of St. Louis encephalitis virus to autogenously developed eggs of Aedes atropalpus mosquitoes. *Journal of the American Mosquito Control Association,* **6**(4), 658–661.

Peng, Z.; Li, H.; and Simons, F. E. (1998). Immunoblot analysis of salivary allergens in 10 mosquito species with worldwide distribution and the human IgE responses to these allergens. *The Journal of Allergy and Clinical Immunology,* **101**(4 Pt 1), 498–505.

Petersen, J. L. (1977). Behavioral differences in two subspecies of *Aedes aegypti* (Diptera L, Culicidae) in East Africa: PhD thesis, University of Notre Dame.

Peyton, E.; Campbell, S. R.; Candeletti, T. M.; Romanowski, M.; and Crans, W. J. (1999). *Aedes (Finlaya) japonicus japonicus* (Theobald), a new introduction into the United States. *Journal of the American Mosquito Control Association,* **15**(2), 238–241.

Pfitzner, W. P.; Lehner, A.; Hoffmann, D.; et al. (2018). First record and morphological characterization of an established population of *Aedes (Hulecoeteomyia) koreicus* (Diptera: Culicidae) in Germany. *Parasites & Vectors* **11,** 662.

Platt, K. B.; Tucker, B. J.; Halbur, P. G.; Tiawsirisup, S.; Blitvich, B. J.; Fabiosa, F. G.; et al. (2007). West Nile virus viremia in eastern chipmunks (Tamias striatus) sufficient for infecting different mosquitoes. *Emerging Infectious Diseases,* **13**(6), 831–837.

Poppert, S.; Hodapp, M.; Krueger, A.; Hegasy, G.; Niesen, W. D.; Kern, W. V.; et al. (2009). Dirofilaria repens infection and concomitant meningoencephalitis. *Emerging Infectious Diseases,* **15**(11), 1844–1846.

Powell, J. R. (2018). Mosquito-borne human viral diseases: Why *Aedes aegypti*? *The American Journal of Tropical Medicine and Hygiene,* **98**(6), 1563–1565. doi:10.4269/ajtmh.17-0866.

Powell, J. R. and Tabachnick, W. J. (2013). History of domestication and spread of *Aedes aegypti* - A review. *Memórias do Instituto Oswaldo Cruz,* **108**, 11–17.

Powell, J. R.; Gloria-Soria, A.; and Kotsakiozi, P. (2018). Recent history of *Aedes aegypti*: Vector genomics and epidemiology records. *Bioscience,* **68**(11), 854–860. doi:10.1093/biosci/biy119.

Pratt, H. D. and Moore, C. G. (1993). Mosquitoes of public health importance and their control, Self-Study Course 3013-G. US Department of Health & Human Services, Atlanta, GA.

Rajavel, A. R.; Natarajan, R.; and Vaidyanathan, K. (2006). Mosquitoes of the mangrove forests of India: Pt VI–Kundapur, Karnataka and Kannur, Kerala. *Journal of the American Mosquito Control Association,* **22**, 582–585.

Ramalingan, S. (1968). The epidemiology of filarial transmission in Samoa and Tonga. *Annals of Tropical Medicine and Parasitology,* **62**, 305–324.

Ramirez, J. L.; et al. (2014). Chromobacterium Csp_P reduces malaria and dengue infection in vector mosquitoes and has entomopathogenic and in vitroanti-pathogen activities. *PLOS Pathogens,* **10**, e1004398.

Reinert, J. F. (2000). New classification for the composite genus *Aedes* (Diptera: Culicidae: Aedini), elevation of subgenus *Ochlerotatus* to generic rank, reclassification of the other subgenera, and notes on certain subgenera and species. *Journal of the American Mosquito Control Association,* **16**(3), 175–188.

Reinert, J. F. (2000). Description of Fredwardsius, a new subgenus of *Aedes* (Diptera: Culicidae). *European Mosquito Bulletin,* **6,** 1–7.

Reinert, J. F.; Harbach, R. E.; and Kitching, I. J. (2004). Phylogeny and classification of Aedini (Diptera: Culicidae), based on morphological characters of all life stages. *Zoological Journal of the Linnean Society,* **142**, 289–368.

Reinert, J. F.; Harbach, R. E.; and Kitching, I. J. (2006). Phylogeny and classification of Finlaya and allied taxa (Diptera: Culicidae : Aedini) based on morphological data from all life stages. *The Zoological Journal of the Linnean Society*, **148**(1), 1–101.

Reiter, P. (1984). *Aedes albopictus* in Memphis, Tennessee: an achievment of modern transportation? *Mosquito News*, **44**, 396–399.

Reiter, P. (1998). *Aedes albopictus* and the world trade in used tires: 1988–1995 the shape of things to come? *Journal of the America Mosquito Control Association*, **14**, 83–94.

Reiter, P. and Darsie, R. F. (1984) *Aedes albopictus* in Memphis, Tennessee (USA): an achievement of modern transportation? *Mosquito News*, **44**(3), 296–399.

Reno, H. E. and Novak, R. J. (2005). Characterization of apyrase-like activity in *Ochlerotatus triseriatus*, *Ochlerotatus hendersoni*, and *Aedes aegypti*. *The American Journal of Tropical Medicine and Hygiene*, **73**(3), 541–545.

Restifo, R. A. and Lanzaro, G. C. (1980). The occurrence of *Aedes atropalpus* (Coquillett) breeding in tires in Ohio and Indiana. *Mosquito News*, **40**, 292–294.

Rezza, G. (2012). Aedes albopictus and the reemergence of Dengue. *BMC Public Health*, **12**, 72.

Ribeiro, J. M.; Nussenzveig, R. H.; and Tortorella, G. (1994). Salivary vasodilators of *Aedes triseriatus* and *Anopheles gambiae* (Diptera: Culicidae). *Journal of Medical Entomology*, **31**(5), 747–753.

Riccardo, F.; Monaco, F.; Bella, A.; Savini, G.; Russo, F.; Cagarelli, R.; et al. (2018). An early start of West Nile virus seasonal transmission: the added value of one heath surveillance in detecting early circulation and triggering timely response in Italy, June to July 2018. *Eurosurveillance*, **23**(32). doi: 10.2807/1560-7917.ES.2018.23.32.1800427.

Richard, V.; Paoaafaite, T.; and Cao-Lormeau, V.-M. (2016). Vector competence of *Aedes aegypti* and *Aedes polynesiensis* populations from French Polynesia for chikungunya virus. *PLoS Neglected Tropical Diseases*, **10**(5), e0004694.

Ritchie, S. A. (2014). Dengue vector bionomics: why *Aedes aegypti* is such a good vector, in *Dengue and Dengue Hemorrhagic Fever*, 2nd edition (Gubler, D., ed.). Oxfordshire, United Kingdom: CAB International, 455–480.

Rizzo, F.; et al. (2014). Molecular characterization of flaviviruses from field-collected mosquitoes in northwestern Italy, 2011–2012. *Parasites & Vectors*, **7**, 395.

Roiz, D.; Vázquez, A.; Seco, M. P.; Tenorio, A.; and Rizzoli, A. (2009). Detection of novel insect flavivirus sequences integrated in *Aedes albopictus* (Diptera: Culicidae) in Northern Italy. *Virology Journal*, **6**, 93.

Romi, R.; Sabatinelli, G.; Savelli, L. G.; Raris, M.; Zago, M.; and Malatesta, R. (1997). Identification of a North American mosquito species, *Aedes atropalpus* (Diptera: Culicidae), in Italy. *Journal of the American Mosquito Control Association*, **13**(3), 245–246.

Rosen, L. (1955). Observations on the epidemiology of human filariasis in French Oceania. *American Journal of Hygiene*, **61**, 219–248.

Rosen, L.; Roseboom, L. E.; Gubler, D. J.; Lien, J. C.; and Chaniotis, B. N. (1985). Comparative susceptibility of mosquito species and strains to oral and parenteral infection with dengue and Japanese encephalitis viruses. *The American Journal of Tropical Medicine and Hygiene*, **34**, 603–615.

Rudnick, A. and Chan, Y. C. (1965). Dengue Type 2 virus in naturally infected *Aedes albopictus* mosquitoes in Singapore. *Science*, **149**, 638–639.

Rueda, L. M.; Kim, H. C.; Klein, T. A.; Pecor, J. E.; Li, C.; Sithiprasasna, R.; et al. (2006). Distribution and larval habitat characteristics of Anopheles Hyrcanus group and related mosquito species (Diptera: Culicidae) in South Korea. *Journal of Vector Ecology*, **31**(1), 198–205.

Russel, R. C.; et al. (2005). *Aedes (Stegomyia) albopictus:* A dengue threat for southern Australia? *Communicable Diseases Intelligence*, **29**(3), 296–298.

Russell, R. C.; Webb, C. E.; and Davies, N. (2005). *Aedes aegypti* (L.) and *Aedes polynesiensis Marks* (Diptera: Culicidae) in Moorea, French Polynesia: A study of adult population structures and pathogen (Wuchereria bancrofti and Dirofilaria immitis) infection rates to indicate regional and seasonal epidemiological risk for dengue and filariasis. *Journal of Medical Entomology*, **42**(6), 1045–1056.

Saleh, F.; Kitau, J.; Konradsen, F.; Alifrangis, M.; Lin, C.-H.; Juma, S.; Mchenga, S. S.; Saadaty, T.; and Schiøler, K. L. (2018). Habitat characteristics for immature stages of *Aedes aegypti* in Zanzibar city, Tanzania. *Journal of the American Mosquito Control Association*, **34**(3), 190–200.

Salgueiro, P.; Serrano, C.; Gomes, B.; et al. (2019). Phylogeography and invasion history of *Aedes aegypti*, the Dengue and Zika mosquito vector in Cape Verde islands (West Africa). *Evolutionary Applications*, **12**, 1797–1811.

Samarawickrema, W. A.; Sone, F.; Kimura, E.; Self, L. S.; Cummings, R. F.; and Paulson, G. S. (1993). The relative importance and distribution of *Aedes polynesiensis* and *Ae. aegypti* larval habitats in Samoa. *Medical and Veterinary Entomology*, **7**, 27–36.

Sardelis, M. R. and Turell, M. J. (2001). *Ochlerotatus j. japonicus* in Frederick County, Maryland: discovery, distribution, and vector competence for West Nile virus. *Journal of the American Mosquito Control Association,* **17**, 737–147.

Sardelis, M. R.; Dohm, D. J.; Pagac, B.; Andre, R. G.; and Turell, M. J. (2002). Experimental transmission of eastern equine encephalitis virus by *Ochlerotatus j. japonicus* (Diptera: Culicidae). *Journal of Medical Entomology,* **39**, 480–484.

Sardelis, M. R.; Turell, M. J.; and Andre, R. G. (2002). Laboratory transmission of La Crosse virus by *Ochlerotatus j. japonicus* (Diptera: Culicidae). *Journal of Medical Entomology,* **39**, 635–639.

Sardelis, M. R.; Turell, M. J.; Andre, R. G. (2003). Experimental transmission of St. Louis encephalitis virus by *Ochlerotatus j. japonicus. Journal of the American Mosquito Control Association,* **19**, 159–162.

Schaffner, F.; Kaufmann, C.; Hegglin, D.; and Mathis, A. (2009). The invasive mosquito *Aedes japonicus* in Central Europe. *Medical and Veterinary Entomology,* **23**(4), 448–451.

Schaffner, F.; Medlock, J. M.; and Van Bortel, W. (2013). Public health significance of invasive mosquitoes in Europe. *Clinical Microbiology and Infection,* **19**, 685–692.

Schaffner, F.; Vazeille, M.; Kaufmann, C.; Failloux, A. B.; and Mathis, A. (2011). Vector competence of *Aedes japonicus* for chikungunya and dengue viruses. *The Journal of the European Mosquito Control Association,* **29**, 141–142.

Scholte, E. J.; Den Hartog, W.; Braks, M.; Reusken, C.; Dik, M.; and Hessels, A. (2009). First report of a North American invasive mosquito species *Ochlerotatus atropalpus* (Coquillett) in the Netherlands, 2009. *Eurosurveillance,* **14**(45), 19400.

Scott, J. J.; VfcNetty, J. R.; and Crans, W. J. (1999). *Aedes japonicus* overwinters in New Jersey. *Vector Ecol Newsl,* **30**(2), 6–7.

Service, M. W. (1970). Studies on the biology and taxonomy of Aedes (Stegomya) vittatus (Bigot) (Diptera: Culicidae) in northern Nigeria. *Transactions of the Royal Entomological Society of London,* **122**, 101–143.

Service, M. W. (1974). Survey of the relative prevalence of potential yellow fever vectors in northwest Nigeria. *Bulletin of the World Health Organization,* **50**, 487–494.

Shields, S. E. (1938). Tennessee Valley mosquito collections. *Journal of Economic Entomology,* **311**, 426–480.

Silaghi, C.; Beck, R.; Capelli, G.; Montarsi, F.; and Mathis, A. (2017). Development of *Dirofilaria immitis* and *Dirofilaria repens* in *Aedes japonicus* and *Aedes geniculatus*. *Parasites & Vectors,* **10**, 94.

Singh, K. R. P. (1967). Cell cultures derived from larvae of *Aedes albopictus* (Skuse) and *Aedes aegypti* (L.). *Current Science,* **36**(19), 506–508.

Sinkins, S. P.; Braig, H. R.; and O'Neill, S. L. (1995). Wolbachia superinfections and the expression of cytoplasmic incompatibility. *Proceedings of the Royal Society B: Biological Sciences,* **261**, 325–330.

Skuse, F. A. A. (1894). The banded mosquito of Bengal. *Indian Museum Notes,* **3**(5), 20.

Slonchak, A.; Hussain, M.; Torres, S.; Agari, S.; and Khromykh, A. A. (2014). Expression of mosquito micro RNA Aae-miR-2940-5p downregulated in response to West Nile virus infection to restrict viral replication. *Journal of Virology,* **88** (15), 8457–8467.

Smith, C. E. (1956). The history of dengue in tropical Asia and its probable relationship to the mosquito *Aedes aegypti*. *The American Journal of Tropical Medicine and Hygiene,* **59**(10), 243–251.

Sudeep, A. B. and Shil, P. (2017). Aedes vittatus (Bigot) mosquito: An emerging threat to public health. *Journal of Vector Borne Diseases,* **54**, 295–300.

Suh, E.; et al. (2009). Pathogenicity of life-shortening Wolbachia in *Aedes albopictus* after transfer from Drosophila melanogaster. *Applied and Environmental Microbiology,* **75**, 7783–7788.

Symes, C. B. (1961). A note on vectors of filariasis in the South Pacific. World Health Organization, WHO/FIL29.

Symes, C. B. and Mataika, J. U. (1959). Observations on Microfilaria fijiensis from fruits bats in Fiji. *Journal of Heminthology,* **33**, 223–232.

Szymczak, L. J. and Rai, K. S. (1987). Genetic differentiation in the *Aedes atropalus* complex. II. Chromosomal divergence between *Ae. atropalus* and *Ae. epactius*. *Journal of Genetics,* **66**(1), 33–34.

Tabachnick, W. J. (1991). Evolutionary genetics and arthropod-borne disease: the yellow fever mosquito. *American Entomologist,* **37**, 14–24.

Tabachnick, W. J.; Connelly, C. R.; and Smartt, C. C. (2006). Blood feeding insect series: yellow fever. *EDIS*. http://edis.ifas.ufl.edu/in659 (27 February 2017).

Tabachnick, W. J.; Munstermann, L. E.; and Powell, J. R. (1978). Genetic distinctness of sympatric forms of *Aedes aegypti* in East Africa. *Evolution,* **33**, 287–295.

Takashima, I. and Rosen, L. (1989). Horizontal and vertical transmission of Japanese encephalitis virus by *Aedes japonicus* (Diptera: Culicidae). *Journal of Medical Entomology,* **26**, 454–458.

Takashima, I. and Rosen, L. (1989). Horizontal and vertical transmission of Japanese encephalitis virus by *Aedes japonicus* (Diptera: Culicidae). *Journal of Medical Entomology,* **26**(5), 454–458.

Tanaka, K.; Mizusawa, K.; and Saugstad, E. S. (1979). A revision of the adult and larval mosquitoes of Japan (including the Ryukyu Archipelago and the Ogasaware Islands) and Korea (Diptera: Culicidae). *Contributions of the American Entomological Institute,* **16**, 1–987.

Tanaka, K.; Mizusawa, K.; and Saugstad, E. S. (1979). A revision of the adult and larval mosquitoes of Japan (including the Ryukyu Archipelago and the Ogasawara islands) and Korea (Diptera: Culicidae). *Contributions of the American Entomological Institute*, **16**, 1–987.

Tedjou, A. N.; Kamgang, B.; Yougang, A. P.; Njiokou, F.; and Wondji, C. S. (2019). Update on the geographical distribution and prevalence of *Aedes aegypti* and *Aedes albopictus* (Diptera: Culicidae), two major arbovirus vectors in Cameroon. *PLOS Neglected Tropical Diseases,* **13**(3), e0007137.

Tesh, R. B. and Gubler, D. J. (1975). Laboratory studies of transovarial transmission of La Crosse and other arboviruses by *Aedes albopictus* and Culex fatigans. *The American Journal of Tropical Medicine and Hygiene,* **24**, 876–880.

Tewari, S. C.; Thenmozhi, V.; Katholi, C. R.; Manavalan, R.; Munirathinam, A.; and Gajanana, A. (2004). Dengue vector prevalence and virus infection in a rural area in south India. *Tropical Medicine & International Health,* **9**(4), 499–507.

Theobald, F. V. (1901). *A Monograph of the Culicidae or Mosquitoes.* London: British Museum (Natural History). Vol. 1: 385; http://www.mosquitocatalog.org/files/pdfs/131700-32.PDF Archived 2016-03-25 at the Wayback Machine.

Thompson, P. H. and Dicke, R. J. Sampling studies with *Aedes vexans* and some other Wisconsin *Aedes* (Diptera: Culcidae). *Annals of the Entomological Society of America,* **58**(6), 927–930.

Tippelt, L.; Werner, D.; and Kampen, H. (2019). Tolerance of three *Aedes albopictus* strains (Diptera: Culicidae) from different geographical origins towards winter temperatures under field conditions in northern Germany. *PLoS One*, **14**(7), e0219553.

Turell, M. J.; Byrd, B. D.; and Harrison, B. A. (2013). Potential for populations of *Aedes j. japonicus* to transmit Rift Valley fever virus in the USA. *Journal of the American Mosquito Control Association,* **29**, 133–137.

Turell, M. J.; Dohm, D. J.; Sardelis, M. R.; Oguinn, M. L.; Andreadis, T. G.; and Blow, J. A. (2005). An update on the potential of north American mosquitoes (Diptera: Culicidae) to transmit West Nile virus. *Journal of Medical Entomology,* **42**(1), 57–62.

Turell, M. J.; Dohm, D. J.; Sardelis, M. R.; Oguinn, M. L.; Andreadis, T. G.; and Blow, J. A. (2005). An update on the potential of north American mosquitoes (Diptera: Culicidae) to transmit West Nile Virus. *Journal of Medical Entomology,* **42**(1), 57–62.

Turell, M. J.; O'Guinn, M. L.; Dohm, D. J.; and Jones, J. W. (2001). Vector competence of North American mosquitoes (Diptera: Culicidae) for West Nile virus. *Journal of Medical Entomology,* **38**(2), 130–134.

Veronesi, E.; Paslaru, A.; Silaghi, C.; Tobler, K.; Glavinic, U.; Torgerson, P.; et al. (2018). Experimental evaluation of infection, dissemination, and transmission rates for two West Nile virus strains in European *Aedes japonicus* under a fluctuating temperature regime. *Parasitology Research,* **117**, 1925–1932.

Versteirt, V.; De Clercq, E. M.; Fonseca, D. M.; Pecor, J.; Schaffner, F.; Coosemans, M.; et al. (2012). Bionomics of the established exotic mosquito species *Aedes koreicus* in Belgium, Europe. *Journal of Medical Entomology,* **49**(6), 1226–1232.

Wagner, S.; Mathis, A.; Schönenberger, A. C.; Becker, S.; Schmidt-Chanasit, J.; Silaghi, C.; et al. (2018). Vector competence of field populations of the mosquito species *Aedes japonicus japonicus* and *Culex pipiens* from Switzerland for two West Nile virus strains. *Medical and Veterinary Entomology,* **32**, 121–124.

Walker, N. (1992). The eastern treehole mosquito, *Aedes triseriatus. Wing Beats,* **3**(2), 17.

Walker, T.; et al. The wMel Wolbachia strain blocks dengue and invades caged *Aedes aegypti* populations. *Nature,* **476**, 450.

Walter Reed Biosystematics Unit: *Aedes polynesiensis.* Archived from the original on 2010-06-24.

Watson, M. S. (1967). *Aedes* (*Stegomyia*) *albopictus*: a literature review. Archived 22 October 2014 at the Wayback Machine. Dep. Army, Ft. Detrick, MD, *Miscellaneous Publications,* **22**, S.1–S.38.

Watts, D. M.; Pantuwatana, S.; DeFoliart, G. R.; Yuill, T. M.; and Thompson, W. H. (1973). Transovarial transmission of La Crosse virus (California encephalitis group) in the mosquito, *Aedes triseriatus*. *Science*, **182**, 1140–1141.

Watts, D. M.; Thompson, W. H.; Yuill, T. M.; DeFoliart, G. R.; and Hanson, R. P. (1974). Overwintering of La Crosse virus in Aedes triseriatus. *The American Journal of Tropical Medicine and Hygiene*, **23**, 694–700.

Weetman, D.; Kamgang, B.; Badolo, A.; Moyes, C. L.; Shearer, F. M.; et al. (2018). *Aedes* mosquitoes and *Aedes*-borne arboviruses in Africa: current and future threats. *International Journal of Environmental Research and Public Health*, **15**(2), E220.

Werner, D.; Zielke, D. E.; and Kampen, H. (2016). First record of *Aedes koreicus* (Diptera: Culicidae) in Germany. *Parasitology Research*, **115**, 1331–1334.

White, D. J. and White, C. P. (1980). *Aedes atropalpus* breeding in artificial containers in Suffolk County, New York. *Mosquito News*, **40**, 106–110.

Wilkerson, R.; Linton, Y.-M.; Fonseca, D.; Schultz, T.; Price, D.; and Strickman, D. (2015). Making mosquito taxonomy useful: A stable classification of tribe Aedini that balances utility with current knowledge of evolutionary relationships. *PLoS One*, **10**, e0133602.

Williams, D. D.; MacKay, S. E.; Verdonschot, R. C.; and Tacchino, P. J. (2007). Natural and manipulated populations of the treehole mosquito, *Ochlerotatus triseriatus*, at its northernmost range limit in southern Ontario, Canada. *Journal of Vector Ecology*, **32**(2), 328–335.

Williges, E.; Farajollahi, A.; Scott, J. J.; Mccuiston, L. J.; Crans, W. J.; and Gaugler, R. (2008). Laboratory colonization of *Aedes japonicus japonicus*. *Journal of the American Mosquito Control Association*, **24**, 591–593.

Wu, J. Y.; Lun, Z. R.; James, A. A.; and Chen, X. G. (2010). Dengue fever in mainland China. *The American Journal of Tropical Medicine and Hygiene*, **83**(3), 664–671.

Xi, Z., Ramirez, J. L., and Dimopoulos, G. (2008). The *Aedes aegypti* toll pathway controls dengue virus infection. *PLOS Pathogens*, **4**, e1000098.

Xi, Z.; et al. (2006). Interspecific transfer of Wolbachia into the mosquito disease vector *Aedes albopictus*. *Proceedings of the Royal Society B: Biological Sciences*, **273**, 1317–1322.

Yang, F.; Chan, K.; Marek, P. E.; Armstrong, P. M.; Liu, P.; Bova, J. E.; et al. (2018). Cache Valley virus in *Aedes japonicus japonicus* mosquitoes, Appalachian region, United States. *Emerging Infectious Diseases*, **24**, 553–557.

Yee, D. A. (2008). Tires as habitats for mosquitoes: a review of studies within the eastern United States. *Journal of Medical Entomology*, **45**(4), 581–593.

Zahouli, J. B. Z.; Utzinger, J.; Adja, M. A.; Müller, P.; Malone, D.; Tano, Y; et al. (2016). Oviposition ecology and species composition of *Aedes* spp. and *Aedes aegypti* dynamics in variously urbanized settings in arbovirus foci in southeastern Côte d'Ivoire. *Parasites & Vectors,* **9**, 523.

Zhang, G.; Hussain, M.; and Agari, S. (2014). Regulation of arginine methyl transferase 3 by a Wolbachia-induced microRNA in Aedes aegypti and its effect on Wolbachia and dengue virus replication. *Insect Biochemistry and Molecular Biology*, **53**, 81–88.

Zielke, D. E.; Ibáñez-Justicia, A.; Kalan, K.; Merdić, E.; Kampen, H.; and Werner, D. (2015). Recently discovered *Aedes japonicus japonicus* (Diptera: Culicidae) populations in The Netherlands and northern Germany resulted from a new introduction event and from a split from an existing population. *Parasites & Vectors,* **8**, 40.

Chapter 3

Viral Pathogens: A General Account

Vinod Joshi,[a] Bennet Angel,[a] Annette Angel,[b] Neelam Yadav,[c] and Jagriti Narang[d]

[a]*Amity Institute of Virology and Immunology, Amity University, Sector 125, Noida, India*

[b]*Division of Zoonosis, National Centre for Disease Control, 22 Sham Nath Marg, Civil Lines, Delhi, India*

[c]*Centre for Biotechnology, Maharshi Dayanand University, Rohtak, India*

[d]*Department of Biotechnology, Jamia Hamdard University, New Delhi, India*

annetteangel_15@yahoo.co.in, vinodjoshidmrc@gmail.com, bennetangel@gmail.com

The book, as its title *Small Bite, Big Threat: Deadly Infections Transmitted by* Aedes *Mosquitoes* suggests, focuses on the two main components of disease transmission, the mosquito *Aedes* and the pathogen that it carries. The first two chapters throw light on the various types of mosquitoes that inhabit Earth and distribution of different species of *Aedes* mosquitoes across the globe. To proceed further, it is important to have a general understanding of the different viral pathogens. Therefore, this chapter presents a brief account of viruses spread by arthropods, in which insects, specifically mosquitoes, play a major role. It presents a general introduction on

Small Bite, Big Threat: Deadly Infections Transmitted by Aedes *Mosquitoes*
Edited by Jagriti Narang and Manika Khanuja

ISBN 978-981-4800-86-0 (Hardcover), 978-1-003-00329-8 (eBook)
www.jennystanford.com

viral pathogens and focuses mainly on the second component, i.e., transmission of pathogens, especially viruses, by *Aedes* mosquitoes.

3.1 Introduction

Infectious diseases are a major cause of concurrent epidemics, pandemics, and seasonal outbreaks throughout the globe. With many diseases, mortality and morbidity are being reported everywhere along with fluctuating disability-adjusted life year (DALY) (nearly 30% of 149 billion DALYs are lost every year due to this) (Taylor and Latham, 2001; WHO, 2005). Studies have reported occurrence of many infectious diseases for the past three decades, which have been found to be viral in nature and more precise to be of zoonotic origin (Cleaveland et al., 2001; Dikid et al., 2013; Wolfe et al., 2007).

Animals occupied the major portion of land on this planet millions of years ago. With the concept of housing and family living, we humans, who are not even 10-million-year old, have built dwellings for ourselves encroaching animal habitats even in peripheral dense forests. A fact that goes unseen here is the interference of humans in the lives of animals, which has unknowingly led to an increase in the diseases of zoonotic origin, some of which are fatal, such as Hantavirus, Crimean–Congo hemorrhagic fever (CCHF), Zika virus, Ebola, and dengue. With continuous increase in population and simultaneous movement toward dense forests for occupying unreached areas, we are entering crucial niches where species conservation has been safeguarded since long, be it macro- or microniches. This can be supplemented with the consideration that names of the diseases mentioned earlier have been kept based on their origin or initial occurrence, such as Ebola (a river in Africa), West Nile (a river in the West Nile area of Africa), and Crimean–Congo (two African regions). Powell and Tabachnick (2013) have rightly referred to this with the extinction of the invaded species due to the encroachment of humans to the native habitat of species which also resulted into evolution of domestication or commensalism over the other.

Even Lounibos (2002) reported the invasiveness of insect vectors. Vectors play a very important role in disease transmission or, in other words, bridging an interacting step of pathogen with the host system (human system).

Viruses form the major group of these disease-causing pathogens, and there are more than 500 viruses listed in the international catalogue of viruses that have been isolated from mosquitoes. Interestingly, a small proportion of these viruses falls in the category of arboviruses. Arboviruses is a jargon used since the early 1940s (Reeves, 2001) for viruses that replicate intracellularly in both vertebrate and invertebrate hosts. As per the World Health Organization (WHO, 1967), the basic requirement for defining an arbovirus is that both viral replication in the phyla of hosts and viral transmission by blood-sucking arthropod to vertebrate hosts should demonstrate viremia. In a later report by the WHO in 1985, the definition was modified to include direct transmission as an alternative mode of transmission (WHO, 1985). So hypothetically, the disease transmission cycle will revolve around three essential components: the virus, the vector, and the vertebrate host. Many researchers, including medical entomologists, epidemiologists, and virologists alike, have raised fundamental questions ranging from the advantages of such a complicated mode of transmission to its impact on the genetics of viruses (Kuno and Chang, 2005).

For species, especially insects, depending on vertebrate hosts for blood meal, the colonization and population density of the human host seem preferable and beneficial to the earlier dependent one (i.e., the animal host). Some have even referred to this phenomenon as switching from zoophagy to anthropophagy. Disease profiles seen in the case of vector-borne diseases are also commonly seen to infect some animal population, specifically of our ancestral origin.

"When to this is added the fact that the more closely a mosquito is associated with man, the more is it the subject of prejudice and misconception, it follows that the prevailing conception of *A. aegypti* in the minds of the general run of entomologists may well be more remote from reality than in the case of most other mosquitoes" (Mattingly, 1957).

The listing of infectious diseases has been growing by and large, categorizing them into emerging and re-emerging diseases (1407 species reported, of which 177 species are headed under emerging and re-emerging category). Of this, more than 30 new infectious agents have been reported from different parts of the world in the last three decades only; 60% of which are zoonotic in origin (Bhatia et al., 2012; Dikid et al., 2013; Fauci, 2001; Taylor and Latham, 2001; WHO, 2005).

This chapter discusses only the vector-borne viral pathogens and related diseases such as dengue, chikungunya, Zika, West Nile fever, and yellow fever. All these are caused by viruses and transmitted through arthropod vectors of zoonotic origin. A brief discussion on their origin and distribution is presented here.

3.2 Viruses Transmitted by Arthropods

Arthropod-borne diseases are transmitted through arthropods, including mosquitoes [dengue, chikungunya, Zika, Japanese encephalitis (JE), Venezuelan equine encephalitis (VEE)], flea (bubonic plague), mites (scabies, scrub typhus), ticks (Kyasanur Forest disease, CCHF, encephalitis), louse (endemic typhus, relapsing fever), sand flies (Rift Valley fever), etc. (Fig. 3.1). Of these, insects act as the major pathogen-transmitting tool and also mainly viral pathogens (designated as arboviruses). The WHO has defined arboviruses as "Viruses that are maintained in nature principally, or to an important extent, through biological transmission between susceptible vertebrate host by hematophagous arthropods or through transovarial and possibly venereal transmission in arthropoda."

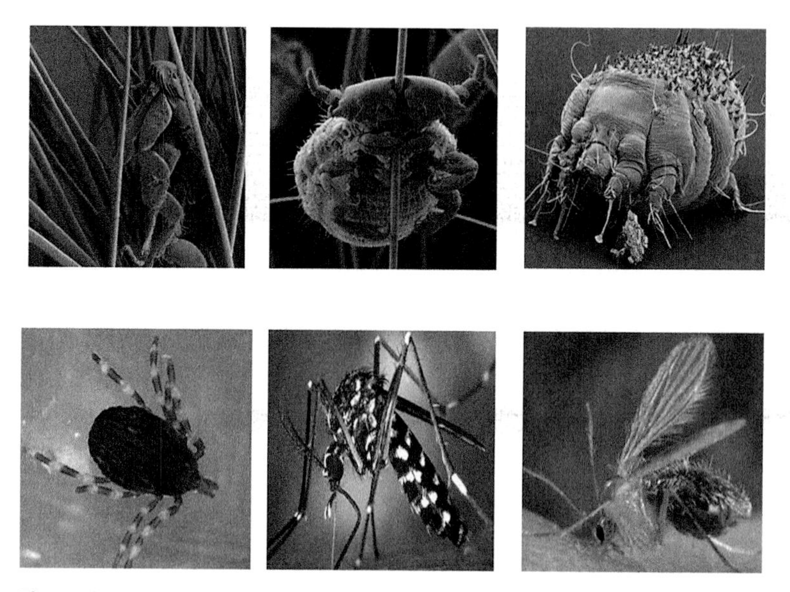

Figure 3.1 Vectors involved in virus transmission: fleas, lice, mites, ticks, mosquitoes, and sandflies. Figure taken from www.pinterest.com.

These arboviruses belong to three families:

1. Togaviruses, which cause chikungunya, eastern equine encephalitis, western equine encephalitis, VEE, etc.
2. Bunyaviruses, which cause sandfly fever, Rift Valley fever, CCHF, etc.
3. Flaviviruses, which cause yellow fever, dengue, JE, West Nile fever, Zika fever, etc.

For some of the pathogens, reservoir hosts have also been identified, while for some, the reservoir hosts remain unknown; for example, monkeys (chikungunya virus, yellow fever virus, and in some reports dengue virus), pigs and birds (JE virus), birds (Sindbis virus), rodents (VEE virus), etc.

3.2.1 Togaviruses

According to the Baltimore classification of viruses, Togaviruses fall under Group IV, which consists of single-stranded (+ve) RNA viruses. The family Togaviridae includes two genera: Alphaviruses and Rubiviruses. Their genome is linear and non-segmented and approximately 10,000–12,000 nucleotides long. The virus has an envelope and a nucleocapsid. It is icosahedral in shape constituted by 240 monomers. Glycoprotein spikes are found throughout the surface, which help to attach efficiently to the host cell surface (Fig. 3.2).

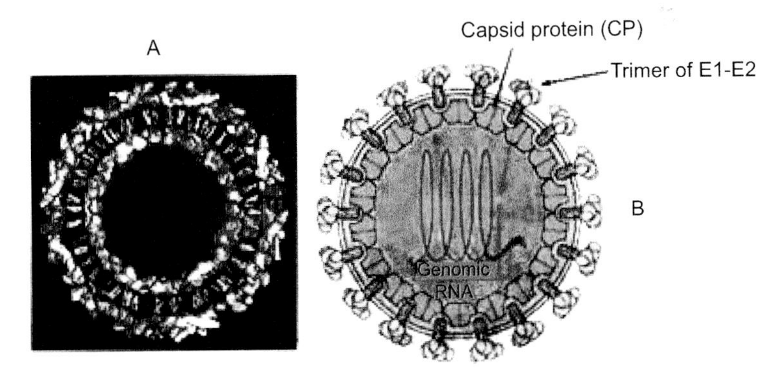

Figure 3.2 Structure of a Togavirus. Figure taken from www.biologydiscussion.com/viruses/animal-viruses/togaviruses-structure-and-replication-microbiology/65689.

Viruses that fall under this family are as follows:

3.2.1.1 Chikungunya virus

The chikungunya virus is named so as people infected by it cannot walk properly or rather walk stooped due to severe joint pain (arthralgia). The word chikungunya means "to become contorted." It is widely spread in the African and South East Asian countries, including India. In 2015, America reported 693,489 cases and 146,914 laboratory confirmed cases of chikungunya (WHO website), of which Brazil reported most cases (Fig. 3.3). The disease is characterized by sudden fever, rashes, pain in joints, and nausea. Though the patients are known to recover, yet some may continue to have muscular pain and feel difficulty in walking for months and even years. *Aedes aegypti* and *Aedes albopictus* are known to transmit this disease in humans. However, some reports suggest birds and monkeys to be reservoir host. At present, there is no chemotherapy or vaccine available against this disease.

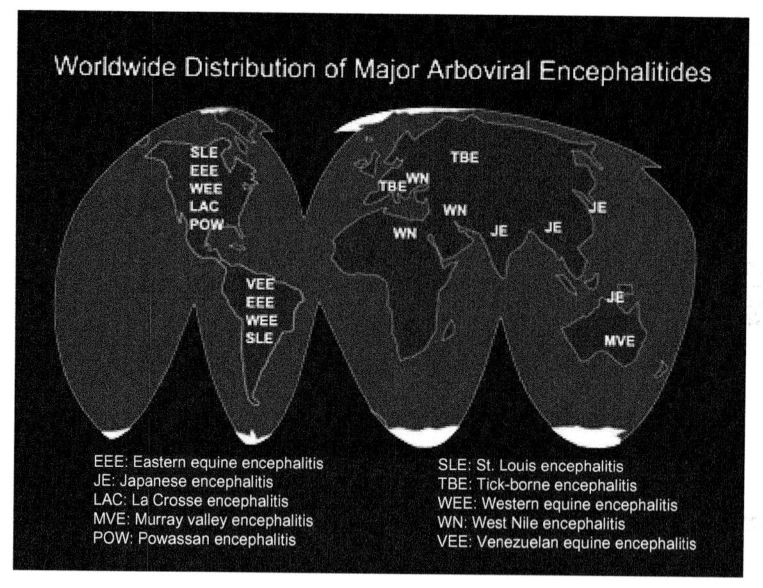

Figure 3.3 Worldwide distribution of the different arboviral encephalitis. Figure taken from http://afludiary.blogspot.com/2011/08/mmwr-arboviral-disease-surveillance.html. Copyright 2006–2019 Michael P. Coston. Awesome Inc. theme.

3.2.1.2 Eastern equine encephalitis virus

Also called EEEV or triple E, this virus was first isolated from horse brain. It caused the death of children during an outbreak in the United States in 1938. High fever, muscular pain, mental alterations, headaches, photophobia, and seizures are associated with the onset of this disease. Since the virus infects the brain, persons who survive this disease may have some physical and mental impairment. The virus is a single-stranded positive-sense RNA genome. The diameter is 60–65 nm. The virus has four lineages of which Group I is the most common followed by Group IIa, IIb, and III.

3.2.2 Bunyaviruses

This is the largest family containing RNA viruses. According to the Baltimore classification of viruses, Bunyaviruses fall under Group V, which consists of single-stranded (–ve) RNA viruses. A total of 330 viruses are under this group (Fig. 3.4). These are enveloped, about 90–100 nm in diameter. The size of the genome is approximately 10.5–22.7 bp. Their genome is tripartite genome made up of large (L), medium (M), and small (S) RNA segments (Guu et al., 2012). The L codes for RNA-dependent RNA polymerase; M codes for glycoproteins, which form the surface projections; and S codes for nucleocapsid. The genome is 10.5 to 22.7 bp. It includes four genera, *Bunyaviruses, Phlebovirus, Nairovirus*, and *Hantavirus*, which constitutes 35 serogroups with at least 304 types and subtypes (Shope, 1996).

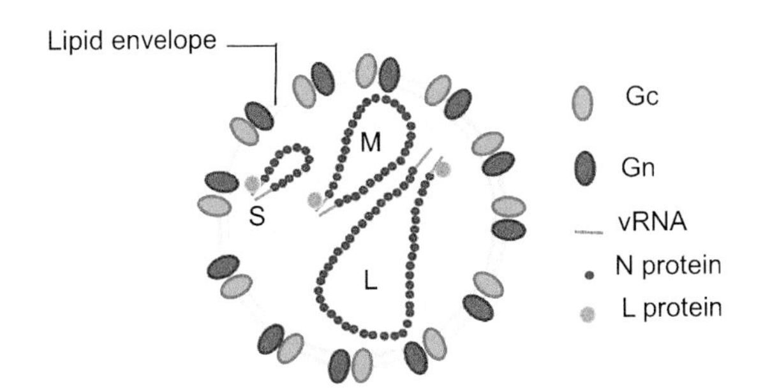

Figure 3.4 Schematic diagram of a Bunyavirus virion. Figure courtesy: Veronica Rezelj, taken from http://cvr.academicblogs.co.uk/bunyaviruses-we-are-one-big-family/.

The viruses that fall under this family are as follows:

3.2.2.1 Crimean–Congo hemorrhagic fever virus

The CCHF virus belongs to *Nairoviruses*, which are basically tick-borne viruses that cause hemorrhagic conditions along with tachycardia, lymphadenopathy, petechial rashes, and ecchymoses, finally leading to pulmonary failure. Usually mortality rate associated with this virus is 30%. The virus is known to be transmitted by small hard ticks (Ixodid) specifically of the genus *Hyalomma*, which are residents of domestic livestock population. So health workers/owners/caretakers of domestic livestock are mainly at risk to the disease caused by this virus, that is, CCHF. The virus is 1000–1500 BC old in origin and was first reported in Tajikistan in the 12th century. In India, the disease was first reported in 2011 (Fig. 3.5).

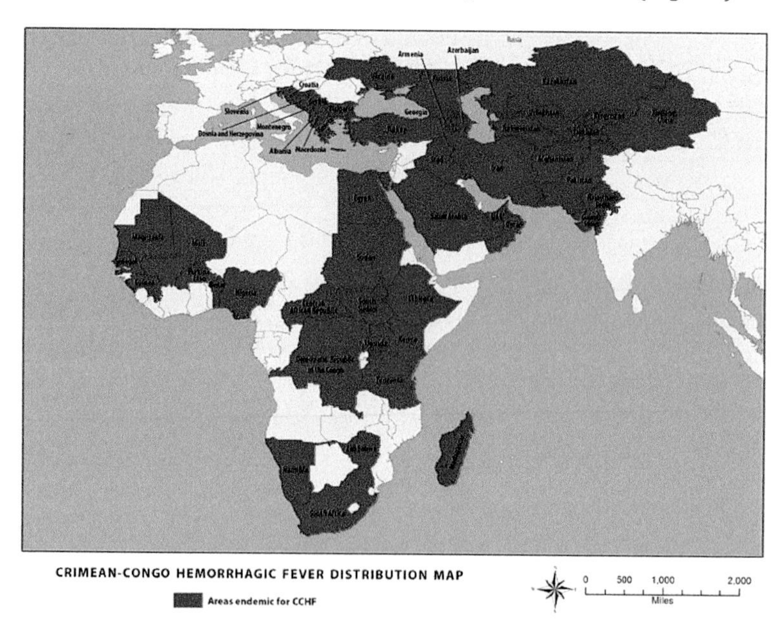

Figure 3.5 Distribution of CCHF outbreaks in different parts of the world. Figure taken from www.cdc.gov/vhf/crimean-congo/outbreaks/distribution-map.html.

3.2.2.2 Rift Valley fever virus

The virus belongs to the genus *Phlebovirus* and is basically a disease of the livestock, which may infect humans in close contact with the infected livestock (with blood and animal tissues, by consuming unpasteurized milk). Though the virus causes mild infection and hepatic liver in humans, 8–10% may report eye disease, encephalitis, hemorrhage conditions, etc. The Rift Valley fever was first reported in Kenya's Rift Valley in 1910 and since then is seen in African areas. Mosquitoes such as *Aedes* and *Culex* are also known to transmit the disease. Since the infection in human lasts only for 2 days, no specific treatment is required. However, researches on vaccines are going on.

3.2.2.3 La Crosse virus

This is an encephalitis virus of the genus *Orthobunyavirus*, discovered from the stored brain and spinal tissue of an infected child (then an unknown virus) in La Crosse of Wisconsin in 1960. The virus leads to a disease referred to as La Crosse encephalitis, leading to headache, coma, paralysis, and brain damage in severe cases. Normally people with no symptoms are also seen. The

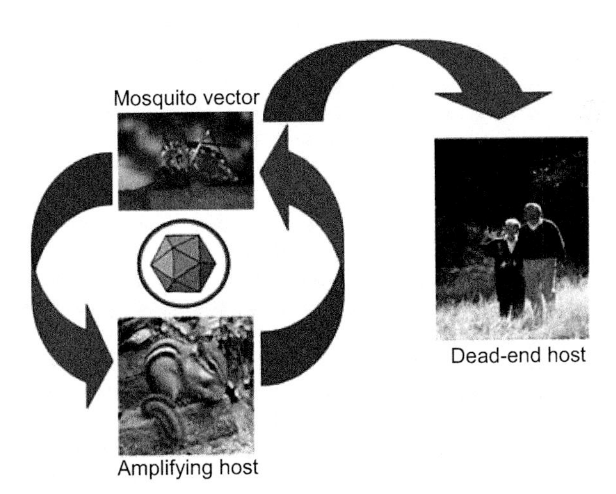

Figure 3.6 Transmission cycle of La Crosse virus. Figure taken from www.CDC.gov.

virus is known to be transmitted through the *Aedes* vector (*Aedes triseriatus*), which is a forest-dwelling species (Fig. 3.6). Since it dwells in forests, its reservoir hosts tend to be chipmunks, tree squirrels, etc. Transovarial transmission has also been reported for this virus. The infection has been reported in children under the age of 16 years as well as in immunocompromised individuals. At present no vaccine or treatment is available. Sometimes the disease goes undiagnosed where medical practitioners think it to be a case of aseptic meningitis (Thompson et al., 1965). In the United States, approximately 63 cases are reported each year due to this infection.

3.2.3 Flaviviruses

The viruses that fall within this group are classified as Group IV +ve ssRNA viruses. The family to which they belong is known as Flaviviridae. Flaviviruses originated in Africa (North Dakota Deptt., 2011), and it is said that Joseph Conrad suffered from yellow fever when he was writing his novel *Heart of Darkness* in 1902. The size of these viruses is approximately 40–65 nm, and they are enveloped in nature. The nucleocapsid is icosahedral in shape. There are approximately 10,000–11,000 bases within their genome, and the genome is non-segmented (Dimmock et al., 2007; Lindenbach et al., 2001). Examples of viruses falling under this category are West Nile virus, yellow fever virus, dengue virus, Zika virus, tick-borne encephalitis virus, etc. Most of these, if we see, are arthropod-borne viruses, that is, they are transmitted by an arthropod vector such as mosquitoes or ticks. So these viruses replicate in both the systems, that is, mosquitoes as well as humans (Fig. 3.7). But in some cases, the virus titer becomes low and cannot re-infect further. In such cases, we call the human system in which the virus is replicating as dead end, such as cases of West Nile virus and encephalitis virus infection. Some of these viruses can also be transmitted through other routes such as during blood transfusion, from mother to progeny and through animal remains.

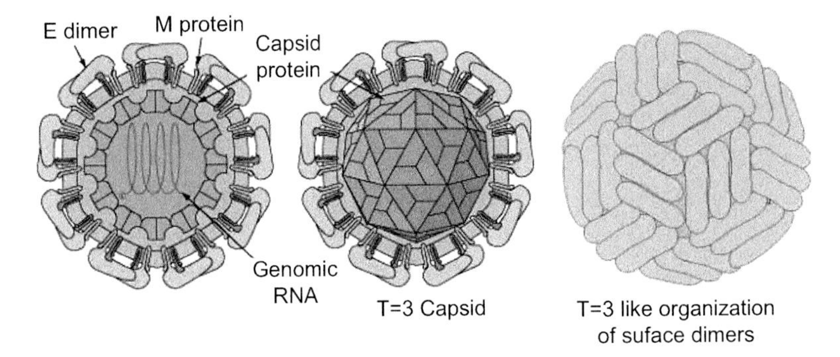

Figure 3.7 A general representation of the flavivirus structure and genome organization. Figure taken from ViralZone: www.expasy.org/viralzone, SIB Swiss Institute of Bioinformatics.

3.2.3.1 Dengue virus

The virus belongs to the family Flaviviridae and is circulating worldwide in the form of four serologically different serotypes named DENV-1, DENV-2, DENV-3, and DENV-4. It is said that these serotypes have emerged from the sylvatic cycle circulating within the forests of South-East Asia (Wang et al., 2000). A fifth variant was discovered in 2013 and named DENV-5. It was identified in a 37-year-old farmer in Sarawak, Malaysia, in 2007 (Mustafa et al., 2015) (Fig. 3.8). Whole genome sequencing of the sample from the patient confirmed it to be different from the existing ones but slightly similar to the DENV-2 strain (Normile, 2013). The virus is also known to have severe life-threatening effects in the form of dengue hemorrhagic fever and dengue shock syndrome. The virus is transmitted by *Aedes* mosquitoes. Different species of *Aedes* mosquitoes are responsible for its spread across countries, such as *Aedes aegypti, Aedes albopictus, Aedes nivalis, Aedes polynesiensis, Aedes scutellaris,* and *Aedes africanus.* Currently, there is no chemotherapy or licensed vaccine available against the dengue virus. Twenty-five candidates are under trials, but still a fully functional one is awaited (Back and Lundkvist, 2013).

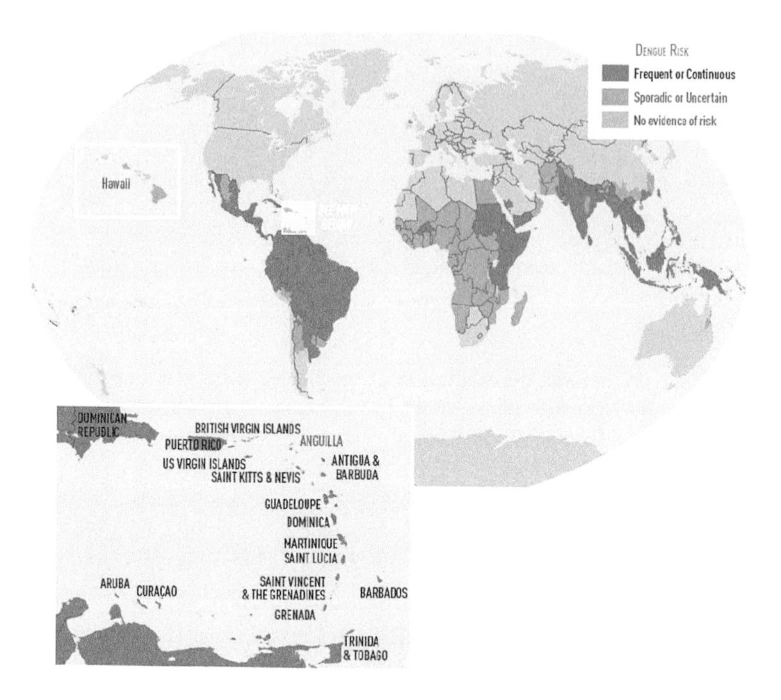

Figure 3.8 World map depicting places of dengue risk. Figure taken from www.cdc.gov/dengue/areaswithrisk/around-the-world.html.

3.2.3.2 Japanese encephalitis virus

This is yet another virus belonging to the family Flaviviridae. It is named so as it was first identified in Japan in 1871 and is characterized by paralysis and encephalitis-like symptoms in severe stages. Japanese encephalitis virus (JEV) is actually one of the viruses falling under the JE serogroup. This group includes viruses causing encephalitis-like West Nile virus, St. Louis encephalitic virus, and Murray Valley encephalitis virus (Gould et al., 2003). The JEV has five genotypes, namely GI to GV, each having a unique distribution pattern. The Indonesia–Malaysia region has all the five genotypes circulating, the Taiwan–Philippines region has GII and GIII circulating, the Thailand–Cambodia–Vietnam region has GI–GIII circulating, the Australia–New Guinea region has GI–II circulating, the Japan–Korea–China region has GI and GIII circulating, while

India–Sri Lanka–Nepal has GIII circulating (Fig. 3.9) (Solomon et al., 2003; Yun et al., 2003).

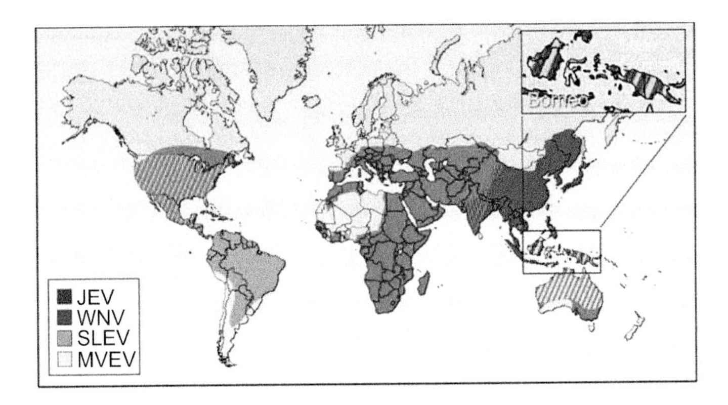

Figure 3.9 The distribution of the different serogroups of JE virus. Reprinted from Yun and Lee, Copyright 2014 Landes Bioscience, under Creative Commons license.

The virus is transmitted to humans through mosquitoes, mainly the Culicine group, which prefers vertebrates hosts such as pigs and birds. Humans get the infection when they come in contact with an infected Culex mosquito. The virus may reach new geographical areas through migratory birds carrying the virus. It is also known that pigs are the main reservoir as they can multiply virus in high titers. Humans, on the other hand, are dead end as the viremia gets too low to transmit further (Rosen, 1983; Solomon and Vaughn, 2002; Vaughn and Hoke, 1992; Weaver and Barrett, 2004; Yun and Lee, 2006). No chemotherapy is available against this virus, but four vaccines have been formulated and are available. These are mouse-brain-derived killed-inactivated, cell-culture-derived live-attenuated, cell-culture-derived killed-inactivated, and genetically engineered live-attenuated chimeric vaccines (Beasley et al., 2008; Halstead and Thomas, 2011; Smith and Halstead, 2011).

3.2.3.3 Yellow fever virus

The virus got its name from the word "yellow" referred due to jaundice-like conditions seen in patients suffering from the yellow fever infection. The virus was first isolated from a patient in Ghana in 1927 and was since then named the Asibi yellow fever strain named after the patient, Asibi. The first evident epidemic was recorded in Yucatan in 1648 (Carter, 1931). Walter Reed, in 1900, showed that the virus is transmitted from mosquitoes. Yellow fever is prevalent in 90% of African regions, and the first epidemic was reported in Kenya (Barnett, 2007). It was Max Theiler who derived the successful yellow fever vaccine, after some modification from the original 17D strain (live-attenuated type). The vaccine was later discontinued in the 1980s (Gardner and Ryman, 2010; Staples and Month, 2008; Smith and Theiler, 1937; Theiler and Smith, 1937). The structure and genomic composition are similar to those of other flaviviruses. It is an enveloped virus with a nucleocapsid and membrane as the structural part, while seven proteins form the nonstructural parts. The EYE (Eliminate Yellow Fever Epidemics) was launched in 2017 by the WHO, UNICEF, and GAVI (the vaccine alliance) and is operating in 40 countries and aims to vaccinate and protect billions of people from the virus by 2026 (https://www.who.int/news-room/fact-sheets/detail/yellow-fever).

3.2.3.4 West Nile virus

As referred earlier, the West Nile virus belongs to the JE serogroup. It was first seen in Uganda in 1937. Encephalitis and meningitis are known to be the serious complications associated with this viral infection. It is also known to occur in horses. Human vaccine is not available, but vaccine for horses is available. Birds are known to be one of the reservoirs specially of the Crow family, and it was isolated from crows in the delta region of Nile and hence the name. There are four to five lineages of this virus. *Culex* mosquitoes are the main vectors responsible for transmission of this virus, though in some parts *Aedes* species have also been reported. The virus causes severe complications in the brain as it is reported to cross the blood–brain barrier efficiently, but the mechanism has not been fully understood yet.

3.3 Conclusion

The chapter discussed all arbovirus infectious diseases spread by mosquitoes. The arboviruses have three families: Togaviruses, which include chikungunya virus, Eastern equine encephalitis virus, western equine encephalitis virus, and Venezuelan equine encephalitis virus; Bunyaviruses, which include sandfly fever virus, Rift Valley fever virus, Crimean–Congo hemorrhagic fever virus; and flaviviruses, which include yellow fever virus, dengue virus, Japanese encephalitis virus, West Nile virus, and Zika virus.

References

Beasley, D. W.; Lewthwaite, P.; and Solomon, T. Current use and development of vaccines for Japanese encephalitis. *Expert Opin. Biol. Ther.* 2008, **8**(1): 95–106.

Bhatia, R.; Narain, J. P.; and Plianbangchang, S. Emerging infectious diseases in East and South-East Asia. In *Public Health in East and South East Asia* (Detels, R., Sullivan, S. G., and Tan C. C., eds.), University of California Press, Berkeley, USA, 2012, pp. 43–78.

Carter, H. R. *Yellow Fever: An Epidemiological and Historical Study of its Place of Origin*. Williams and Wilkins, Baltimore, MD, 1931.

Cleaveland, S.; Laurenson, M. K.; and Taylor, L. H. Diseases of humans and their domestic mammals: Pathogen characteristics, host range and the risk of emergence. *Philos. Trans. R. Soc. Lond. B Biol. Sci.* 2001, **356**(1411): 991–999.

Dikid, T., Jain, S. K.; Sharma, A.; Kumar, A.; and Narain, J. P. Emerging and re-emerging infections in India: An overview. *IJMR* 2013, **138:** 19–31.

Dikid, T.; Jain, S. K.; Sharma, A.; Kumar, A.; and Narain, J. Emerging and re-emerging infections in India: An overview. *Indian J. Med. Res.* 2013, **138**: 19–31.

Dimmock, N. J.; Easton, A. J.; and Leppard, K. N. *Introduction to Modern Virology*, 6th edn., Blackwell Publishing, Massachusetts, 2017, pp. 453–454.

Elizabeth, D. B. Yellow fever: Epidemiology and prevention. *Clin. Infect. Dis.* 2007, **44**(6): 850–856.

Fauci, A. S. Infectious diseases: Considerations for the 21st century. *Clin. Infect. Dis.* 2001, **32**: pp. 675–685.

Gardner, C. L. and Ryman, K. D. Yellow fever: A reemerging threat. *Clin. Lab. Med.* 2010, **30**(1): 237–260.

Gould, E. A.; de Lamballerie, X.; Zanotto, P. M.; and Holmes, E. C. Origins, evolution, and vector/host coadaptations within the genus *Flavivirus*. *Adv. Virus Res.* 2003, **59**: 277–314.

Guu, T. S.; Zheng, W.; and Tao, Y. J. Bunyavirus: Structure and replication. *Adv. Exp. Med. Biol.* 2012, **726**: 245–266. doi: 10.1007/978-1-4614-0980-9_11.

Halstead, S. B. and Thomas, S. J. New Japanese encephalitis vaccines: Alternatives to production in mouse brain. *Expert Rev. Vaccines.* 2011, **10**(3), 355–364.

http://www.who.int/emergencies/zika-virus/timeline/en/.

http://www.who.int/mediacentre/factsheets/fs207/en/.

http://www.who.int/mediacentre/factsheets/fs208/en/.

http://www.who.int/mediacentre/factsheets/fs327/en/.

http://www.who.int/mediacentre/factsheets/fs354/en/.

https://en.wikipedia.org/wiki/Bunyavirales.

https://en.wikipedia.org/wiki/Crimean%E2%80%93Congo_hemorrhagic_fever.

https://en.wikipedia.org/wiki/Eastern_equine_encephalitis.

https://en.wikipedia.org/wiki/Flavivirus.

https://en.wikipedia.org/wiki/History_of_yellow_fever.

https://en.wikipedia.org/wiki/La_Crosse_encephalitis.

https://en.wikipedia.org/wiki/Orthohantavirus.

https://en.wikipedia.org/wiki/Togaviridae.

https://microbewiki.kenyon.edu/index.php/Togaviridae.

https://www.cdc.gov/easternequineencephalitis/tech/transmission.html).

https://www.cdc.gov/hantavirus/index.html.

https://www.cdc.gov/lac/index.html.

https://www.cdc.gov/vhf/crimean-congo/index.html.

https://www.cdc.gov/vhf/rvf/index.html.

Kuno, G. and Chang, G. J. Biological transmission of arboviruses: Re-examination of and new insights into components, mechanisms, and unique traits as well as their evolutionary trends. *Clin. Microbiol. Rev.* 2005, **18**(4): 608–637.

Lindenbach, B. D. and Rice, C. M. Flaviviridae: The viruses and their replication. In *Fields Virology*, 4th edn. (Knipe, D. M. and Howley, P. M., eds.). Lippincott Williams and Wilkins, Philadelphia, 2001, pp. 991–1041.

Lounibos, L. P. Invasions by insect vectors of human disease. *Annu. Rev. Entomol.* 2002, **47**: 233–266.

Mustafa, M. S.; Rasotgi, V.; Jain, S.; and Gupta, V. Discovery of fifth serotype of dengue virus (DENV-5): A new public health dilemma in dengue control. *Med. J. Armed Forces India* 2015, **71**(1): 67–70.

Normile, D. Surprising new dengue virus throws a spanner in disease control efforts. *Science* 2013, **342**: 415.

North Dakota Department of Health. Department of Preparedness and Response flaviviruses fact sheet. North Dakota Department of Health (http://www.ndhealth.gov./EPR/publci/viral/Flavivirusfact.htm), accessed April 2, 2011, 2005.

Powell, J. R. and Tabachnick, W. J. History of domestication and spread of *Aedes aegypti*: A review. *Mem. Inst. Oswaldo Cruz.* 2013, **108** (1): 11–17.

Reeves, W. C. Partners: Serendipity in arbovirus research. *J. Vector Ecol.* 2001, **26**(1): 1–6.

Rosen, L. The natural history of Japanese encephalitis virus. *Annu. Rev. Microbiol.* 1986, **40**: 395–414.

Shope, R. E. Bunyaviruses. In *Medical Microbiology*, 4th edn. (Baron, S., ed.), University of Texas Medical Branch at Galveston, Galveston (TX), 1996.

Smith, H. H. and Theiler, M. The adaptation of unmodified strains of yellow fever virus to cultivation in vitro. *J. Exp. Med.* 1937, **65**(6): 801–808.

Solomon, T. and Vaughn, D. W. Pathogenesis and clinical features of Japanese encephalitis and West Nile virus infections. *Curr. Top Microbiol. Immunol.* 2002, **267:** 171–194.

Solomon, T.; Ni, H.; Beasley, D. W.; Ekkelenkamp, M.; Cardosa, M. J.; and Barrett, A. D. Origin and evolution of Japanese encephalitis virus in southeast Asia. *J. Virol.* 2003, **77**(5): 3091–3098.

Staples, J. E. and Monath, T. P. Yellow fever: 100 years of discovery. *JAMA* 2008, **300**: 960–962.

Taylor, L. H.; Latham, S. M.; and Woolhouse, M. E. Risk factors for human disease emergence. *Philos. Trans. R. Soc. Lond. B Biol. Sci.* 2001, **356**: 983–989.

Theiler, M. and Smith, H. H. The use of yellow fever virus modified by in vitro cultivation for human immunization. *J. Exp. Med.* 1937, **65**(6): 787–800.

Thompson, W. H.; Kalfayan, B.; and Anslow, R. O. (1965). Isolation of California encephalitis virus from a fatal human illness. *Am. J. Epidemiol.* 1965, **81**: 245–253.

Vaughn, D. W. and Hoke, C. H. Jr. The epidemiology of Japanese encephalitis: Prospects for prevention. *Epidemiol. Rev.* 1992, **14**: 197–221.

Weaver, S. C. and Barrett, A. D. Transmission cycles, host range, evolution and emergence of arboviral disease. *Nat. Rev. Microbiol.* 2004, **2**(10): 789–801.

Wilder-Smith, A. and Halstead, S. B. Japanese encephalitis: Update on vaccines and vaccine recommendations. *Curr. Opin. Infect Dis.* 2010, **23:** 426–431.

Wolfe, N. D.; Dunavan, C. P.; and Diamond, J. Origins of major human infectious diseases. *Nature* 2007, **447**(7142): 279–283.

World Health Organization, Regional Office for South East Asia Region. *Combating Emerging Infectious Diseases in the South-East Asia Region*, World Health Organization, WHO SEARO, New Delhi, 2005.

Yun, S. I. and Lee, Y. M. Japanese encephalitis virus: Molecular biology and vaccine development. In *Molecular Biology of the Flavivirus* (Kalitzky, M. and Borowski, P., eds.), Horizon Scientific Press, Norwich, United Kingdom, 2006, pp. 225–271.

Yun, S. I. and Lee, Y. M. Japanese encephalitis: The virus and vaccines. *Hum. Vaccin. Immunother.* 2014, **10**(2): 263–279.

Yun, S. I.; Kim, S. Y.; Choi, W. Y.; Nam, J. H.; Ju, Y. R.; Park, K. Y.; Cho, H. W.; and Lee, Y. M. Molecular characterization of the full-length genome of the Japanese encephalitis viral strain K87P39. *Virus Res.* 2003, **96**(1–2): 129–140.

Chapter 4

Dengue Fever: A Viral Hemorrhagic Fever of Global Concern

Bennet Angel,[a] Neelam Yadav,[b] Jagriti Narang,[c] Annette Angel,[d] and Vinod Joshi[a]

[a]*Amity Institute of Virology and Immunology, Amity University, Sector 125, Noida, India*
[b]*Centre for Biotechnology, Maharshi Dayanand University, Rohtak, India*
[c]*Department of Biotechnology, Jamia Hamdard University, New Delhi, India*
[d]*Division of Zoonosis, National Centre for Disease Control, 22 Sham Nath Marg, Civil Lines, Delhi, India*
jags_biotech@yahoo.co.in, vinodjoshidmrc@gmail.com, bennetangel@gmail.com

Of the many viruses transmitted by *Aedes* mosquitoes, dengue virus is responsible for causing dengue fever (DF). More severe forms of the disease are termed dengue hemorrhagic fever (DHF) and dengue shock syndrome (DSS). Until 1970s, only nine countries were recorded to be affected by the dengue virus, but since then, the disease has spread around the globe (Fig. 4.1). The World Health Organization (WHO) has estimated that each approximately 50–100

Small Bite, Big Threat: Deadly Infections Transmitted by Aedes *Mosquitoes*
Edited by Jagriti Narang and Manika Khanuja

ISBN 978-981-4800-86-0 (Hardcover), 978-1-003-00329-8 (eBook)
www.jennystanford.com

million new dengue patients appear, out of which 2.5–5% of the two billion people are at risk; hundreds of thousands are hospitalized; and 20,000–25,000 mortality cases are reported (Fig. 4.2) (Gubler and Meltzer, 1999; Halstead, 1988; Rigau-Pérez and Gubler, 1999; WHO, 2002). Compared to the last 50 years, countries around the globe are now facing a 30-fold increase in the disease burden (WHO website; http://www.who.int/denguecontrol/disease/en/). This figure is not an exact one because many cases go undetected, resulting in the disease getting under-reported from many parts of the world. India is also among those countries that witness the dengue epidemic every year. The Chinese encyclopedia, which dates back to the Jin Dynasty (265–420 AD), is the first ever record to mention a probable case of dengue fever. During those days, the disease was referred to as "water poison" that was associated with flying insects (http://www.denguevirusnet.com/history-of-dengue.html). The next reference of the disease appeared in 1779 in the form of the first-ever epidemics that occurred in Batavia (present-day Jakarta) and in the city of Cairo (Fig. 4.3) (Henchal and Putnak, 1990; Siler et al., 1926). Soon, similar epidemics started occurring in many countries across the globe, affecting Philadelphia in 1780, Zanzibar in 1823 and 1870, West Indies in 1827, and Hong Kong in 1901.

4.1 Epidemiology

Aedes mosquitoes transmit many viruses, and among them dengue is caused by dengue viruses. Dengue virus results into dengue fever, while its severe forms are dengue hemorrhagic fever (DHF) and dengue shock syndrome (DSS). This disease has spread globally, but in 1970 the disease affected only nine countries (Fig. 4.1). As per reports of the World Health Organization (WHO), approximately 100 million infections appear each year, 5% of the 2 billion people are at risk, hundred thousands are hospitalized, and about 20,000–25,000 mortality cases are reported each year due to this disease (Fig. 4.2) (Gubler and Mertizer, 1999; Halstead, 1988; Riagau-Perez and Gubler, 1999; WHO, 2000). There has been a 30-fold increase in disease burden since the last 50 years (www.who.int/denguecontrol/disease/en/; website WHO). Regardless, many cases are underreported and not diagnosed in many developing countries. The epidemic of dengue is also witnessed every year in India. The first case of dengue was reported in the Chinese encyclopedia, which dates back to 265–420 AD. But at that time, the disease was not designated as dengue; it was named water poisoning related to flying insects (www.denguevirusnet.com/history-of-dengue.html). In Batavia in Jakarta, it was reported in 1779, and the first epidemic that was ever reported was in Cairo (Fig. 4.3) (Henchal and Putnak, 1990; Siler et al., 1926). The epidemic then spread to many countries throughout the world affecting Philadelphia in 1780, Zanzibar in 1823 and 1870, West Indies in 1827, and Hong Kong in 1901.

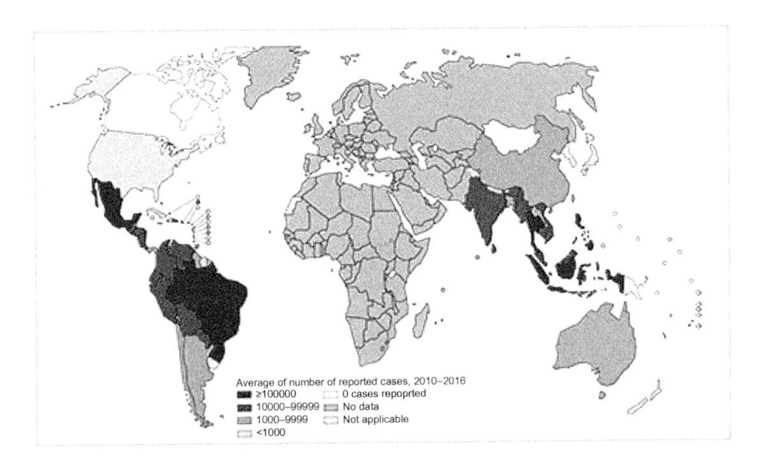

Figure 4.1 Spread of dengue virus (DENV) across the world. Figure taken from http://www/who.int/denguecontrol/epidemology/en.

In India, the first case of dengue was reported and documented in 1780 in Madras (Gupta et al., 2012). Later it was documented in Calcutta in 1824, with consequent epidemics reported in 1853, 1871, 1905, and so on (Chaturvedi and Nagar, 2008; Henchal and Putnak, 1990). The evolution of dengue as DHF and DSS was first documented in Queensland in 1897, in the United States in 1922, and in South Africa (SA) in 1927. Greece was affected in 1928, and Formosa was affected in 1931. The severe form of dengue was documented in the Philippines in 1954 (Halstead, 1966). In India, different strains were also reported in different parts of Kanpur in 1969 (DENV-2 and DENV-4). In Rajasthan, DENV-1 and DENV-3 were reported, and Vellore witnessed all serotypes of dengue in 1968 (Gupta et al., 2012). First only DENV-2 was reported in Delhi, but later all the four serotypes of dengue were documented in 2003. Remarkably, in 2004, a swift change replaced the earlier circulating serotypes of dengue (i.e., DENV-2 with DENV-3), which was considered an important reason for the appearance of severe forms of dengue, that is, DHF and DSS. Replacement with different serotypes of dengue is still prevalent in Delhi; replacement from DENV-2 and DENV-3 to DENV-4 was observed during 2007–2009 (Gupta et al., 2012). A few authors also conducted studies in Rajasthan regarding evolution of various serotypes of dengue (Angel et al., 2014; Joshi et al., 2012; Sharma et al., 2008).

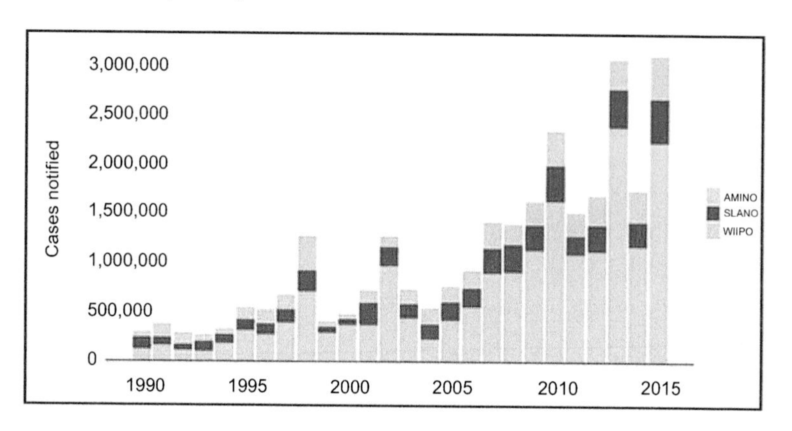

Figure 4.2 Laboratory proven and suspected dengue cases documented by the WHO, 1990–2015. Figure taken from http://www.who.int/denguecontrol/epidemology/en/.

Due to the lack of medication and management of the disease, the rate of mortality and morbidity increased globally. The TDR portfolio of the WHO showed some researchable issues on dengue. Among all researchable issues, vector–virus interaction is yet to be fully discovered. This requires significant effort to figure out the molecular epidemiological aspects of dengue.

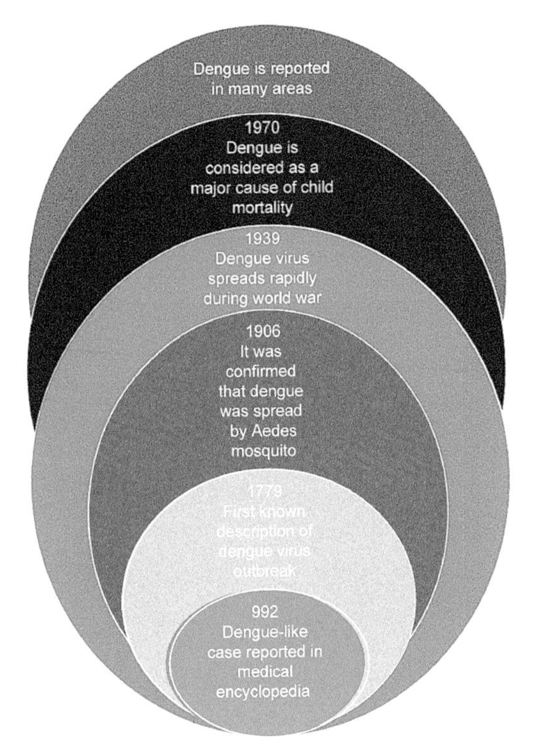

Figure 4.3 Situation scrutiny of origin of dengue virus and recent trend of dengue fever.

4.2 Virus Morphology

According to the virus classification of Baltimore, dengue virus comes under Group IV because it is a +ssRNA virus. The dengue virus belongs to the Arthropod family consisting a large group and to the family Flaviviridae. The size of a single virion is 50 nm and has about 1.23 g/cc density (Rusell et al., 1980). The total genome size is 11,000 bases. The genome of the virus codes for 10 different proteins, namely, three structural proteins and seven nonstructural proteins. Structural proteins are designated as nucleocapsid protein, membrane-linked protein M and envelope protein E (Fig. 4.4), and seven nonstructural proteins are designated as NS1, NS2a, NS2b, NS3, NS4a, NS4b, and NS5 (Henchal and Putnak, 1990). This was first documented by Rice et al. in 1985 and 1986. In viral genome, capping was found at the 5′ end of ssRNA (Cleaves and Dublin, 1979; Wengler et al., 1978), while at the 3″ end, there is absence of poly A tail (Hahn et al., 1988; Irie et al., 1989; Rice et al., 1985). It has a single open reading frame.

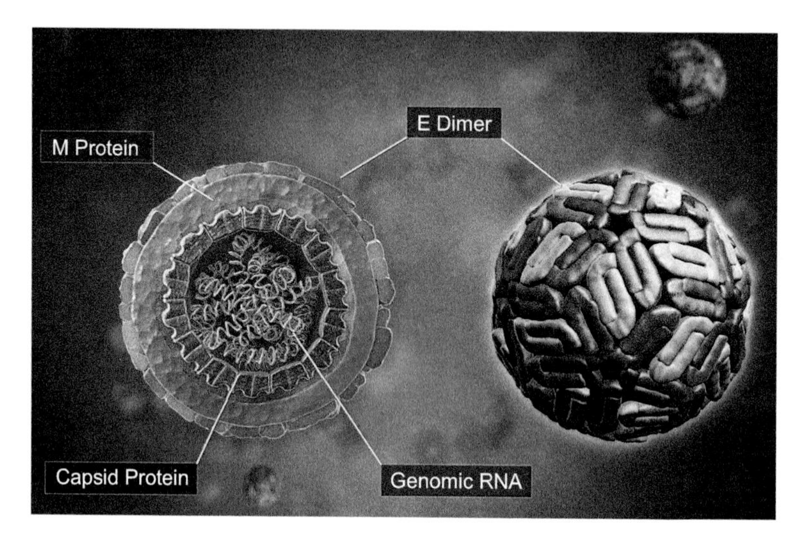

Figure 4.4 Dengue virus (DENV) morphology. Figure taken from https://en.wikipedia.org/wiki/Dengue_virus#/media/File:SAG_Dengue-Virus_160413_01.jpg.

Four strains of dengue virus were reported and are serologically different, namely, DENV-1, DENV-2, DENV-3, and DENV-4. Another serotype, the fifth serotype, was documented in 2013 (Normile, 2013). During 1944–1945, these strains were discovered after experimental studies. A number of studies were conducted by Sabin and his group, and infectious material was inoculated into human volunteers; then only two serotypes were designated, that is, Hawaii and the new Guinea B, C, and D strains (Sabin, 1952). Afterward, advancement in the virology field led to the use of suckling Swiss Albino mice for performing several experiments; the DENV-1 serotype was identified (Hotta, 1969; Kimura and Hotta, 1944; Sabin and Schlesinger, 1945; Sabin, 1952), while DENV-3 and DENV-4 were isolated and investigated during an epidemic of DHF that occurred in the Philippines islands (Hammon et al., 1960).

An immature virion has a pre-membrane protein (PrM), and it is also a precursor of membrane protein, which is present in a mature virion (Shaprio et al., 1972; Wengler and Wengler, 1989), as shown in Fig. 4.5 (Rice et al., 1985; Stollar, 1969). All structural and nonstructural proteins are coded by a single polypeptide chain, as shown in Figs. 4.5 and 4.6 (Bell et al., 1985; Cleaves, 1985; Deubel et al., 1988; Hahn et al., 1988; Henchal and Putnak, 1990; Mason et al., 1987; Osatomi et al., 1988; Rice et al., 1985; Zhao et al., 1986). The membranous structure of virion is shown in Fig. 4.7.

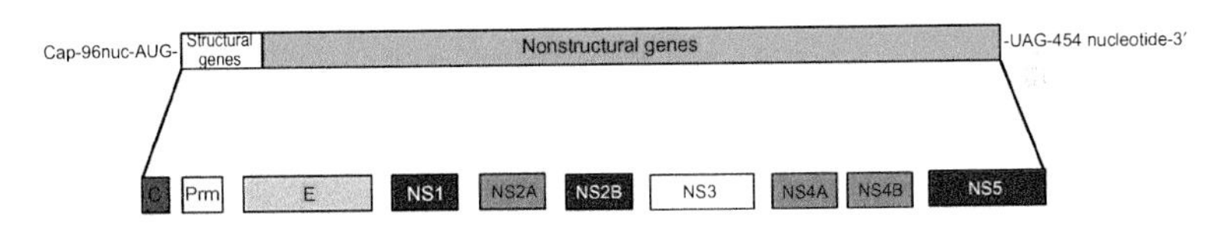

Figure 4.5 Single polypeptide chain code all structural and nonstructural proteins of DENV.

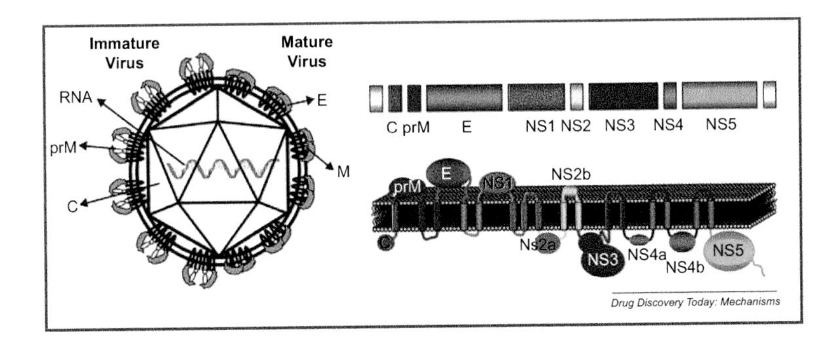

Figure 4.6 The genome of dengue virus (immature and mature virion). Figure reprinted from Hottz et al., 2011, with permission from Elsevier.

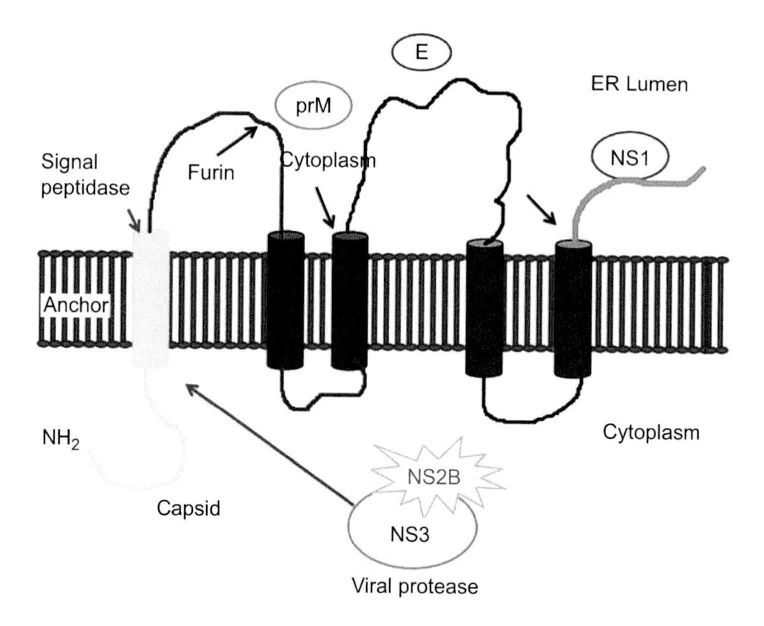

Figure 4.7 DENV membrane with transmembrane helices NS2B protease (represented by cylinders) cleavage site (represented by red arrow), post-translational site cleavage site (represented by dark red color) and site for furin cleavage (represented by black color).

4.2.1 Structural Proteins

4.2.1.1 The nucleocapsid protein

Structural protein involves a C protein or nucleocapsid protein, which has a molecular weight (MW) of approximately 13.5 KDa. Nucleocapsid proteins comprise 100 monomer units and have 26 basic amino acids, which include lysine and arginine and have three acidic amino acids. Due to a fewer number of acidic amino acids, it is a highly basic protein. The N-terminal end acts as an RNA-binding site, which is involved in the formation of viral particles (Samsa et al., 2012), though the C terminal residues act as a transmembrane signal for the pre-membrane protein (PrM) (Fig. 4.8) (Byk and Gamarnik, 2016; Jones et al., 2003; Wang et al., 2004).

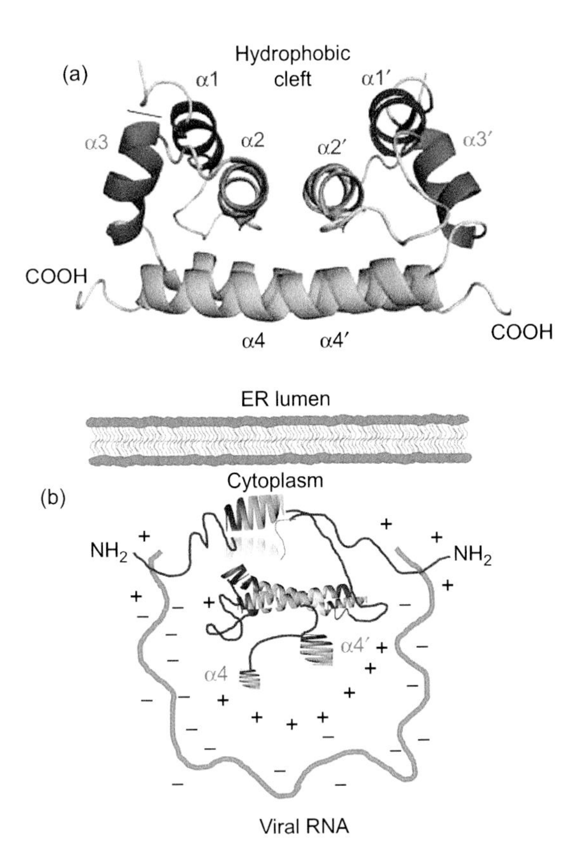

Figure 4.8 Dengue virus and capsid protein structure: (a) Homodimers consist of four helices and are denoted by different colors (such as red color denotes α1, yellow color denotes α2, blue color denotes α3, and green color denotes α4). As it is a homodimer, marking for one of the units is denoted with prime symbols (helices α1′, α2′, α3′, and α4′). The hydrophobic cleft is indicated. Figure reprinted from Bundo and Igarashi, 1985, with permission from Elsevier. (b) Genome of DENV showed interaction with the α4–α4′ and unstructured N region of capsid protein.

4.2.1.2 The membrane protein

After pre-membrane protein or PrM cleavage, there is formation of membrane protein or M protein. The cleavage of 22,000 Da PrM protein leads to the generation of 8000 Da protein (Deubel et al., 1988; Hahn et al., 1988; Mason et al., 1987; Osatomi et al., 1988; Randolf et al., 1990; Rice et al., 1985). The E protein involved in binding the receptors of host molecules has three domains: Domains I, II, and III (Fig. 4.9). Domain I links the other two domains, that is, Domains II and Domains III. The role of Domain II is to form dimerization through Domain III, which is involved in binding the receptors of host cells (Chin et al., 2007; Hidari and Suzuki, 2011; McMinn, 1997; Modis et al., 2003; Modis et al., 2005; Stiansny et al., 2007; Zhang et al., 2003). This protein is composed of 75 amino acids.

4.2.1.3 The envelope protein

The structural E protein of dengue virus is a glycoprotein having MW approximately 51,000 to 60,000 Da. The gene that codes for this protein comprises consensus sequence (as compared to other flaviviruses) containing 12 cysteine amino acids, and these cysteine amino acids are involved in forming six disulfide bridges (Mandal et al., 1989; Nowak and Wengler, 1987). As depicted in Fig. 4.9, three non-overlapping antigenic domains of envelope proteins, that is, A, B, and C, are also present. The E protein helps in determining the assembly of virus, that is, icosahedral, and also acts as a trigger for the fusion of viral and host cell membrane (Harrison, 2008; Lindenbach et al., 2007).

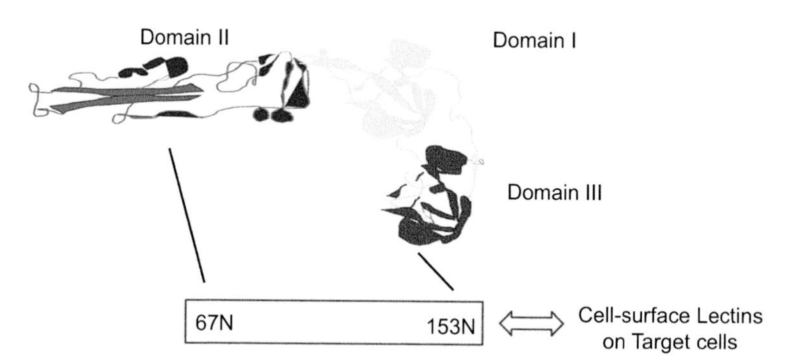

Figure 4.9 The structural E protein involved in binding the receptors of host molecules and has three domains: Domains I represented by yellow color, Domains II represented by blue color, and Domains III represented by dark red color. Domain I links the other two domains, that is, Domains II and Domains III. The role of Domain II is to form dimerization through Domain III, which is involved in binding the receptors of host cells. At 67th and 153th position of nucleotide, N-glycan is present which is involved in the binding of DENV with the host proteins.

Envelope proteins consist of three domains I, II, and III. Further, an S region is present next to domain III, which is juxtamembrane stem, and this is followed by a transmembrane anchor (TM). A mature virion is covered by 90 dimers. The lipid content of the protein also helps in protecting the DENV genome.

4.2.2 Nonstructural Proteins

4.2.2.1 NS1 protein

This is the first protein in line with the E structural protein. It is a nonstructural glycoprotein having an MW of 48 KDa. NS1 protein is responsible for viral genome replication, and it is a marker of virus infection. This protein mainly interacts with the host immune response and is involved in evasion, which results in pathogenesis; nevertheless, its role in the life cycle of virus is yet to be proven (Akey et al., 2014; Avirutnan et al., 2010, 2011; Chung et al., 2003; Krishna et al., 2009; Young et al., 2000). NS1 protein is initially formed as hydrophilic monomer, but afterward it changes into a hydrophobic dimer. But when this protein enters Golgi bodies, two N-glycans of protein are modified to form a complex structure. The hydrophobic dimer has three domains, namely, β-ladder, β-roll, and wing domain, as depicted in Figs. 4.10 and 4.11. Owing to its diverse role, it is nowadays exploited for experiments globally as it is considered a target for therapeutic application.

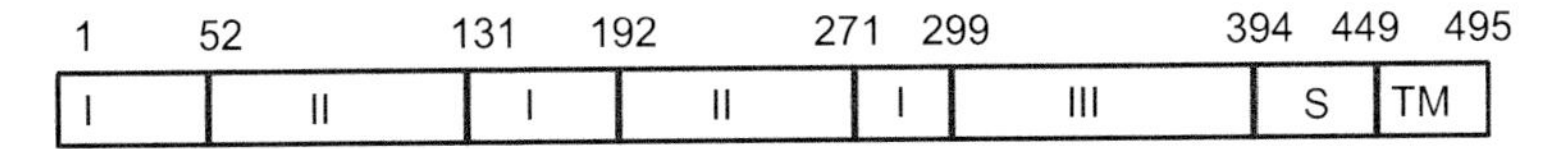

Figure 4.10 Organization and positions of various domains in E protein of DENV.

4.2.2.2 NS2 protein

NS2 protein comprises two proteins NS2a and NS2b having MWs of 20 and 14.5 KDa, respectively. Both these proteins are hydrophobic in nature. NS2a is mainly involved in the proteolytic role in the C terminus of NS1 protein, while the function of NS2b is yet to be documented (Chambers et al., 1989; Falgout et al., 1989; Speight et al., 1988).

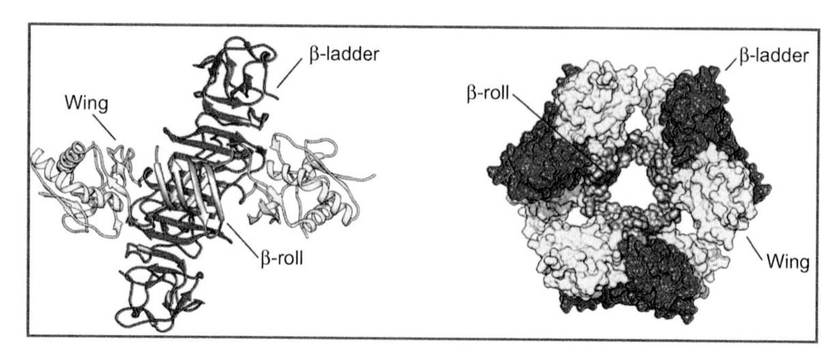

Figure 4.11 DENV structure NS1 protein, dimer, and hexamer forms, respectively. Reprinted from Scaturro et al., 2015, under Creative Commons license.

4.2.2.3 NS3 protein

Another nonstructural dengue virus is NS3 protein, which is a viral protease comprising six beta strands arranged as two β-barrels having 1-180 residues. NS3 protein is a hydrophilic protein unlike NS1 protein, having an MW of 70,000 Da. Their function is mainly like that of an RNA helicase and RTPase/NTPase (Fig. 4.12). In between the two β-barrels, the active sites are present, which involve His-51, Asp-75, and ser-135. The activity of these active sites depends on the presence of the NS2B cofactor. A six-stranded parallel β-sheet is surrounded by four α helices, which result in the formation of three domains, namely, subdomains I to III (Henchal and Putnak, 1990; Perera and Kuhn, 2008).

4.2.2.4 NS4A protein

The NS4 protein of dengue comprises two forms: NS4a and NS4b having MWs of 16 and 27 KDa, respectively. This protein is hydrophobic in nature (Chambers et al., 1989; Speight et al., 1988; Speight and Westaway, 1989). Cell curvature is changed due to this protein, and it also triggers autophagy (Miller et al., 2007). This protein is mainly responsible for viral replication, and it leads to oligomerization (Lee et al., 2015). If mutation occurs in NS4A, then the viral replication gets inhibited because NS4B is not able to form association with NS4A. The interaction of NS4A with NS4B is must for viral replication (Zou et al., 2015) and is also required as a cofactor along with NS5 protein.

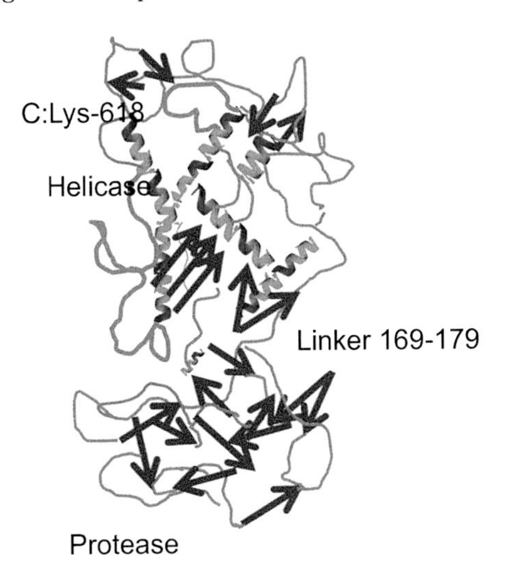

Figure 4.12 3D structure of dengue virus NS3 protein.

4.2.2.5 NS5 protein

The nonstructural protein NS5 of dengue virus comprises 900 residue peptides having RNA-dependent RNA polymerase and located at C terminal (residues 320–900), whereas at the N-terminal, methyltransferase domain is located (residues 1–296) (Perera and Kuhn, 2008). NS5 protein has a molecular weight of 105,000 Da. It is the most conserved protein and shows 67% sequence homology between all serotypes of dengue. The NS5 protein structure is shown in Fig. 4.13; yellow color is denoted by MTase, RdRp fingers are denoted by green color, palm is blue colored, and thumb is salmon colored. The domains are linked by a helix (residues 263–267) and are designated by orange color, whereas sticks show GTP and cofactor SAH. Zinc ions are represented by spheres.

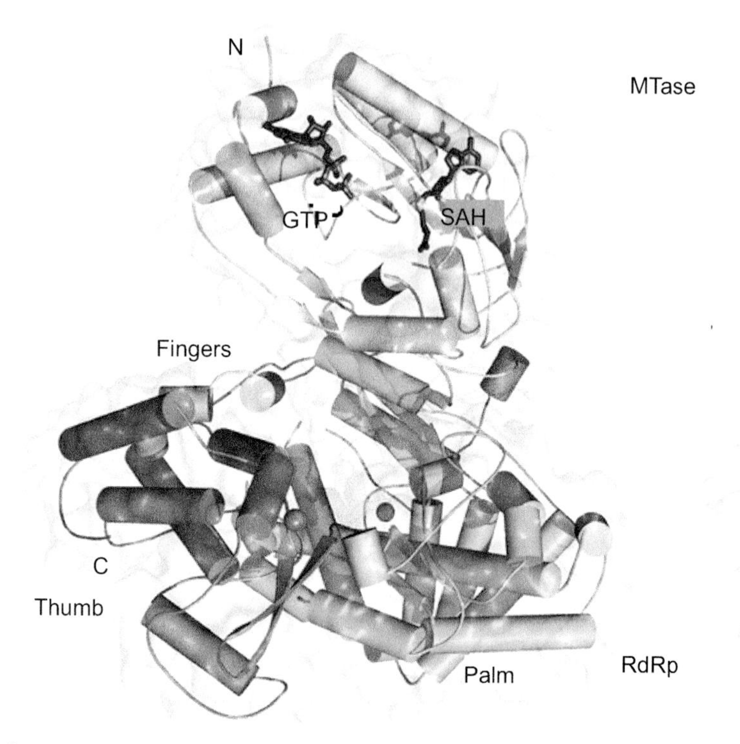

Figure 4.13 3D structure of NS5 protein of dengue virus. Figure reprinted from Zhao et al., 2015, under Creative Commons license.

Table 4.1 shows the structure and function of the 10 viral proteins contained by dengue virus-2, and Fig. 4.14 shows the pictorial representation of the virus structure and position and functional details of the virus.

The four strains of DENV are yet to be classified into many genetic groups or genotypes based on their sequence diversity. Earlier genotype was defined on the basis of 6% sequence divergence within the 240-nucleotide region of DENV-1 and DENV-2 E/NS1 junction (Ricco-Hesse, 1990). After the introduction of tools of bioinformatics, the genotypes are designated based on phylogenetic analysis (Ong, 2010). Tables 4.2–4.4 show the various genotypes and their original distribution (adapted from Ong, 2010).

Table 4.1 Dengue virus-2 proteins: formation and function

Name	Glycosylated	N-terminal cleavage[a]	Number of amino acids in mature protein	Function
C	No	M|NNQ (aminopeptidase)	99	Nucleocapsid
PrM	Yes	VMA|FHL (signalase)	166	M precursor
M	No	EXR|SVA (dibasic)	75	Membrane protein
E	Yes[b]	SMT|MRC (signalase)	495	Envelope protein
NS1	Yes	VQA|DSG (signalase)	352	Virus assembly?
NS2a	No	VTA|GHG (unknown)	218	NS1 processing?
NS2b	No	KKR|SWP (dibasic)	130	Unknown
NS3	No	KQR|AGV (dibasic)[c]	618	Protease/ NTPase?
NS4a	No	GKR|SLT (dibasic)	150	Unknown
NS4b	No	TMA|NEM (signalase)	248	Unknown
NS5	No	TRR|GTG (dibasic)	900	RNA polymerase?

[a]Cleavage site for amino acid at N terminus. The potential protease which is responsible for cleavage is shown in parentheses.
[b]DENV contains glycosylated E protein, but it is not in the case of all flaviviruses.
[c]Dibasic amino acid specific protease are similar to the motif.

Source: Henchal and Putnak, 1990.

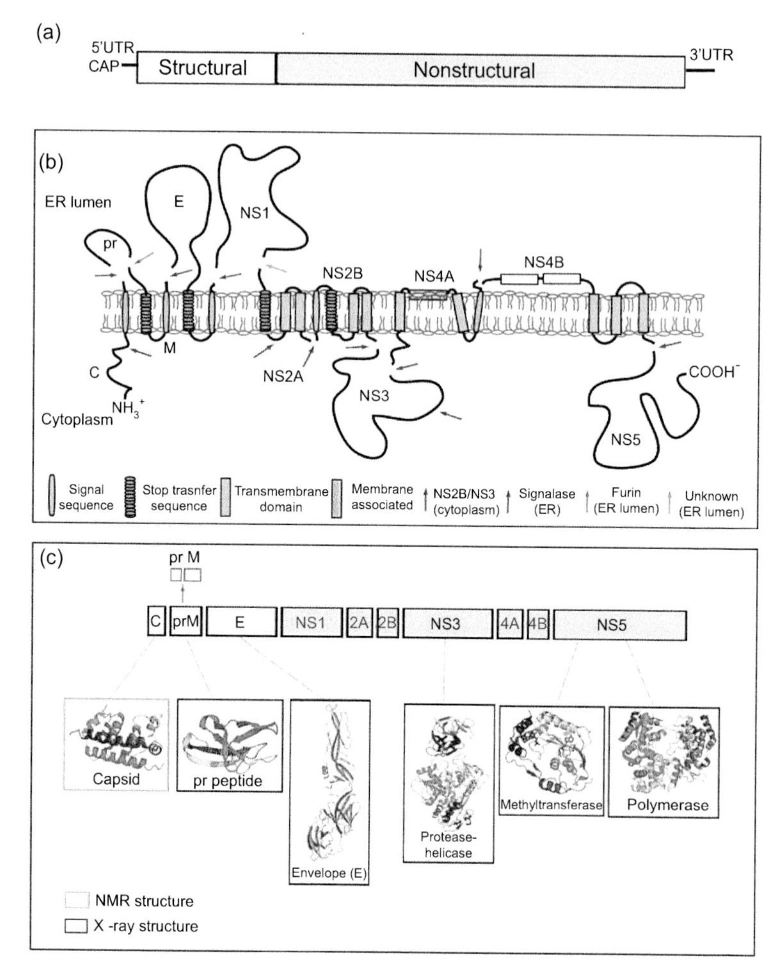

Figure 4.14 Structural and functional study of DENV by NMR and X-rays. Figure reprinted from Perera and Kuhn, 2008, with permission from Elsevier.

Table 4.2 Genotype of dengue virus-1 according to Goncalvez et al. (2002)

Genotypes	Original known distribution
I	Japan, Hawaii in the 1940s (the prototype strains), China, Taiwan, and South Ease Asia
II	Thailand in the 1950s and 1960s
III	Sylvatic source in Malaysia
IV	Nauru, Australia, Indonesia, and the Philippines
V	Africa, South East Asia, and the Americas

Source: Ong, 2010.

Table 4.3 DENV-2 genotypes according to Twiddy et al. (2002)

Genotypes	Original known distribution
American	Formerly known as Subtype V. Found in Latin America, old strains from India (1957), the Caribbean, and the Pacific Islands between 1950 and 1970s.
American/ Asian	Formerly known as Subtype III. Found in China, Vietnam, Thailand, and in Latin America since the 1980s.
Asian I	Thailand, Myanmar, and Malaysia
Asian II	Formerly known as Subtypes I and II. Found in China, the Philippines, Sri Lanka, Taiwan, and Vietnam. Includes the New Guinea C prototype strain.
Cosmopolitan	Formerly known as genotype IV. Wide distribution, including Australia, the Pacific Islands, South East Asia, the Indian subcontinent, Indian Ocean Islands, Middle East, and both East and West Africa.
Sylvatic	Isolated from nonhuman primates in West Africa and Malaysia.

Source: Ong, 2010.

Table 4.4 DENV-3 genotypes accruing to Lanciotti et al. (1994) and the known distribution of the genotypes prior to 1993

Genotypes	Original known distribution
I	Indonesia, Malaysia, Thailand, Burma, Vietnam, the Philippines, and the South Pacific Islands (French, Polynesis, Fiji, and New Caledonia). Includes the H87 prototype strain.
II	Thailand, Vietnam, and Bangladesh
III	Singapore, Indonesia, South Pacific Islands, Sri Lanka, India, Africa, and Samoa

Source: Ong, 2010.

The genotypes of DENV-4 were initially given by Lanciotti et al. (1997) as two genotypes, but later two more genotypes were described by Klungthong et al. in 2004.

4.3 Transmission Route

As mentioned in the previous chapters, the dengue virus is transmitted through *Aedes* mosquitoes, acting as the vector. These mosquitoes are known to transmit the dengue virus in two ways, horizontally and vertically. In horizontal transmission, the dengue virus is transmitted from an infected individual to a healthy person, through the *Aedes* mosquito and, in vertical transmission, transovarial transmission (TOT) of virus takes place, which means that the virus is transmitted from the parent mosquito to their progeny. Many researchers have documented the mechanism of TOT and its effect in maintaining DENV through the inter-epidemic periods without involvement of any humans and other vertebrate hosts (Francy et al., 1981; Hull et al., 1984; Joshi et al., 1996, 2002; Khin and Than, 1983; Leake, 1992; Rosen et al., 1983). Both horizontal and TOT routes occur in nature concurrently in a specific region and in particular settings, which is the major reason behind the epidemics or outbreaks. Therefore, facts on both horizontal and TOT routes are vital to recognize the disease condition in a certain part whether it is rural or urban and is also important to recognize the places wherever vertical transmission occurs during the inter-epidemic period.

Besides the transmission cycle that happens between mosquitoes and human hosts, there is one more cycle known as the sylvatic cycle or the peri-urban or the mosquito–monkey cycle (Figs. 4.15 and 4.16). The virus was primarily known to be a virus of primates (DeSilva et al., 1999; Gubler, 1992; Nathan et al., 2001). Since the peri-urban sites are isolated, viruses circulating in these niches are thought to be undisturbed and crude, which may induce a severe manifestation of symptoms when human hosts enter these areas. Tree holes are the main reservoirs here where the *Aedes* mosquito prefers to lay eggs, and if the eggs are possessed with virus, then these may be an unchallenged virus form. If strains contained by mosquitoes of peri-urban foci happen to be different than the urban strain, mixing of the two strains may be causal for DHF (Joshi et al., 2006).

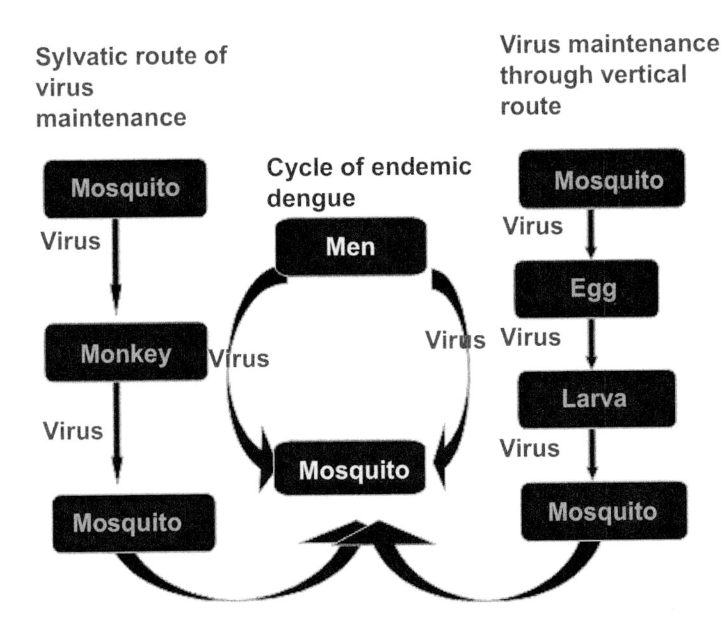

Figure 4.15 Transmission routes of dengue virus.

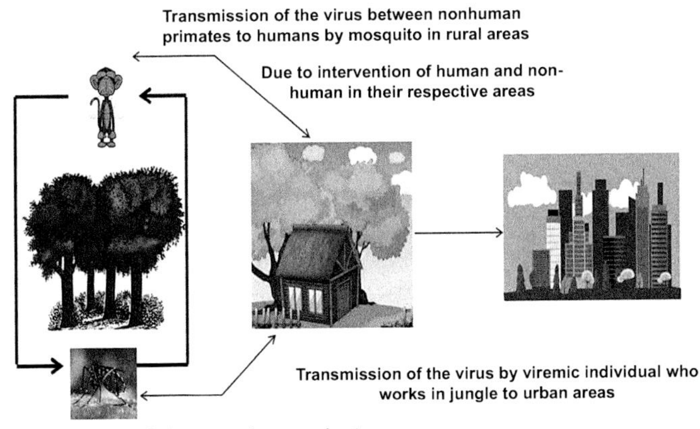

Figure 4.16 A pictorial representation of the transmission cycle of dengue infection. Figure adapted from https://www.cdc.gov/dengue/educationtraining/index.html.

The molecular and cellular steps involved in the viral replication and dissemination within a single cell or mosquito or human host cell should be fully comprehended in order to understand the dengue virus transmission.

4.3.1 Replication of Dengue Virus within Systems: The Extrinsic System

The dengue virus has been found intracellularly in two systems. Among them, one is within the mosquitoes, which is the "extrinsic form."

When a DENV-infected female *Aedes* mosquito bites a person, blood along with the virus particles is ingested from the proboscis into the gut of the mosquito. Therefore, midgut is the first site of virus replication where DENV cells can efficiently survive and increase their population. In addition, the DENV reaches the hemolymph, where the virus particles access other mosquito tissues such as ovaries, nervous system, fat bodies, body cavity, and salivary glands. (Fig. 4.17) (Leake, 1992). It has been reported that some bacterial species inside the mosquito gut also transmit dengue infection (Charan et al., 2016).

Tissues such as muscle, diverticula, hindgut, and Malpighian tubules are not infected by DENV, which shows selective tissue tropism in insects as it has been reported in humans (Linthicum et al., 1996; Sriurairatna and Bhamarapravati, 1997). The well known mechanism of DENV entry is not clear yet due to the diversity in DENV cellular tropism both in vitro and in vivo. However there may be possibility that DENV enters inside the host cells by using receptor present at the surface of affected host cell class (Acosta et al., 2008). In 1989, Hase and his coworkers reported that DENV virions can penetrate the plasma membrane of the mosquito cells by interrupting the membrane formed at the adsorption sites. According to a study by Salas-Benito and del Angel, DENV-4 binds to two glycoproteins having MWs of 40,000 and 45,000 Da located at C6/C36 cell surface (Salas-Benito and del Angel, 1997). Angel and Joshi (2008) and Angel et al. (2014) explained the mechanism of association of 200 kDa protein found in the ovarian tissues of *Aedes* mosquito with the DENV. Presence of this ovarian protein inhibits virus multiplication and circulates through the TOT route (Angel and Joshi, 2008; Angel et al., 2014). Transmission of DENV is on C6/C36 cells, and titer of particles of virus has been investigated by infectious center plaque titration or infectious center assays on BHK-15 cells (Diamond et al., 2000; Lambeth et al., 2005; Roche et al., 2000). In genus *Aedes*, it has been demonstrated that DENV and its receptors are present in different tissues of mosquito in diverse stages of its life cycle (Angel, 2012; Frier and Grimstad, 1983; Khin and Than, 1983; Rosen et al., 1983, 1987; Sriurairatna and Bhamarapravati, 1977).

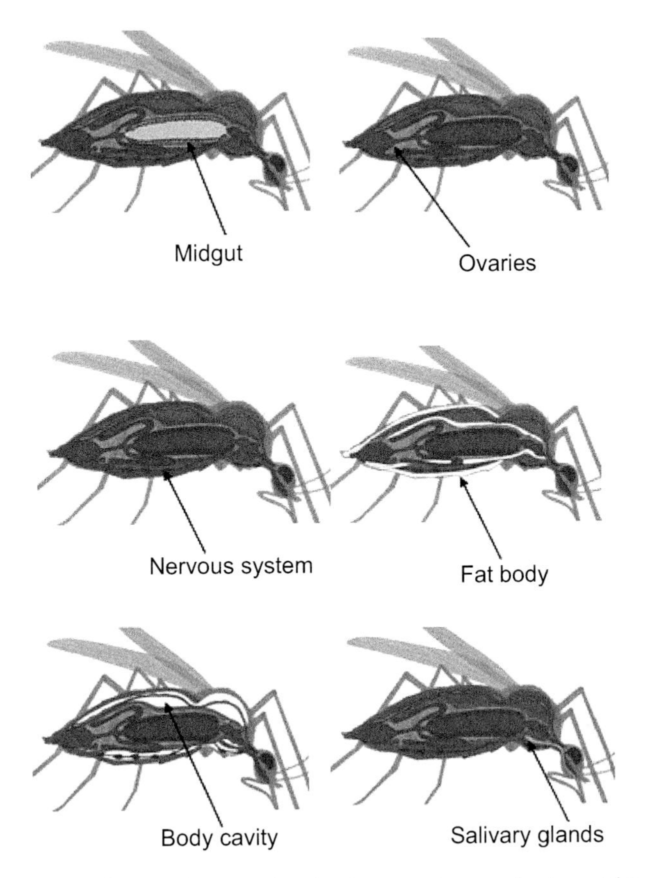

Figure 4.17 Different tissues wherein dengue virus replicates within the mosquito system. Figure taken from https://www.who.int/tdr/diseases-topics/dengue/en/.

4.3.2 Replication of Dengue Virus within Systems: The Intrinsic System

The second mechanism of virus replication is through humans, which is known as the "intrinsic form." It is characterized by symptoms such as extremely high fever, headache, and nausea. Human beings are infected by DENV-carrying female *Aedes* mosquito during a blood meal. As the mosquito penetrates the skin by using proboscis, the virus is released into the blood vessel along with some salivary proteins. Now DENV from the peripheral blood vessel circulates in the blood system and infects tissues such as lymphatic system, liver, spleen, and thymus, as shown in Fig. 4.18.

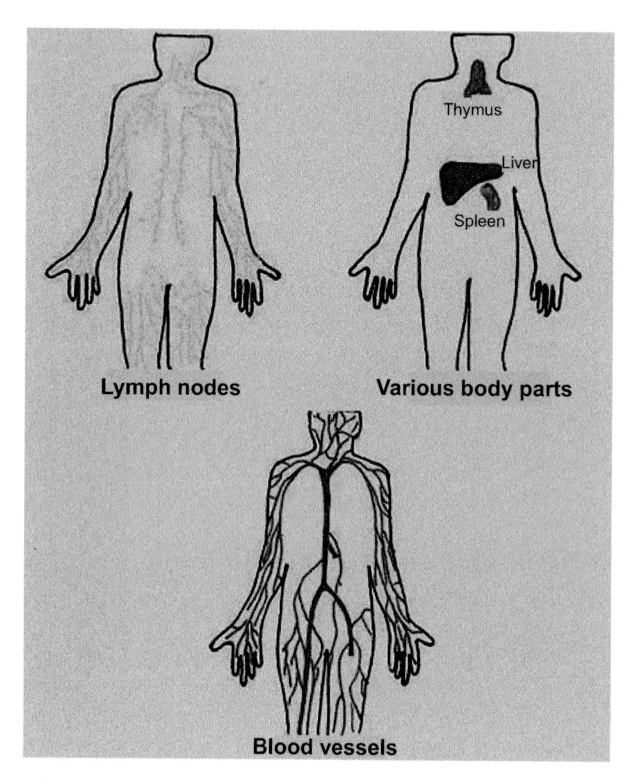

Figure 4.18 Target organs of dengue virus replication in human hosts.

It has been reported that transmission of DENV in humans is same as in the case of mosquito. First, the virus enters the peripheral blood monocytes through plasma or pinocytic membranes of vacuoles (Hase et al., 1989). C6/C36 cells have been used for virus attachment and its penetration in human monocytes (Barth, 1992; Hase et al., 1989). It has been assumed that some molecules act as DENV receptors present on distinct host cells like a highly sulfated type of heparan (HS) present on Vero and BHK cells (Chen et al., 1997; Hung et al., 1999), and two proteins ranging from 40 to 45 kDa and 70 to 75 kDa, respectively, are localized in the myelomonocyte cell line HL60 and non-Epstein–Barr virus transformed B cells (Bielefeldt-Ohmann, 1998). Therefore, dengue virus uses precise molecules for attachment as well as entry into distinct cell lines.

4.3.3 Intracellular Replication of Dengue Virus

Since in the earlier chapters, different tissues of the mosquito as well as the human system being infected by the dengue virus have been discussed, it is important to study the replication strategy of the virus at the intracellular level in tissues of both the host and the vector. Virion is the infected form of dengue virus. When this infected form penetrates the host cell, receptors on the surface of the virion bind with the host cell receptors and are internalized by endocytosis to form endosome within the host cell. This endosome consists of invaginated cell membrane of human host cell, inner layer of envelope of the invaginated virion, and double-layered nucleocapsid with single-stranded RNA in the core region. Decreasing pH mediates the fusion of cell membrane with the envelope of virion and thereby release of nucleocapsid into the cytoplasm of host cell (www.who.int/tdr).

The DENV RNA attaches to the ribosomes of rough endoplasmic reticulum (rER) and forms single 3000 amino acid poly proteins and undergoes post-translational modification of host and viral proteases in the cytoplasm of host cell. The DENV replicates inside the nucleus; consequently, the positive-sense RNA transcribes into the negative-sense RNA, which is used for the generation of positive-sense daughter RNA. Daughter nucleocapsids are formed by the interaction of newly synthesized genomes with the capsid proteins of parent virion in cytoplasm. Now nucleocapsids are surrounded with a membrane that contains envelope proteins followed by generation of immature virions and then diffused to Golgi bodies (GBs) where sugar moieties are added. Mature virus particles are released outside by the process of exocytosis in the form of secretory vesicles (www.who.int/tdr). Figure 4.19 depicts the internalization and subsequent intracellular multiplication of dengue virus (Angel, 2008).

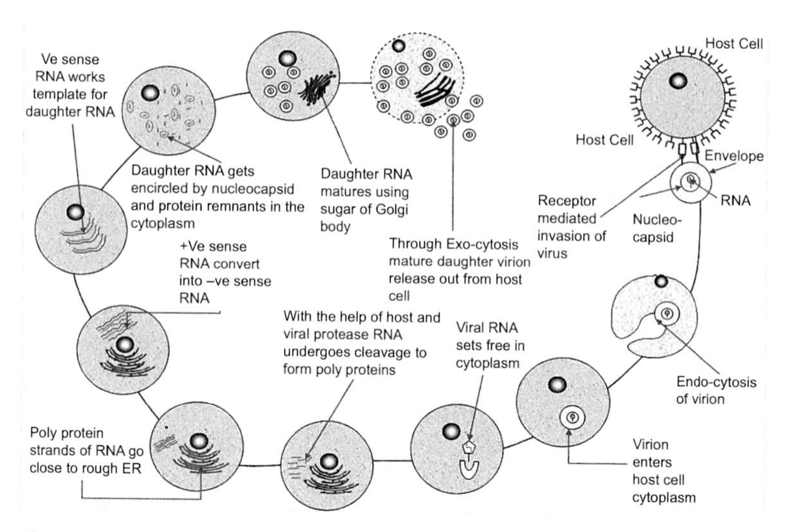

Figure 4.19 Pictorial expression of replication pathway of dengue virus.

4.4 Pathogenicity

Till date there is no well-known mechanism for the pathogenicity caused by DENV, and severe forms of dengue are DHF and DSS. Efforts are going on for studying pathogenesis from the viewpoint of either host or virus, or type of vector species. In the 1780s, Benjamin Rush first described the disease as "bilious remitting fever" or "breakbone fever" (Rush, 1789). The disease is characterized as symptomatic and asymptomatic (Harris et al., 2000; Henchal and Putnak, 1990; Martina et al., 2009). In symptomatic diseases, patients suffer from elevated fever (103–106°F), stinging at the back of eyes, headaches, myalgia, back pain, inflammations, nausea, and vomiting. On the other hand, asymptomatic diseases are estimated to constitute approximately three-quarters of the total dengue infections (Bhatt et al., 2013; Burke et al., 1988; Duong et al., 2015; Endy et al., 2011; Mammen et al., 2008).

DHF and DSS are the severe forms of DENV infection that show symptoms such as coagulopathy, elevated vascular fragility, permeability, leucopenia, occasional thrombocytopenia, while DSS is characterized by hypovolemic shock. The WHO has classified DHF into four categories (Table 4.5); however, the categorization is being further modified (Fig. 4.20) (Martina et al., 2009; WHO Guidelines, 2009). Patients suffering from DENV fever have any of the four distinct serotypes of DENV. Studies are going on to investigate the severity of DENV infection. However, researchers are facing major problems in understanding the mechanism of DENV infection due to the lack of appropriate animal models.

Table 4.5 The criterial of the World Health Organization for characterizing DHF patients

Grades of disease	Signs and symptoms
I	Fever accompanied by non-specific constitutional symptoms with a positive tourniquet test as the only hemorrhagic manifestation
II	Same as grade I, except with spontaneous hemorrhagic manifestation
III	Circulatory failure manifested by rapid, weak pulse pressure (<20 mmHg) or hypotension
IV	Profound shock with undetectable blood pressure and pulse

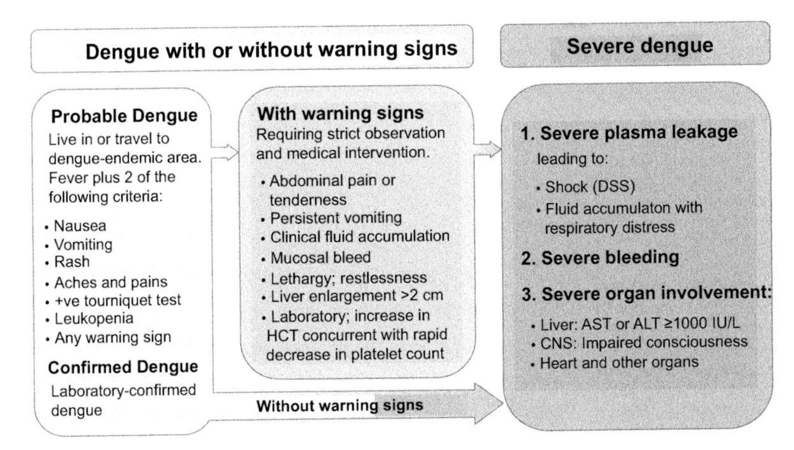

Figure 4.20 Classification of dengue disease by the World Health Organization (WHO, 2009).

The DHF form of dengue infection occurs due to antibody-dependent augmentation of the host system because a person infected with one strain of DENV gets infected again with another strain of virus (Halstead, 1970, 1980, 1988) (Fig. 4.21). DENV has exhibited great variability during their replication inside the host system which can be the cause of DHF as it is confirmed by researchers. Therefore, some DENV strains might have shown more outbreak possibility (Gubler et al., 1978; Gubler, 1988; Rosen, 1977). According to Chaturvedi and Nagar (2008), the DHF form may be due to the result of cascade events of cytokines. T-cell activation during secondary infection may be the promising basis of DHF (Mongkolsapaya et al., 2003). One more study has suggested that during the primary phase of infection (primary case), the circulation of multiple strains of DENV within *Aedes* mosquito systems in outbreak region leads to DHF infection (Angel et al., 2015).

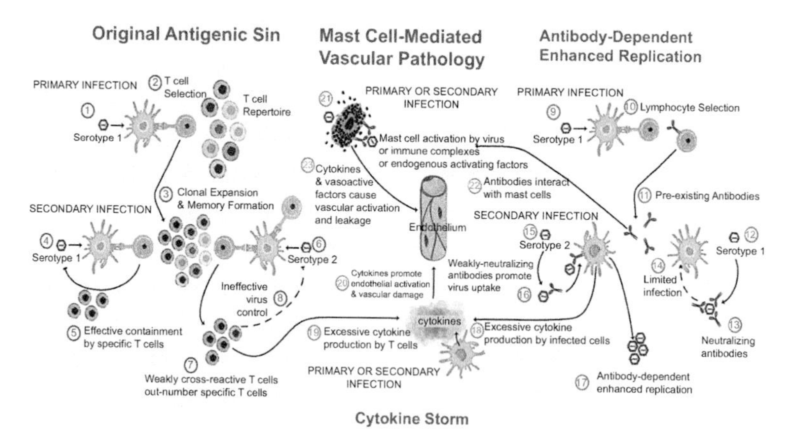

Figure 4.21 Diagrammatic representation of the virus theories for dengue virus pathogenicity. Reprinted from St. John, 2015, under Creative Commons Attribution License.

Although extreme blood loss has not been reported in DHF and DSS patients, if DENV infects the gastrointestinal tract, it leads to anoxia, cell death, and bleeding in the gastrointestinal tract. During viral infection in the immune system, endothelial cells (ECs) lining the blood vessels and liver get activated. Langerhans cells (epidermal dendritic cells) are the primary cells to interact with DENV during infection followed by keratinocytes present in the dermis and epidermis regions of the skin. Now, DENV diffuses into lymph nodes and encounters monocytes and macrophages, which cause primary viremia. Furthermore, spleen and liver macrophages and myeloid dendritic cells are also infected (Fig. 4.22) (Boonnak et al., 2008; Durbin et al., 2008; Ho et al., 2004; Jessie et al., 2004; Kou et al., 2008; Limon-Flores et al., 2005; Martina et al., 2009; Wu et al., 2000). In blood, virus exists for 2–7 days, and during this time, a person develops fever. After an incubation period of 4–10 days, the bite from an infected mosquito shows symptoms within 2–7 days (www.who.int/denguecontrol/human/en/) (Figs. 4.22 and 4.23).

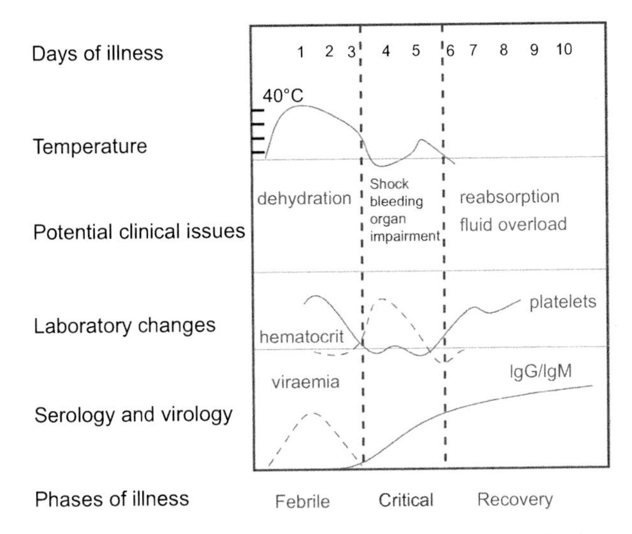

Figure 4.22 The clinical course of dengue (Courtesy: WHO, 2009).

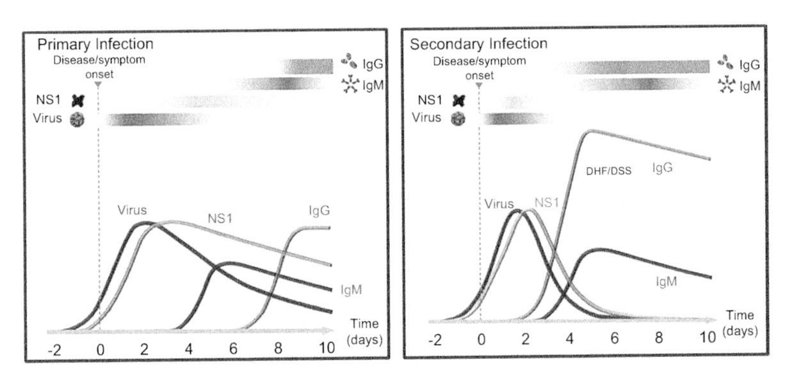

Figure 4.23 Graphical representation of primary and secondary infection of dengue. Figure reprinted from Muller et al., 2017, with permission from Oxford University Press.

4.5 Treatment and Diagnostics

4.5.1 Dengue Vaccine Update

Unfortunately, dengue disease has no reported vaccine and chemotherapy in the market till date (Blanc et al., 1929; Simmons et al., 1931). However, scientists have synthesized an inactive strain of DENV-1 using mice brain (Sabin and Sclesinger, 1952). Scientists are putting extra efforts to develop inactivated virus, attenuated live antigens, subunit-based vaccines, and vectored and DNA vaccines. However, the major problem that arises is the defense mechanism as it should be against all dengue serotypes (Whitehead et al., 2007). A tetravalent vaccine was developed against all serotypes of dengue, and the experiment was done in the kidney cells of dogs. The developed vaccine had a great shelf life of about 15 months to 45 years, but its response declined, i.e., 36–63% in adult cells (Edelman et al., 2003; Simasathein et al., 2008; Sun et al., 2009; Watanaveeradej et al., 2011). A recombinant DNA generated by ligating segments from different sources, termed chimeric YF17D-DENV tetravalent dengue vaccine (CYD TDV), has been synthesized (McArthur et al., 2013; Sabchareon et al., 2012). Another vaccine, DENVax, has been produced by CDC, Atlanta (Osaorio et al., 2011). A tetra-Vax DV vaccine was also developed employing site-directed mutagenesis by the National Institute of Allergy and Infectious Disease (NIAID), United States (Blaney et al., 2004; Murphy and Whitehead, 2011).

Additionally, there are a number of recombinant subunit vaccines under pre-clinical and primary trial. The developed vaccines are prepared using envelope or capsid protein of DENV as it is present in all serotypes of dengue and expression systems, namely, *Saccharomyces cerevisiae*, *Escherichia coli*, *Pichia pastoris*, Chinese hamster ovary DHFR system, baculovirus or Drosophila S2 cells (Coller et al., 2007). Virus-like particle (VLP)-based vaccines are also in the trial stages using adenovirus type 5 (AdV5) vector (Khanam et al., 2011; Rollier et al., 2011). DNA-based tetravalent vaccines have been prepared using four distinct plasmids encoding PrM/E genes of all the DENV serotypes (Danko et al., 2011).

4.5.2 Disease Management

The WHO have laid down certain instructions that should be advised to patients and which the primary and secondary health centers should follow (WHO, 2009 guidelines, WHO website). Extensive efforts are being made for the management of the disease as the WHO has formulated a pro forma regarding information on dengue patients, including family history of the disease (WHO, 2009 guidelines, WHO website). The general information asked in the aforesaid pro forma includes medical details, traveling in the dengue-affected regions within or outside the country, and area details. Various dengue-related tests should be done by taking blood sample from the patient. Patients should be advised appropriately on diet such as intake of plenty of fluid, ORS, and proper rest. As no specific drug against dengue is available in the market, patients are advised to take antipyretic drugs, namely, paracetamol to reduce fever. However, nonsteroidal anti-inflammatory drugs and aspirin or ibuprofen should not be prescribed as these induce internal bleeding. Patients should be under proper medical surveillance if there is severe condition and should be given blood transfusion, platelet transfusion, and intravenous fluid therapy should be advised and recommended (WHO guidelines, 2009). Hence, regular efforts have been supported globally to lessen the disease problems. Still more efficient preventive measures should be considered to control the disease risk.

4.5.3 Laboratory Diagnosis of Dengue Virus

4.5.3.1 Virus isolations

Intracerebral inoculation in mice: The infected sera from patients were inoculated into suckling mice for possible infection. This method was used conventionally for the isolation of dengue virus, as shown in Fig. 4.24. This method takes almost 1–2 days for intracerebral inoculation and then monitored for possible dengue symptoms (Henchal and Putnak, 1990; Kimura and Hotta, 1944; Meiklejohn et al., 1952; Sabin, 1952; Sabin and Schlesinger, 1945; Schlesinger, 1977).

Figure 4.24 Intracerebral inoculation procedure. Figure taken from https://commons.wikimedia.org/wiki/File:Intracerebral_inoculation_of_infectious_material_into_suckling_mice.jpg.

Intrathoracic inoculation in mosquitoes:

There is one more conventional approach in which DENV is intrathoracically injected in *Aedes* mosquitoes, as shown in Fig. 4.25 (Gubler and Rosen, 1976; Kuberski and Rosen, 1977). Afterward there is DENV multiplication within 4 to 5 days in the range of 10^6 to 10^7 MID_{50} depending on the incubation temperature (Paula and Fonseca, 2008).

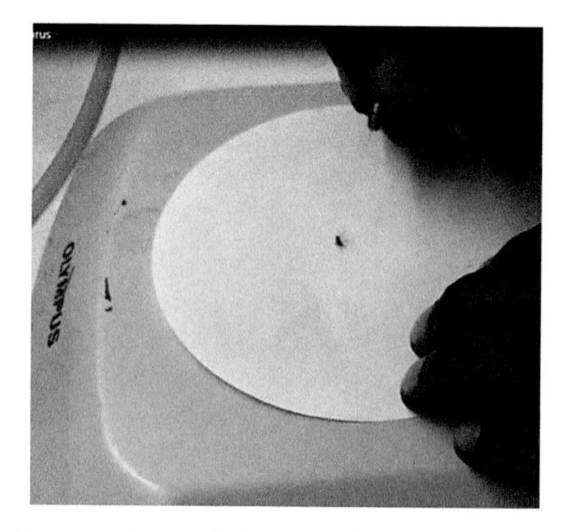

Figure 4.25 Intrathoracic inoculation procedure.

Mammalian cell line culture: Various cell lines such as Vero cell lines (monkey kidney cell lines; LLCMK2) and baby hamster kidney cell lines (BHK-21) are exploited for the isolation of DENV, as shown in Fig. 4.26. However, the most exploited cell line is the Vero cell line, but better results could not be attained (Diercks, 1959; Guzman et al., 1984; Halstead et al., 1964; Hotta and Evan, 1956; Miles and Austin, 1963; Rosen and Gubler, 1974; Sukhavachna et al., 1966; Yuill et al., 1968).

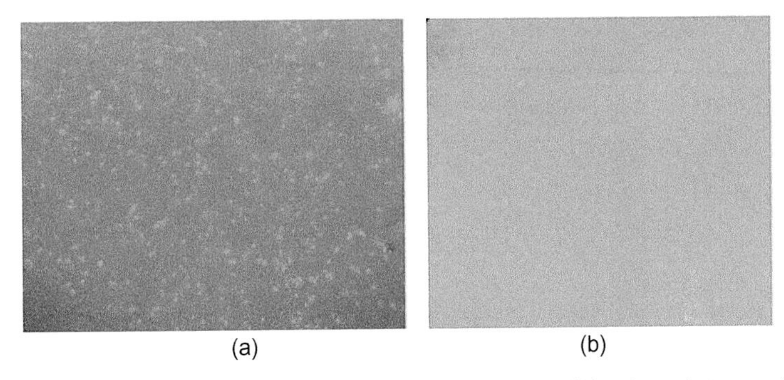

Figure 4.26 (a) Vero cell line inoculated with DENV. (b) Infected Vero cells showing cytopathic effect.

Mosquito cell line culture: Singh and Paul (1969) have used the first instar larval extract of *Aedes albopictus* for the preparation of mosquito cell lines. Other mosquito cell lines that have also been documented include AP61 (*Aedes pseudoscutellaris*) (Hebert et al., 1980; Thongcharoen et al., 1993; Varma et al., 1974), Tra-284 (*Toxorynchites amboinensis*) (Kuno et al., 1985), C636 (*Aedes albopictus)* (Igarashi, 1978; Tesh, 1979), AP64 (clone of *Aedes pseudoscutellaris)* (Morier et al., 1991), and CLA-1 (clone of *Aedes pseudoscutellaris)* cell lines (Morier and Castillo, 1992; Morier et al., 1993). It has been found that the mosquito cell line is more sensitive for all the four dengue serotypes. These mosquito cell lines were easy to culture and can survive and grow at room temperature (Gubler et al., 1984; Guzman and Kouri, 1996; Igarashi, 1978; Kuno et al., 1985; Race et al., 1979; Tesh, 1979).

The immunofluorescence technique has been used to check the growth and multiplication of all serotypes of DENV, as shown in

Fig. 4.27 (Kuberski and Rosen, 1977). The aforesaid technique enables indirect exploration using polyclonal or monoclonal forms of DENV (Henchal et al., 1983). The structural changes in cell line culture can be easily seen under an inverted microscope, as shown in Fig. 4.26.

Figure 4.27 IFAT test positivity as observed under fluorescence microscope.

4.5.3.2 Serological diagnostic tests

Hemagglutination inhibition test (HI): The technique employs antiviral antibodies to inhibit the growth and multiplication of DENV. It is a widely used technique. Erythrocytes from goose and Turkey sources are being utilized for HI assay. However, the major limitation associated with this method is non-specificity (Clarke and Casala, 1958; Gubler, 1989; Vordam and Kuno, 1997).

Plaque reduction neutralization test (PRNT): The test enables detection of antibodies in the patient's serum sample. The antibodies are allowed to react and then inoculated into Vero cell lines. The obtained suspension was then covered by a layer of carboxy methyl cellulose (CMC), which eventually leads to the formation of plaque and is counted under microscope. Staining can also be done in order to expediate the cell counting process (Guzman and Kouri, 1996; Russell et al., 1967).

Enzyme-linked immuno-sorbent assay (ELISA): ELISA is a highly specific (100%) and sensitive (92%) technique. Various kits are available in the market for performing the test. The test relies on the antigen–antibody interaction. Antibody formed in the acute stage is IgM, while antibody formed in the improving phase of disease is IgG. The symptoms appear within 2–5 days, while the DENV-specific IgM antibody appears within 5 days of infection. Both direct and indirect assays are performed in order to detect the DENV (Bundo and Igarashi, 1985; Chungue et al., 1989; Innis et al., 1989; Saluzzo et al., 1986).

4.5.3.3 Molecular diagnostic assays

Reverse transcriptase-polymerase chain reaction (RT-PCR): RT-PCR is a well-exploited technique for the detection of RNA-based viruses. The first step is the isolation of RNA from the DENV by commercially available kits (Qiagen, Invitrogen, Ambion, etc.) or the TRIazol method (Chomczynski–Sacchi) (DePaula et al., 2002; Deubel et al., 1990; Deubel and Pierre, 1994; Figueiredo et al., 1998; Fulop et al., 1993; Lanciotti et al., 1996; Suk-Yin et al., 1994), and then it is amplified using primers. Reverse transcriptase enzyme helps in the conversion of RNA into DNA template. Three authors from the chapter have designed indigenous set of primers for all DENV serotypes (Joshi et al., 2018). DENV can be tested by using different types of PCR such as semi-nested PCR (Gomes et al., 2007), NASBA technique (Usawattankul et al., 2002), and RT-LAMP (Sahni et al., 2013).

Genome sequencing of dengue virus: The entire genetic information of all serotypes of dengue (Angel et al., 2016; Anoop et al., 2012; Chao et al., 2005; Dayaraj et al., 2011; Ong et al., 2008; Patil et al., 2011; Schreiber et al., 2009; Sharma et al., 2011; Shin et al., 2013) and partial sequences (Angel et al., 2016; Domingo et al., 2006; Kukreti et al., 2008, 2009) have been investigated. Table 4.6 depicts the results of protein sequencing. The genetic material of DENV is RNA and, thus, mutates at a fast rate (Drake, 1993; Holmes and Burch, 2000; Bennett et al., 2003). Chin-Inmanu et al. (2012) have used pyro-sequencing strategy for sequencing the viral genome. The Broad Institute Massachusetts, USA, and many pioneer institutes have contributed to huge data of DENV genome (Ong, 2010).

Web portals and consortiums have sequenced the genome, and other pertinent information can be made available on a single platform in a convenient way. Portals such as DengueInfo (Schreiber et al., 2007), Flavitrack (Misra and Schein, 2007), NCBI's Virus Variation Resources (Resch et al., 2009), DENVirDB (Asnet et al., 2014), and Flavivirus Toolkit provide genome-related information on DENV.

Table 4.6 Length of the DENV proteins derived from complete gene sequence from NCBI database

Proteins	DENV-I	DENV-II	DENV-III	DENV-IV
C	114	114	113	113
PrM/M	166	166	166	166
E	495	495	493	495
NS1	352	352	352	352
NS2a	218	218	218	218
NS2b	130	130	130	130
NS3	619	618	619	618
NS4a	150	150	150	150
NS4b	249	248	248	245
NS5	899	900	900	900
Length of CDS	3392	3391	3390	3387

Source: Ong, 2010.

4.5.3.4 Nanotechnology-based detection methods

Biosensor-based methods: Conventional methods show high specificity, sensitivity, and reliability, but there are still some issues; for example, the earlier methods do not exhibit portability, which limits their use for point-of-care applications. However, biosensors can be an alternative to the conventional methods as they exhibit many characteristic features such as high analytical performance, cost effectiveness, facile preparation, fast response, and potential for miniaturization and point-of-care analysis (Mascini and Palchetti, 2011; Sin et al., 2014).

Electrochemical biosensor: An electrochemical biosensor produces a current signal upon interaction of biological recognition element and analyte. The redox reaction produces changes in the chemical signals, which consequently change the electrical signal in the form of currents or voltage or resistance. The change in response means increased or decreased response, which is related to the concentration of analyte in the biological sample (Perumal and Hashim, 2014). Some biosensors are reportedly employed for the detection of dengue viruses. Some of which are summarized as follows:

Cratylia Mollis designed a device that can recognize the glycoprotein of DENV. The developed design has the capability to identify three different types of dengue serotypes in the real samples. The developed sensor has the potential to identify specific glycoproteins of dengue, and the sensitivity of the developed sensor is also increased. In this approach, lectin is employed to identify glycoprotein (Andrade et al., 2011; Oliveria et al., 2009). An immunosensor for the detection of NS1 antigen was designed by Cavalcant et al. (2012). The developed immunosensor showed better sensitivity, specificity, and a wide linear range. In addition, the developed sensor is also cost effective. Nanomaterials and oriented captured antibody were immobilized to enhance the sensitivity of the developed sensor (Bhattacharya et al., 2011). Immobilized antibodies are very specific toward the specific serotypes of DENV, which further enhanced the sensitivity of biosensor (Lee et al., 2013). Biosensors are also employed for the detection of NS1 antigens in real samples (Luna et al., 2014). In addition, a device based on Con A and lipid membrane was employed for the detection of DENV in patients (Avelino et al., 2014; Luna et al., 2014). Avelino et al. (2014) have constructed a lectin-based biosensor for the differentiation of glycoproteins from the sera of patients infected with dengue.

An electrochemical biosensor based on antibody immobilization by streptavidin and biotin linkage is available for the detection of glycoprotein in real samples (Prakash et al., 2014). Some researchers isolated IgY antibodies from the egg yolk and considered a highly specific antibody and can be considered biological recognition element for further biosensor applications (Figueiredo et al., 2015). Another sensor was fabricated, which was based on alumina-modified platinum electrode for the detection of dengue serotype. The developed sensor showed lower detection limit and wide linear range, and response time was also very fast. The sensor was fabricated to detect the RNA of DENV (Cheng et al., 2012) (Fig. 4.28).

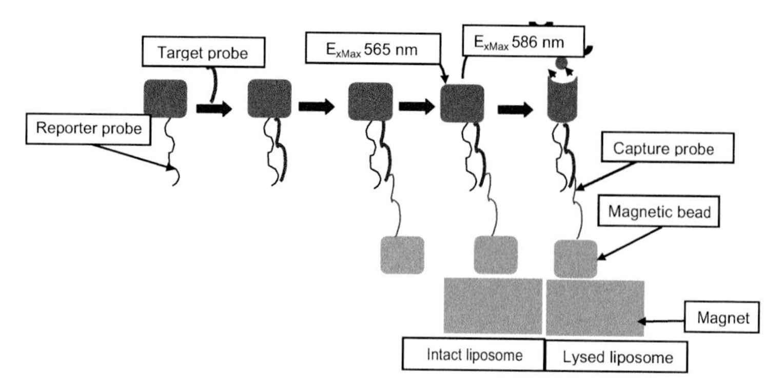

Figure 4.28 Schematic representation of a biosensor based on DNA/RNA hybridization, magnetic bead complex formation and fluorescence detection of RNA-specific complexes via intact (a) and lysed (b) liposomes. Reprinted from Kwakye et al., 2006, with permission from Elsevier.

A DNA biosensor based on pencil graphite electrode was constructed to detect the nucleic acid sequence of dengue serotype I. The developed sensor showed many characteristic features such as high sensitivity, low limit of detection and specificity (Souza et al., 2011). Another microfluidic biosensor was constructed on the basis of HRP-labeled antibody against DENV Ag using polymethacrylate microspheres (Lieberzeit et al., 2016). The sensitive dengue-DNA-based biosensor was constructed on the nanocomposite framework, that is, gold nanoparticles and polyaniline (PANI) (Nascimento et al., 2011). Another DNA-based biosensor was constructed and was employed for the detection of dengue serotype I and III (Rai et al., 2012).

- **Advantages:** Electrochemical biosensors offer many advantageous features, which make them most exploited in biosensing applications. Various features are high sensitivity, simplistic approach, cost effective, easy to handle, and tunable elasticity.
- **Disadvantages:** These electrochemical biosensors take more time to detect the labeled analyte of interest.

Piezoelectric biosensors: Piezoelectric biosensors are those biosensors that rely on the change in mass as they depict change in mass upon interaction of biological recognition element and analyte. These biosensors exploit the piezoelectric material as a transducing element such as quartz. After interaction of biological recognition element and analyte, the piezoelectric material resonates at specific frequency and controlled by the potential applied, which results into generation of an oscillating electric field. After surface interactions or reaction of recognition element and analyte, there is change in the value of current due to distinct resonance frequency by crystal microbalance (QCM) (Perumal and Hashim, 2014).

A piezoelectric biosensor was fabricated to detect dengue virus in the earlier phase of infection (Su et al., 2013). The sensor was found to have good sensitivity in real samples as well. Another antibody- and antigen-based piezoelectric biosensor was fabricated for the detection of dengue viruses. The biosensor was employed for the detection of envelope protein and NS1 protein. The limit of detection was found to be good enough to detect the virus in real samples (Wu et al., 2005).

Another piezoelectric biosensor was fabricated for the detection of nonstructural protein antigen. The biosensor was designed by exploiting thin molecular imprinted polymer, which was specific for the detection of nonstructural protein of dengue virus. The polymer was anchored onto the QCM and formed amino acid cavities, which brought specificity to the biosensor (Tai et al., 2006). The developed biosensor did not require any pre-treatment steps. Furthermore, Chen et al. (2009) also constructed another piezoelectric biosensor, which exploited gold nanoparticles into circulating flow QCM for the detection of dengue serotype II.

- **Advantages:** Piezoelectric biosensors offer many advantageous features such as high specificity, ease of operation, sensitivity, and economy as well (Chen et al., 2009).
- **Disadvantages:** Piezoelectric biosensors take more time to detect the change.

Optical biosensors: Optical biosensors rely on the principle of change in absorption upon interaction of biological recognition element and analyte. The intensity of absorption and fluorescent signals can be detected by the fluorescent detector (Abdulhalim et al., 2007). Various optical biosensors were fabricated for the investigation and detection of various serotypes of dengue. An optical biosensor based on super-magnetic beads was designed for the detection of RNA of dengue serotype III. The developed biosensor utilizes capture and reporter probes. The capture probe showed interaction with the RNA of DENV, while the reporter probe is associated with the encapsulated liposome, which is utilized for fluorescent signal. After the interaction of probe and analyte, a complex bead associated liposome complex like structure forms, which gives fluorescence and can be easily detected by a fluorescence microscope. The detection limit was also in picomoles (Kwakye and Baeumner, 2003).

Another microfluidic biosensor was fabricated for the specific detection of different serotypes of dengue virus. The aforementioned biosensor exhibited unique sensitivity with low limit of detection for both intact and lysed liposome system (0.125 nM and 50 pM, respectively) and response time 20 min (Zaytseva et al., 2005).

For the detection of specific DENV serotypes, Baeumner et al. (2002) fabricated a genosensor based on isothermal nucleic acid sequence-based amplification (NASBA). This optical genosensor exhibited advantages such as simplicity, portability, and cost effectiveness.

A marker-free optical biosensor involving serological assays has also been used for the investigation of DENV. IgM antibodies specific for dengue have been detected by using surface plasmon resonance. The approach for detection of dengue virus was based on two immunoassay techniques such as direct immunoassay for detecting dengue-specific IgM antibody and indirect competitive inhibition for detecting dengue in acute dengue phase of infection. In direct immunoassay, antibody from sera was adsorbed at the surface of recognition element confined with dengue antigen that elevates the resonance angle. On the other hand, indirect immunoassay involves the incubation of monoclonal antibody specific for DENV followed by the addition of antigen-incubated solution on the sensor confined with DENV antigen. As a result, decline in resonance angle indicates the restricted interaction. This assay can be useful for the investigation of acute and late dengue infection (Kumbhat et al., 2010).

A chemiluminescent optical fiber immunosensor (OFIS) was developed by Atias et al. (2009) for the investigation of IgM antibody specific for dengue. This OFIS exhibited enhanced sensitivity and low limit of detection (Atias et al., 2009).

- **Advantages:** The aforementioned optical biosensors have shown high selectivity, sensitivity, independence of marker, and reliable results. These biosensors can be used to screen multiple samples simultaneously.
- **Disadvantages:** In these biosensors, a microscope equipped with fluorescence filters has been used. Consequently, detection of dengue becomes very expensive with these devices.

4.6 Risk Factors

Virus, vectors, and human hosts are considered risk factors linked with dengue fever. The components that pose a risk include the fast mutating behavior of the dengue virus; the other components are vector and human hosts, which are important from the viewpoint of management and feasibility. The various factors that manifest the risk of DF/DHF/DSS include housing patterns, socioeconomic and environmental conditions. The socioeconomic conditions rely on the profile or status of the people living in a society such as low, medium, and high profile people. Mass housing pattern, water storage devices for household chores, living in huts, throwing waste materials near houses, pits for water lodging, etc. are risk factors that provide ample sites for mosquito breeding. Although high socioeconomic areas have better water storage, garbage disposal facility, good housing design, better hygiene, and sanitation, peri-domestic containers such as pots for gardening and animal feeding are kept outside and have been surveyed to be positive sites for larval breeding. In India, religious practice and traditions require the placing of such containers in peri-domestic areas. Small and big pots kept outside or inside the house for gardening are also considered breeding sites for mosquito larvae (Joshi et al., 2006, 2012, 2013; Sharma et al., 2008). As *Aedes* mosquitoes bite in daytime, housewives and babies are more at risk compared to persons who move in and out of residential areas (Velaso-Salas et al., 2104). Integrated vector management (IVM) strategies should be employed as suggested and stressed by the WHO so that different forms of vector species can be controlled. Recently our group surveyed approximately 130,525 containers in the state of Rajasthan and observed that only 1.09% containers are positive for *Aedes* larval breeding, which suggests that small efforts and measures if applied to small niches may help in management of disease by controlling the vector (Fig. 4.29) (Angel et al., 2016).

Figure 4.29 To identify the reservoirs that can be potential sites for larval breeding, epidemiological and entomological parameters were investigated. (a) Bird bath kept outside house as sites of *Aedes* larval breeding. (b) Plant pots containing lodged water indicate sites of larval breeding. (c) Waste tires kept

(*Continued*)

outside homes are sites for collection of rainwater. (d) Resting sites for *Aedes* adults in one of the city gardens located at the periphery of the urban area. (e) Water collection in the garden situated at the periphery of urban areas for storing water for plants. (f) Nurseries act as sites of water collection and lodging. (g) Peri-domestic cement water containers kept for cow/cattle drinking. (h) *Aedes* eggs laid on the surface of the cement container. (i) Indoor water containers are sites for larval breeding. (j) Rural housing pattern and water-storing containers are sites for larval breeding. (k) Rural area showing necessity of water collection due to insufficient water supply. (l) Cement water containers kept at the peri-domestic site for animal drinking. (m) Collection of water by villagers from tank made available by municipal department and water lodging nearby becomes efficient sites for larval breeding. (o) Surveying water containers kept outside a house adjacent to a primary school. (p) Checking of water tank for larval breeding. (q) Investigation of "kunda" (a form of earthen vessel) kept below the rounded earthen pots filled with water. These are rarely cleaned, so they become sites of larval breeding. (r) Investigation and questioning of the past and recent episodes of illness in the house of an infected patient.

Public places such as zoos, parks, and gardens also provide ample environmental conditions for breeding of mosquitoes. It is also known that the sylvatic cycle is responsible for circulating the DENV strains from forest areas to urban areas. If the strain is different from the already prevailing strains in urban areas, then it may pose a risk to persons visiting the areas and they may bring the new strain to urban areas (Moncayo et al., 2004; Wolfe et al., 2001).

Visitors arriving from dengue endemic areas pose a serious risk and increase the chance of DENV circulation (Semenza et al., 2014). Studies also concluded that age (Angel et al., 2016, 2017), seasonal distribution (Angel and Joshi, 2008, 2009), and change in climate and global warmings are also key factors in disease transmission (Goud and Higgs, 2009; Wu et al., 2009). Increased urbanization is also considered a risk factor as mobility and transportation increase (Sungmo et al., 2016).

A research was performed in order to elucidate the linkage between the risk of DHF and patients suffering from hypertension, diabetes, and allergic diseases. The study concluded that people having diabetes or allergy toward steroids were at two and half times more risk of DHF as steroids tend to release pro-inflammatory cytokines due to inflammation, which causes capillary fragility (Brown et al., 2008; Trivedi and Llyod, 2007). The study also concluded that in type II diabetes, physiological and anatomical stability of capillary endothelium gets disturbed, which ultimately causes release of pro-inflammatory cytokines such as IFN-γ and TNF-α and leads to capillary fragility (Dandona et al., 2004; Hsueh et al., 2004). However, people with hypertension were not linked with the high risk of DHF (Figueiredo et al., 2010). A case study by our group found that when fresh blood (instead of stocked blood, containing live WBCs) was transferred to patients with severe clinical manifestation, an increase in IFN-γ levels was found (Joshi et al., 2012).

4.7 Conclusion

In the 1970s, only nine countries reported dengue; since then the DENV infection has spread around the globe. DENV belongs to the Flaviviridae family. It has three structural proteins, namely, nucleocapsid protein C, membrane-linked protein M, and envelope protein E, and seven nonstructural proteins, namely, NS1, NS2a, NS2b, NS3, NS4a, NS4b, and NS5. The structural protein plays a role in determining the icosahedral virion assembly, while non-structural proteins are crucial for DNV replication. The severe forms of the disease are DHF and DSS. The DHF form is characterized by coagulopathy, increased vascular fragility, permeability, leucopenia, and occasional thrombocytopenia. On the other hand, DSS is characterized by hypovolemic shock.

References

Abdulhalim, I.; Zourob, M.; and Lakhtakia, A. Overview of optical biosensing techniques. In *Handbook of Biosensors and Biochips*, Wiley: Hoboken, NJ, USA, 2007.

Acosta, E. G.; Talaric, L. B.; and Damonte, E. B. Cell entry of dengue virus. *Future Virology*, 2008, **3**(5): 471–479.

Akey, D. L.; Brown, W. C.; Dutta, S.; Konwerski, J.; Jose, J.; Jurkiw, T. J.; DelProposto, J.; Ogata, C. M.; Skiniotis, G.; Kuhn, R. J.; and Smith, J. L. Flavivirus NS1 structures reveal surfaces for associations with membranes and the immune system. *Science*, 2014, **343**(6173): 881–885.

Andrade, C. A.; Oliveira, M. D.; De Melo, C. P.; Coelho, L. C.; Correia, M. T.; Nogueira, M. L.; Singh, P. R.; and Zeng, X. Diagnosis of dengue infection using a modified gold electrode with hybrid organic–inorganic nanocomposite and *Bauhinia monandra* lectin. *Journal of Colloid and Interface Science*, 2011, **362**(2): 517–523.

Angel, A. Study of macromolecular factors to detect dengue transmission or blocking proteins in *Aedes* mosquitoes. PhD Thesis. Jai Narain Vyas University, Jodhpur, 2012.

Angel, A.; Angel, B.; and Joshi, V. Rare occurrence of natural transovarial transmission of dengue virus and elimination of infected foci as a possible intervention method. *Acta Tropica*, 2016, **155**: 1–5.

Angel, A.; Angel, B.; Bohra, N.; and Joshi, V. Structural study of mosquito ovarian proteins participating in transovarial transmission of dengue viruses. *International Journal of Current Microbiology and Applied Sciences*, 2014, **3**(4): 565–572.

Angel, A.; Angel, B.; Joshi, A. P.; Baharia, R. J.; Rathore, S.; and Joshi, V. First study of complete genome of Dengue-3 virus from Rajasthan, India: Genomic characterization, amino acid variations and phylogenetic analysis. *Virology Reports*, 2016, **6**: 32–40.

Angel, A.; Angel, B.; Joshi, V.; and Kucheria, K. Sequential emergence of serological responses of human hosts in a progressive outbreak of dengue and dengue hemorrhagic fever in Rajasthan, India. *Indian Journal of Applied Research*, 2014, **4**(5), 24–26.

Angel, A.; Angel, B.; Yadav, K.; Sharma, N.; Joshi, V.; Thanvi, I.; and Thanvi, S. Age of initial cohort of dengue patients could explain the origin of disease outbreak in a setting: A case control study of Rajasthan, India. *Virus Diseases*, 2017, **28**(2): 205–208.

Angel, B and Joshi, V. Distribution of dengue virus types in *Aedes aegypti* in dengue endemic districts of Rajasthan, India. *Indian Journal of Medical Research*, 2009, **129**: 665–668.

Angel, B. and Joshi, V. Distribution and seasonality of vertically transmitted dengue viruses in *Aedes* mosquitoes in arid and semi-arid areas of Rajasthan, India. *Journal of Vector Borne Diseases*, 2008, **45**: 56–59.

Angel, B. Proteomics of *Aedes* mosquitoes of Rajasthan for development of molecular markers of vector competence for dengue viruses. PhD Thesis. Jai Narain Vyas University, Jodhpur, 2008.

Angel, B.; Angel, A.; and Joshi, V. 2015. Multiple dengue viruses harboured by individual mosquitoes. *Acta Tropica*, 2015, **150**: 107–110.

Angel, B.; Sharma, K.; and Joshi, V. 2008. Association of ovarian proteins with transovarial transmission of dengue viruses by *Aedes* mosquitoes in Rajasthan, India. *Indian Journal of Medical Research,* 2008, **128**: 181–184.

Anoop, M.; Mathew, A. J.; Jayakumar, B.; Issac, A.; Nair, S.; Abraham, R.; Anupriya, M. G.; and Sreekumar, E. Complete genome sequencing and evolutionary analysis of dengue virus serotype 1 isolates from an outbreak in Kerala, South India. *Virus Genes*, 2012, **45**: 1–13.

Asnet, M. J.; Rubia, A. G. P.; Ramya, G.; Nagalakshmi, R. N.; and Shenbagarathai, R. DENVirDB: A web portal of dengue virus sequence information on Asian isolates. *Journal of Vector Borne Diseases*, 2014, **51**: 82–85.

Atias, D.; Liebes, Y.; Chalifa-Caspi, V.; Bremand, L.; Lobel, L.; Marks, R. S.; and Dussart, P. Chemiluminescent optical fiber immunosensor for the detection of IgM antibody to dengue virus in humans. *Sensors and Actuators B: Chemical*, 2009, **140**(1): 206–215.

Avelino, K. Y. P. S.; Andrade, C. A. S.; De Melo, C. P.; Nogueira, M. L.; Correia, M. T. S.; Coelho, L. C. B. B.; and Oliveira, M. D. L. Biosensor based on hybrid nanocomposite and Cramoll lectin for detection of dengue glycoproteins in real samples. *Synthetic Metals,* 2014, **194**: 102–108.

Avirutnan, P.; Fuchs, A.; Hauhart, R. E.; Somnuke, P.; Youn, S.; Diamond, M. S.; and Atkinson, J. P. Antagonism of the complement component C4 by flavivirus nonstructural protein NS1. *The Journal of Experimental Medicine*, 2010, **207**: 793–806.

Avirutnan, P.; Hauhart, R. E.; Somnuke, P.; Blom, A. M.; Diamond, M. S.; and Atkinson, J. P. Binding of flavivirus nonstructural protein NS1 to C4b binding protein modulates complement activation. *Journal of Immunology,* 2011, **187**: 424–433.

Baeumner, A. J.; Schlesinger, N. A.; Slutzki, N. S.; Romano, J.; Lee, E. M.; and Montagna, R. A. Biosensor for dengue virus detection: Sensitive, rapid, and serotype specific. *Analytical Chemistry,* 2002, **74**(6): 1442–1448.

Bell, J. R.; Kinney, R. W.; Trent, D. W.; Lenches, E. M.; Dalgarno, L.; and Strauss, J. H. Amino-terminal amino acid sequences of structural proteins of three flaviviruses. *Virology*, 1985, **143**: 224–229.

Bennett, S. N.; Holmes, E. C.; Chirivella, M.; Rodriguez, D. M.; Beltran, M.; Vorndam, V.; Gubler, D. J.; and Mc Millan, W. O. Selection-driven evolution of emergent dengue virus. *Molecular Biology and Evolution,* 2003, **20**: 1650–1658.

Bhatt, S.; Gething, P. W.; Brady, O. J.; Messina, J. P.; Farlow, A. W.; Moyes, C. L.; Drake, J. M.; Brownstein, J. S.; Hoen, A. G.; Sankoh, O.; Myers, M. F.; George, D. B.; Jaenisch, T.; Wint, G. R.; Simmons, C. P.; Scott, T. W.; Farrar,

J. J.; and Hay, S. I. The global distribution and burden of dengue. *Nature*, 2013, **496**(7446): 504–507.

Bhattacharya, M.; Hong, S.; Lee, D.; Cui, T.; and Goyal, S. M. Carbon nanotube based sensors for the detection of viruses. *Sensors and Actuators B: Chemical*, 2011, **155**(1): 67–74.

Bielefeldt-Ohmann, H.; Meyer, M.; Fitzpatrick, D. R.; and Mackenzie, J. S. Dengue virus binding to human leukocyte cell lines: Receptor usage differs between cell types and virus strains. *Virus Research*, 2001, **73**: 81–89.

Blanc, G. and Caminopetros, J. Contribution a l'etude de la vaccination contre la dengue. *Bulletin de L'Académie Nationale de Médecine*, 1929, **102**(24): 1–4.

Blaney, J. E. Jr.; Hanson, C. T.; Firestone, C. Y.; Hanley, K. A.; Murphy, B. R.; and Whitehead, S. S. Genetically modified, live attenuated dengue virus type 3 vaccine candidates. *American Journal of Tropical Medicine and Hygiene*, 2004, **71**(6): 811–821.

Blaney, J. E. Jr.; Hanson, C. T.; Hanley, K. A.; Murphy, B. R.; and Whitehead, S. S. Vaccine candidates derived from a novel infectious cDNA clone of an American genotype dengue virus type 2. *BMC Infectious Diseases*, 2004, **4**: 39.

Boonnak, K.; Slike, B. M.; Burgess, T. H.; Mason, R. M.; Wu, S. J.; Sun, P.; Porter, K.; Rudiman, I. F.; Yuwono, D.; Puthavathana, P.; and Marovich, M. A. Role of dendritic cells in antibody-dependent enhancement of dengue virus infection. *Journal of Virology*, 2008, **82**: 3939–3951.

Brown, J. M.; Wilson, T. M.; and Metcalfe, D. D. The mast cell and allergic diseases: Role in pathogenesis and implications for therapy. *Clinical and Experimental Allergy*, 2008, **38**(1): 4–18.

Bundo, K. and Igarashi, A. Antibody-capture ELISA for detection of immunoglobulin M antibodies in sera from Japanese encephalitis and dengue hemorrhagic fever patients. *Journal of Virology Methods*, 1985, **11**: 15–22.

Burke, D. S.; Nisalak, A.; Johnson, D. E.; and Scott, R. M. A prospective study of dengue infections in Bangkok. *American Journal of Tropical Medicine and Hygiene*, 1998, **38**(1): 172–180.

Byk, L. A. and Gamarnik, A. V. Properties and functions of the dengue virus capsid protein. *Annual Reviews in Virology*, 2016, **3**(1): 263–281.

Cavalcanti, I. T.; Guedes, M. I.; Sotomayor, M. D.; Yamanaka, H.; and Dutra, R. F. A label-free immunosensor based on recordable compact disk chip for early diagnostic of the dengue virus infection. *Biochemical Engineering Journal*, 2012, **67**: 225–230.

Cavalcanti, I. T.; Silva, B. V. M.; Peres, N. G.; Moura, P.; Sotomayor, M. D. P. T.; Guedes, M. I. F.; and Dutra, R. F. A disposable chitosan-modified carbon fiber electrode for dengue virus envelope protein detection. *Talanta*, 2012, **91**, 41–46.

Chambers, T. J.; McCourt, D. W.; and Rice, C. Yellow fever virus proteins NS2a, NS2b, and NS4b: Identification and partial N-terminal amino acid sequence analysis. *Virology*, 1989, **169**: 100–109.

Chao, D. Y.; King, C. C.; Wang, W. K.; Chen, W. J.; Wu, H. L.; and Chang, G. J. Strategically examining the full genome of dengue virus type 3 in clinical isolates reveals its mutation spectra. *Virology Journal*, 2005, **2**: 1–10.

Charan, S.; Pawar, K.; Gavhale, S.; Tikhe, C. V.; Charan, N.; Angel, B.; Joshi, V.; Patole, M.; and Shouche, Y. Comparative analysis of midgut bacterial communities of three *Stegomyia* mosquito species from dengue-endemic and non-endemic areas of Rajasthan, India. *Medical and Veterinary Entomology*, 2016, **30**(3): 264–277.

Chaturvedi, U. C. and Nagar, R. Dengue and dengue haemorrhagic fever: Indian perspective. *Journal of Bioscience*, 2018, **33**(4): 429–441.

Chen, S. H.; Chuang, Y. C.; Lu, Y. C.; Lin, H. C.; Yang, Y. L.; and Lin, C. S. A method of layer-by-layer gold nanoparticle hybridization in a quartz crystal microbalance DNA sensing system used to detect dengue virus. *Nanotechnology*, 2009, **20**(21): 215501.

Chen, Y.; Maguire, T.; Hileman, R. E.; Fromm, J. R.; Esko, J. D.; Linhardt, R. J.; and Marks, R. M. Dengue virus infectivity depends on envelope protein binding to target cell heparan sulfate. *Nature Medicine*, 1997, **3**: 866–871.

Cheng, M. S.; Ho., J. S.; Tan, C. H.; Wong, J. P. S.; Ng, L. C.; and Toh, C. S. Development of an electrochemical membrane-based nanobiosensor for ultrasensitive detection of dengue virus. *Analytica Chimica Acta*, 2012, **725**: 74–80.

Chin, J.-F.; Chu, J.-J.; and Ng, M. L. The envelope glycoprotein domain III of dengue virus serotypes 1 and 2 inhibit virus entry. *Microbes and Infection*, 2017, **9**: 1–6.

Chin-inmanu, K.; Suttitheptumrong, A.; Sangsrakru, D.; Tangphatsornruang, S.; Tragoonrung, S.; Malasit, P.; Tungpradabkul, S.; Suriyaphol, P. Feasibility of using 454 pyrosequencing for studying quasispecies of the whole dengue viral genome. *BMC Genomics*, 2012, **13**(7): S7 1–8.

Chung, K. M.; Liszewski, M. K.; Nybakken, G.; Davis, A. E.; Townsend, R. R.; Fremont, D. H.; Atkinson, J. P.; and Daimond, M. S. West Nile virus nonstructural protein NS1 inhibits complement activation by binding the regulatory protein factor H. *Proceedings of National Academy of Sciences, USA*, 2006, **103**: 19111–19116.

Chungue, E.; Boutin, J. P.; and Roux, J. Significance of IgM titration by an immunoenzyme technic for the serodiagnosis and epidemiological surveillance of dengue in French Polynesia. *Research in Virology*, 1989, **140**: 229–240.

Clarke, D. H. and Casals, J. Techniques for hemagglutination and hemagglutination-inhibition with arthropod borne viruses. *American Journal of Tropical Medicine and Hygiene*, 1958, **7**: 561–573.

Cleaves, G. R. and Dubin, D. T. Methylation status of intracellular dengue type 2 40S RNA. *Virology*, 1979, **96**: 159–165.

Cleaves, G. R. Identification of dengue type 2 virus specific high molecular weight proteins in virus-infected BHK cells. *Journal of General Virology*, 1985, **66**: 2767–2771.

Coller, B. A.; Clements, D. E.; Bett, A. J.; Sagar, S. L.; and Ter Meulen, J. H. The development of recombinant subunit envelope-based vaccines to protect against dengue virus induced disease. *Vaccine*, 2011, **29**(42): 7267–7275.

Dandona, P.; Alijada, A.; Chaudhuri, A.; and Mohanty, P. Endothelial dysfunction, inflammation and diabetes. *Reviews in Endocrine and Metabolic Disorders*, 2004, **5**: 189–197.

Danko, J. R.; Beckett, C. G.; and Porter, K. R. Development of dengue DNA vaccines. *Vaccine*, 2011, **29**(42): 7261–7266.

Dayaraj, C.; Kakade, M. B.; Bhagat, A. B.; Vallentyne, J.; Singh, A.; Patil, J. A.; Todkar, S. M.; Varghese, S. B.; and Shah, P. S. Detection of dengue-4 virus in Pune, western India after an absence of 30 years: Its association with two severe cases. *Virology Journal*, 2011, **8**: 46.

De Paula, S. O. and Fonseca, B. A. L. Dengue: A review of the laboratory tests a clinician must know to achieve a correct diagnosis. *The Brazilian Journal of Infectious Diseases*, 2004, **8**(6): 390–398.

De Paula, S. O.; Pires Neto, R. J.; Corrêa, J. A. C. T.; Assumpcao, S. R.; Costa, M. L. S.; Lima, D. M.; and Fonseca, B. A. L. The use of reverse transcription-polymerase chain reaction (RT-PCR) for the rapid detection and identification of dengue virus in an endemic region: A validation study. *Transactions of Royal Society of Tropical Medicine and Hygiene*, 2012, **96**: 266–269.

De Silva, A. M.; Dittus, W. P. J.; Amerasinghe, P. H., and Amerasinghe, F. P. Serologic evidence for an epizootic dengue virus infecting toque Macaques (*Macaca sinica*) at Polonnaruwa, Sri Lanka. *American Journal of Tropical Medicine and Hygiene*, 1999, **60**(2): 300–306.

Deubel, V. and Pierre, V. Molecular techniques for rapid and more sensitive detection and diagnosis of flavivirus. In *Rapid Methods and Automation in Microbiology and Immunology* (R.C. Spencer, E. P. Wright, and S. W. B. Newsom, Eds.), Intercept, Andover, United Kingdom, 1994, pp. 227–237.

Deubel, V.; Kinney, R. M.; and Trent, D. W. Nucleotide sequence and deduced amino acid sequence of the nonstructural proteins of dengue type 2 virus, Jamaica genotype: Comparative analysis of the full-length genome. *Virology*, 1998, **165**: 234–244.

Deubel, V.; Laille, M.; Hugnot, J. P.; Chungue, E.; Guesdon, J. L.; Drouet, M. T.; Bassot, S.; and Chevrier, D. Identification of dengue sequences by genomic amplification: Rapid diagnosis of dengue virus serotypes in peripheral blood. *Journal of Virological Methods*, 1990, **30**: 41–54.

Diamond, M. S.; Edgil, D.; Roberts, T. G.; Lu, B.; and Harris, E. Infection of human cells by dengue virus is modulated by different cell types and viral strains. *Journal of Virology*, 2000, **74**: 7814–7823.

Diercks, F. H. Isolation of a type 2 dengue virus by use of hamster kidney cell culture. *American Journal of Tropical Medicine and Hygiene*, 1959, **8**: 488–491.

Domingo, C.; Palacios, G.; Jabado, O.; Reyes, N.; Niedrig, M.; Gasco, J.; Cabreirizo, M.; Lipkin, W. I.; and Tenorio, A. Use of a short fragment of the C-terminal E gene for detection and characterization of two new lineages of dengue virus 1 in India. *Journal of Clinical Microbiology*, 2006, **44**: 1519–1529.

Drake, J. W. Rates of spontaneous mutation among RNA viruses. *Proceedings of National Academy of Science, U.S.A*, 1993, **90**: 4171–4175.

Duong, V.; Lambrechts, L.; Paul, R. E.; Ly, S.; Lay, R. S.; Long, K. C.; Huy, R.; Tarantola, A.; Scott, T. W.; Sakuntabhai, A.; and Buchy, P. Asymptomatic humans transmit dengue virus to mosquitoes. *Proceedings of National Academy of Sciences, USA*, 2015, **112**(47): 14688–14693.

Durbin, A. P.; Vargas, M. J.; Wanionek, K.; Hammond, S. N.; Gordon, A.; Rocha, C.; Balmaseda, A.; and Harris, E. Phenotyping of peripheral blood mononuclear cells during acute dengue illness demonstrates infection and increased activation of monocytes in severe cases compared to classic dengue fever. *Virology*, 2008, **376**: 429–435.

Edelman, R.; Wasserman, S. S.; Bodison, S. A.; Putnak, R. J.; Eckels, K. H.; Tang, D.; Kanesa-Thasan, N.; Vaughn, D. W.; Innis, B. L.; and Sun, W. Phase I trial of 16 formulations of a tetravalent live-attenuated dengue vaccine. *American Journal of Tropical Medicine and Hygiene*, 2003, **69**(6): 48–60.

Endy, T. P.; Anderson, K. B.; Nisalak, A.; Yoon, I. K.; Green, S.; Rothman, A. L.; Thomas, S. J.; Jarman, R. G.; Libraty, D. H.; Gibbons, R. V. Determinants of inapparent and symptomatic dengue infection in a prospective study of primary school children in Kamphaeng Phet, Thailand. *PLoS Neglected Tropical Diseases*, 2011, **5**(3): 1–10.

Falgout, B.; Chanock, R.; and Lai, C. J. Proper processing of dengue virus nonstructural glycoprotein NS1 requires the N-terminal hydrophobic signal sequence and the downstream nonstructural protein, NS2a. *Journal of Virology*, 1989, **63:** 1852–1860.

Figueiredo, A.; Vieira, N. C.; Dos Santos, J. F.; Janegitz, B. C.; Aoki, S. M.; Junior, P. P.; and Guimaraes, F. E. Electrical detection of dengue biomarker using egg yolk immunoglobulin as the biological recognition element. *Scientific Reports*, 2015, **5**: 7865.

Figueiredo, L. T.; Batista, W. C.; Kashima, S.; and Nassar, E. S. Identification of Brazilian flaviviruses by a simplified reverse transcription-polymerase chain reaction method using *flavivirus* universal primers. *American Journal of Tropical Medicine and Hygiene*, 1998, **59**(3): 357–362.

Figueiredo, M. A. A.; Rodrigues, L. C.; Barreto, M. L.; Lima, J. W. O.; Costa, M. C. N.; Morato, V.; Blanton, R.; Vasconcelos, P. F. C.; Nunes, M. R. T.; and Teixeira, M. G. Allergies and diabetes as risk factors for dengue hemorrhagic fever: Results of a case control study. *PLoS Neglected Tropical Diseases*, 2010, **4**(6): 1–6.

Francy, D. B.; Rush, W. A.; Montoya, M.; Inglish, D. S.; and Bolin, R. A. Transovarial transmission of St. Louis encephalitis virus by *Culex pipens* complex mosquitoes. *American Journal of Tropical Medicine and Hygiene*, 1981, **30**: 699–705.

Frier, J. E. and Grimstad, P. R. Transmission of dengue virus by orally infected *Aedes triseriatus*. *American Journal of Tropical Medicine and Hygiene*, 1983, **32**: 1429–1434.

Fulop, L.; Barrett, D. T.; Phillpotts, R.; Martin, L.; Leslie, D.; and Titball, R. W. Rapid identification of flavivirus based on conserved NS5 gene sequences. *Journal of Virological Methods*, 1993, **44**: 179–188.

Gomesa, A. L. V.; Silvab, A. M.; Cordeirob, M. T.; Guimarãesb, G. F.; Marques, E. T. A.; and Abath, F. G. C. [in memorium]. Single-tube nested PCR using immobilized internal primers for the identification of dengue virus serotypes. *Journal of Virological Methods*, 2007, **145**(1): 76–79.

Goncalvez, A. P.; Escalante, A. A.; Pujol, F. H.; Ludert, J. E.; Tovar, D.; Salas, R. A.; and Liprandi, F. Diversity and evolution of the envelope gene of dengue virus type 1. *Virology*, 2002, **303**: 110–119.

Gould, E. A. and Higgs, S. Impact of climate change and other factors on emerging arbovirus diseases. *Transactions of Royal Society of Tropical Medicine and Hygiene*, 2009, **103**: 109–121.

Gubler, D. J. and Meltzer, M. Impact of dengue/dengue hemorrhagic fever on the developing world. *Advances in Virus Research*, 1999, **53**: 35–70.

Gubler, D. J. and Rosen, L. A simple technique for demonstrating transmission of dengue virus by mosquitoes without the use of vertebrate hosts. *American Journal of Tropical Medicine and Hygiene*, 1976, **25**: 146–150.

Gubler, D. J. Dengue and dengue hemorrhagic fever. *Clinical Microbiology Reviews*, 1998, **11**(3): 480–496.

Gubler, D. J. Dengue. In *Epidemiology of Arthropod-Borne Viral Diseases* (T. P. Monath, ed.), CRC Press, Inc., Boca Raton, Fla. 1988, pp. 223–260.

Gubler, D. J.; Kuno, G.; Sather, G. E.; Velez, M.; and Oliver, A. Mosquito cell cultures and specific monoclonal antibodies in surveillance for dengue viruses. *American Journal of Tropical Medicine and Hygiene*, 1984, **33**: 158–165.

Gubler, D. J.; Reed, D.; Rosen, L.; and Hitchcock, J. C. J. Epidemiologic, clinical and virological observations on dengue in the Kingdom of Tonga. *American Journal of Tropical Medicine and Hygiene*, 1978, **27**: 581–589.

Gubler, D. J.; Suharyono, W.; Tan, R.; Abidin, M.; and Sie, A. Viremia in patients with naturally acquired dengue infection. *Bulletin W.H.O*, 1981, **59**: 623–630.

Gupta, N.; Srivastava, S.; Jain, A.; and Chaturvedi, U. C. Dengue in India. *Indian Journal of Medical Research*, 2012, **136**(3): 373–390.

Guzman, M. G. and Kouri, G. Advances in dengue diagnosis. *Clinical and Diagnostic Laboratory Immunology*, 1996, **3**(6): 621–627.

Guzman, M. G.; Kourı, G.; Soler, M.; Morier, L.; and Va´zquez, S. Aislamiento del virus dengue 2 en sueros de pacientes utilizando el rato´n lactante y cultivo de ce´lulas LLCMK2. *Revista Cubana de Medicina Tropical*, 1984, **36**: 4–10.

Hahn, Y. S.; Galler, R.; Hunkapillar, T.; Dalrymple, J. M.; Strauss, J. H.; and Strauss, E. G. Nucleotide sequence of dengue 2 RNA and comparison of the encoded proteins with those of other flaviviruses. *Virology*, 1988, **162**: 167–180.

Halstead, S. B. Dengue hemorrhagic fever: Public health problem and a field for research. *Bulletin W.H.O.*, 1980, **58**: 1–21.

Halstead, S. B. Mosquito-borne hemorrhagic fevers of South and Southeast Asia. *Bulletin W.H.O.*, 1966, **35**: 3–15.

Halstead, S. B. Observations related to pathogenesis of dengue hemorrhagic fever. VI. Hypotheses and discussion. *Yale Journal of Biology and Medicine*, 1970, **42**: 350–362.

Halstead, S. B. Pathogenesis of dengue: challenges to molecular biology. *Science*, 1988, **239**: 476–481.

Halstead, S. B.; Sukhavachana, P.; and Nisalak, A. Assay of mouse adapted dengue viruses in mammalian cell culture by an interference method. *Proceedings of Society of Experimental Biology and Medicine*, 1964, **115**: 1062–1068.

Hammon, W. M.; Rudnick, A.; and Sather, G. E. Viruses associated with epidemic hemorrhagic fevers of the Philippines and Thailand. *Science*, 1960, **131**: 1102–1103.

Harris, E.; Videa, E.; Perez, L.; Sandoval, E.; Tellez, Y.; Perez, M. L.; Cuadra, R.; Rocha, J.; Idiaquez, W.; Alonso, R. E.; Delgado, M. A.; Campo, L. A.; Acevedo, F.; Gonzalez, A.; Amador, J. J.; and Balmaseda, A. Clinical, epidemiologic, and virologic features of dengue in the 1998 epidemic in Nicaragua. *American Journal of Tropical Medicine and Hygiene*, 2000, **63**: 5–11.

Harrison, S. C. Viral membrane fusion. *Nature Structure and Molecular Biology*, 2008, **15**: 690–698.

Hebert, S. J.; Bowman, K. A.; Rudnick, A.; and Burton, J. J. S. A rapid method for the isolation and identification of dengue viruses employing a single system. *Malaysian Journal of Pathology*, 1980, **3**: 67–68.

Henchal, E. A. and Putnak, R. J. The dengue viruses. *Clinical Microbiology Reviews*, 1990, **3**(4): 376–396.

Henchal, E. A.; McCown, J. M.; Seguin, M. C.; Gentry, M. K.; and Brandt, W. E. (1983). Rapid identification of dengue virus isolates by using monoclonal antibodies in an indirect immunofluorescence assay. *The American Journal of Tropical Medicine and Hygiene*, **32**(1), 164–169.

Hidari, K. I. P. J. and Suzuki, T. Dengue virus receptor. *Tropical Medicine and Health*, 2011, **39**(4): 37–43.

Ho, L. J.; Shaio, M. F.; Chang, D. M.; Liao, C. L.; and Lai, J. H. Infection of human dendritic cells by dengue virus activates and primes T cells towards Th0-like phenotype producing both Th1 and Th2 cytokines. *Immunological Investigations*, 2004, **33**: 423–437.

Holmes, E. C. and Burch, S. S. The causes and consequences of genetic variation in dengue virus. *Trends in Microbiology*, 2000, **8**: 74–77.

Holmes, E. C. and Twiddy, S. S. The origin, emergence and evolutionary genetic of dengue virus. *Infection, Genetics and Evolution*, 2003, **3**: 19–28.

Hotta, S. and Evan, C. A. Cultivation of mouse-adapted dengue virus (type 1) in rhesus monkey tissue culture. *Journal of Infectious Disease,* 1956, **98**: 88–97.

Hotta, S. *Dengue and Related Hemorrhagic Diseases.* Warren H. Green, Inc., St. Louis. 1969, pp. 36–44.

Hsueh, W. A.; Lyon, C. J.; and Quiñones, M. J. Insulin resistance and the endothelium. *The American Journal of Medicine,* 2004, **117**: 109–117.

http://www.denguevirusnet.com/history-of-dengue.html accessed on 10th December 2018.

http://www.who.int/mediacentre/factsheets/fs117/en/print.html, accessed 26 September 2006.

https://www.cdc.gov/dengue/entomologyecology/index.html.

Hull, B.; Tikasingh, E.; De Souza, M.; and Martinez, R. Natural transovarial transmission of dengue 4 virus in *Aedes aegypti* in Trinidad. *American Journal of Tropical Medicine and Hygiene,* 1984, **33**: 1248–1250.

Hung, S. L.; Lee, P. L.; Chen, H. W.; Chen, L. K.; Kao, C. L.; and King, C. C. Analysis of the steps involved in dengue virus entry into host cells. *Virology,* 1999, **25**: 156–167.

Igarashi, A. Isolation of a Singh's *Aedes albopictus* cell clone sensitive to dengue and chikungunya virus. *Journal of General Virology*, 1978, **40**: 531–544.

Innis, B. L.; Nisalak, A.; Nimmannitya, S.; Kusalerdchariya, S.; Chongwasdi, V.; Suntayakorn, S.; Puttisri, P.; and Hoke, C. H. An enzyme-linked immunosorbent assay to characterize dengue infections where dengue and Japanese encephalitis co-circulate. *American Journal of Tropical Medicine and Hygiene,* 1989, **40**: 418–427.

Irie, K.; Mohan, P. M.; Sasaguri, Y.; Putnak, R.; and Padmanabhan, R. Sequence analysis of cloned dengue virus type 2 genome (New Guinea-C strain). *Gene*, 1989, **74**: 197–211.

Je, S.; Bae, W.; Kim, J.; Seok, S. H.; and Hwang, E. S. Epidemiological characteristics and risk factors of dengue infection in Korean travelers. *Journal of Korean Medical Science*, 2016, **31**: 1863–1873.

Jessie, K.; Fong, M. Y.; Devi, S.; Lam, S. K.; and Wong, K. T. Localization of dengue virus in naturally infected human tissues, by immunohistochemistry and in situ hybridization. *Journal of Infectious Diseases,* 2004, **189**: 1411–1418.

Jones, C. T.; Ma, L.; Burgner, J. W.; Groesch, T. D.; Post, C. B.; and Kuhn, R. J. Flavivirus capsid is a dimeric α-helical protein. *Journal of Virology*, 2003, **77**: 7143–7149.

Joshi, A. P.; Angel, A.; Angel, B.; Baharia, R. K.; Rathore, S.; Sharma, N.; Yadav, K.; Thanvi, S.; Thanvi, I.; and Joshi, V. In-silico designing and testing of primers for Sanger genome sequencing of dengue virus types of Asian origin. *Journal of Genomics*, 2018, **6**: 34–40.

Joshi, V.; Angel, A.; Angel, B.; and Kucheria, K. Egg laying sites of *Aedes aegypti* and their elimination as the crucial etiological intervention to prevent dengue transmission in Western Rajasthan, India. *International Journal of Scientific Research*, 2013, **2**(12): 468–469.

Joshi, V.; Angel, B.; Chauhan, R.; Bohra, N.; Angel, A.; Singhi, M.; Mathur, A.; and Solanki, A. Immunity in dengue hemorrhagic fever patients could be sensitized by fresh blood transfusion. *Indian Journal of Medical Case Reports*, 2012, **1**(1): 13–14.

Joshi, V.; Angel, B.; Purohit, A.; Singhi, M.; Angel, A.; and Bohra, N. Studies on dengue outbreak in an arid town Jodhpur, Rajasthan. *Journal of Communicable Diseases*, 2012, **44**(2): 109–113.

Joshi, V.; Mourya, D. T.; and Sharma, R. C. Persistence of dengue-3 virus through transovarial transmission passage in successive generations of *Aedes aegypti* mosquitoes. *American Journal of Tropical Medicine and Hygiene*, 2002, **67**: 158–161.

Joshi, V.; Sharma, R. C.; Sharma, Y.; Adha, S.; Sharma, K.; Singh, H.; Purohit, A.; and Singhi, M. Importance of socio-economic status and tree hole distribution in *Aedes* mosquitoes (Diptera: Culicidae) in Jodhpur, Rajasthan, India. *Journal of Medical Entomology*, 2006, **43**(2): 330–336.

Joshi, V.; Singhi, M.; and Choudhary, R. C. Transovarial transmission of dengue virus by *Aedes aegypti*. *Transactions of Royal Society of Tropical Medicine and Hygiene*, 1996, **90**: 643–644.

Khanam, S.; Pilankatta, R.; Khanna, N.; and Swaminathan, S. An adenovirus type 5 (AdV5) vector encoding an envelope domain III-based tetravalent antigen elicits immune responses against all four dengue viruses in the presence of prior AdV5 immunity. *Vaccine*, 2009, **27**(43): 6011–6021.

Khin, M. M. and Than, K. A. Transovarial transmission of dengue-2 virus by *Aedes aegypti* in nature. *American Journal of Tropical Medicine and Hygiene*, 1983, **32**(3): 590–594.

Kimura, R. and Hotta, S. On the inoculation of dengue virus into mice. *Nippon Lgakku*, 1944, **3379**: 629–633.

Klein, D. E.; Choi, J. L.; and Harrison, S. C. Structure of a dengue virus envelope protein late-stage fusion intermediate. *Journal of Virology*, 2013, **87**(4): 2287–2293.

Klungthong, C.; Zhang, C.; Mammen, M. P. Jr.; Ubol, S.; and Holmes, E. C. The molecular epidemiology of dengue virus serotype 4 in Bangkok, Thailand. *Virology*, 2004, **329**: 168–179.

Kou, Z.; Quinn, M.; Chen, H.; Rodrigo, W. W.; Rose, R. C.; Schlesinger, J. J.; and Jin, X. Monocytes, but not T or B cells, are the principal target cells for dengue virus (DV) infection among human peripheral blood mononuclear cells. *Journal of Medical Virology*, 2008, **80**: 134–146.

Kouri, G. P.; Guzman, M. G.; and Bravo, J. R. Why dengue hemorrhagic fever in Cuba? 2. An integral analysis. *Transaction of Royal Society of Tropical Medicine and Hygiene*, 1987, **81**(5): 821–823.

Krishna, V. D.; Rangappa, M.; and Satchidanandam, V. Virus-specific cytolytic antibodies to nonstructural protein 1 of Japanese encephalitis virus effect reduction of virus output from infected cells. *Journal of Virology*, 2009, **83**: 4766–4777.

Kuberski, T. T. and Rosen, L. A simple technique for the detection of dengue antigen in mosquitoes by immunofluorescence. *American Journal of Tropical Medicine and Hygiene*, 1977, **26**: 533–537.

Kuberski, T. T. and Rosen, L. Identification of dengue viruses using complement fixing antigen produced in mosquitoes. *American Journal of Tropical Medicine and Hygiene*, 1977, **26**: 538–543.

Kukreti, H.; Chaudhary, A.; Rautela, R. S.; Anand, R.; Mittal, V.; Chhabra, M.; Bhattacharya, D.; Lal, S.; and Rai, A. Emergence of an independent lineage of dengue virus type 1 (DENV-1) and its co-circulation with predominant DENV-3 during the 2006 dengue fever outbreak in Delhi. *International Journal of Infectious Disease*, 2008, **12**, 542–549.

Kukreti, H.; Dash, P. K.; Parida, M.; Chaudhary, A.; Saxena, P.; Rautela, R. S.; Mittal, V.; Chhabra, M.; Bhattacharya, D.; Lal, S.; Rao, P. V.; and Rai, A. Phylogenetic studies reveal existence of multiple lineages of a single genotype of DENV-1 (genotype III) in India during 1956–2007. *Virology Journal*, 2009, **6**: 1.

Kumbhat, S.; Sharma, K.; Gehlot, R.; Solanki, A.; and Joshi, V. Surface plasmon resonance based immunosensor for serological diagnosis of dengue virus infection. *Journal of Pharmaceutical and Biomedical Analysis*, 2010, **52**(2): 255–259.

Kuno, G.; Gubler, D. J.; Velez, M.; and Oliver, A. Comparative sensitivity of three mosquito cell lines for isolation of dengue viruses. *Bulletin of the World Health Organization*, 1985, **63**: 279–286.

Kwakye, S. and Baeumner, A. A microfluidic biosensor based on nucleic acid sequence recognition. *Analytical and Bioanalytical Chemistry*, 2003, **376**(7): 1062–1068.

Kwakye, S.; Goral, V. N.; and Baeumner, A. J. Electrochemical microfluidic biosensor for nucleic acid detection with integrated minipotentiostat. *Biosensors and Bioelectronics*, 2006, **21**(12): 2217–2223.

Lambeth, C. R.; White, L. J.; Johnston, R. E.; and de Silva, A. M. Flow cytometry-based assay for titrating dengue virus. *Journal of Clinical Microbiology*, 2005, **43**: 3267–3272.

Lanciotti, R. S.; Calisher, C. H.; Gubler, D. J.; Chang, G. J.; and Vorndam, A. V. Rapid detection and typing of dengue viruses from clinical samples by using reverse transcriptase-polymerase chain reaction. *Journal of Clinical Microbiology*, 1992, **30**: 545–551.

Lanciotti, R. S.; Gubler, D. J.; and Trent, D. W. Molecular evolution and phylogeny of dengue-4 viruses. *Journal of General Virology*, 1997, **78**: 2279–2286.

Lanciotti, R. S.; Lewis, J. G.; Gubler, D. J.; and Trent, D. W. Molecular evolution and epidemiology of dengue-3 viruses. *Journal of General Virology*, 1994, **75**: 65–75.

Leake, C. J. Arbovirus-vector interactions and vector specificity. *Parasitology Today*, 1992, **8**: 123–127.

Lee, J. E.; Seo, J. H.; Kim, C. S.; Kwon, Y.; Ha, J. H.; Choi, S. S.; and Cha, H. J. A comparative study on antibody immobilization strategies onto solid surface. *Korean Journal of Chemical Engineering*, 2013, **30**(10): 1934–1938.

Lieberzeita, P. A.; Chuntaa, S.; Navakulb, K.; Sangmab, C.; and Jungmanna, C. Molecularly imprinted polymers for diagnostics: Sensing high density lipoprotein and dengue virus. *Procedia Engineering*, 2016, **168**: 101–104.

Limon-Flores, A. Y.; Perez-Tapia, M.; Estrada-Garcia, I.; Vaughan, G.; Escobar-Gutierrez, A.; Calderon-Amador, J.; Herrera-Rodriguez, S. E.; Brizuela-Garcia, A.; Heras-Chavarria, M.; Flores-Langarica, A.; Cedillo-Barron, L.; and Flores-Romo, L. Dengue virus inoculation to human skin explants: An effective approach to assess in situ the early infection and the effects on cutaneous dendritic cells. *International Journal of Experimental Pathology*, 2005, **86**: 323–334.

Lindenbach, B. D.; Thiel, H.-J.; and Rice, C. M. Flaviviridae: The viruses and their replication. In *Fields Virology*, 5th ed., vol. 1 (D. M. Knipe, P. M. Howley, D. E. Griffin, R. A. Lamb, M. A. Martin, B. Roizman, and S. E. Straus, eds.), Lippincott Williams and Wilkins, Philadelphia, PA, 2007.

Linthicum, K. L.; Platt, K.; Myint, K. S.; Lerdthusnee, K.; Innis, B. L.; and Vaughn, D. W. Dengue-3 virus distribution in the mosquito *Aedes aegypti*: An immunocytochemical study. *Medical Veterinary Entomology,* 1996, **10**: 87–92.

Luna, D. M.; Oliveira, M. D.; Nogueira, M. L.; and Andrade, C. A. Biosensor based on lectin and lipid membranes for detection of serum glycoproteins in infected patients with dengue. *Chemistry and Physics of Lipids*, 2014, **180**: 7–14.

Luo, D.; Xu, T.; Hunke, C.; Grüber, G.; Vasudevan, S. G.; and Lescar, C. Crystal structure of the NS3 protease-helicase from dengue virus. *Journal of Virology*, 2008, **82**(1): 173–183.

Mammen, M. P.; Pimgate, C.; Koenraadt, C. J.; Rothman, A. L.; Aldstadt, J.; Nisalak, A.; Jarman, R. G.; Jones, J. W.; Srikiatkhachom, A.; Ypil-Butac, C. A.; Getis, A.; Thammapalo, S.; Morrison, A. C.; Libraty, D. H.; Green, S.; and Scott, T. W. Spatial and temporal clustering of dengue virus transmission in Thai villages. *PLoS Medicine,* 2018, **5**(11): 1–12.

Mandl, C. W.; Guirakhoo, F.; Holzmann, H.; Heinz, F. X.; and Kunz, C. Antigenic structure of the flavivirus envelope protein E at the molecular level, using tick-borne encephalitis virus as a model. *Journal of Virology*, 1989, **63**: 564–571.

Martina, B. E. E.; Koraka, P.; and Osterhaus, A. D. M. E. Dengue virus pathogenesis: An integrated view. *Clinical Microbiology Reviews*, 2009, 564–581.

Mascini, M. and Palchetti, I. (eds.). *Nucleic Acid Biosensors for Environmental Pollution Monitoring*. Royal Society of Chemistry, 2011.

Mason, P. W.; McAda, P. C.; Mason, T. L.; and Fournier, M. J. Sequence of the dengue-1 virus genome in the region encoding the three structural proteins and the major nonstructural protein NS1. *Virology*, 1987, **161**: 262–267.

McArthur, M. A.; Sztein, M. B.; and Edelman, R. Dengue vaccines: Recent developments, ongoing challenges and current candidates. *Expert Review of Vaccines*, 2013, **12**(8): 933–953.

McMinn, P. C. The molecular basis of virulence of the encephalitogenic flaviviruses. *Journal of General Virology*, 1997, **78**: 2711–2722.

Meiklejohn, G. B. and Lennette, E. H. Adaptation of dengue virus strains in unweaned mice. *American Journal of Tropical Medicine and Hygiene*, 1952, **1**: 51–58.

Meiklejohn, G.; England, B.; and Lennette, E. H. Adaptation of dengue virus strains in unweaned mice. *American Journal of Tropical Medicine and Hygiene*, 1952, **1**: 51–58.

Miles, J. A. and Austin, F. J. Growth of arboviruses in BHK21 cells. *Australian Journal of Science*, 1963, **25**: 466.

Misra, M. and Schein, C. H. Flavitrack: An annotated database of flavivirus sequences. *Bioinformatics*, 2007, **23**(19): 2645–2647.

Modis, Y.; Ogata, S.; Clements, D.; and Harrison, S. C. A ligand-binding pocket in the dengue virus envelope glycoprotein. *Proceedings of National Academy of Sciences, USA*, 2003, **100**: 6986–6991.

Modis, Y.; Ogata, S.; Clements, D.; and Harrison, S. C. Variable surface epitopes in the crystal structure of dengue virus type 3 envelope glycoprotein. *Journal of Virol*ogy, 2005, **79**: 1223–1231.

Moncayo, A. C.; Fernandez, Z.; Ortiz, D.; Diallo, M.; Sall, A.; Hartman, S.; Davis, C. T.; Coffey, L. L.; Mathiot, C. C.; Tesh, R. B.; and Vasilakis, N. Dengue emergence and adaptation to peridomestic mosquitoes. *Emerging Infectious Diseases*, 2004, **10**(10): 1790–1796.

Mongkolsapaya, J.; Dejnirattisai, W.; Xu, X.; Vasanawathana, S.; Tangthawornchaikul, N.; Chairunsri, A.; Sawasdivron, S.; Duangchinda, T.; Dong, T.; Jones, S. R.; Yenchitosomanas, P.; McMichael, A.; Malasit, P.; and Screaton, G. Original antigenic sin and apoptosis in pathogenesis of dengue haemorrhagic fever. *Nature Medicine*, 2003, **9**(7): 921–927.

Morier, L. and Castillo, A. Obtencio´n de un clono de la lı´nea cellular AP-61. Su utilidad para la multiplicacio´n de los virus dengue 1 y 2. *Revista Cubana de Medicina Tropical*, 1992, **44**: 181–184.

Morier, L.; Aleman, M. R.; Castillo, A.; and Pe´rez, V. Estudio preliminary de la lı´nea celular AP-64 (Aedes pseudoscutellaris) para la multiplicacio ´n de los virus dengue 1 y 2. *Revista Cubana de Medicina Tropical*, 1991, **43**: 156–161.

Muller, D. A.; Depelsenaire, A. C. I.; and Young, P. R. Clinical and laboratory diagnosis of dengue virus infection. *The Journal of Infectious Diseases*, 2017, **215**(2): S89–S95.

Murphy, B. R. and Whitehead, S. S. Immune response to dengue virus and prospects for a vaccine. *Annual Reviews in Immunology*, 2011, **29**: 587–619.

Nascimento, H. P.; Oliveira, M. D.; de Melo, C. P.; Silva, G. J.; Cordeiro, M. T.; and Andrade, C. A. An impedimetric biosensor for detection of dengue serotype at picomolar concentration based on gold nanoparticles-polyaniline hybrid composites. *Colloids and Surfaces B: Biointerfaces*, 2011, **86**(2): 414–419.

Normile, D. Tropical medicine: Surprising new dengue virus throws a spanner in disease control efforts. *Science*, 2013, **342**(6157): 415.

Nowak, T. and Wengler, G. Analysis of disulfides present in the membrane proteins of the West Nile flavivirus. *Virology*, 1987, **156**: 127–137.

Oliveira, M. D.; Correia, M. T.; and Diniz, F. B. A novel approach to classify serum glycoproteins from patients infected by dengue using electrochemical impedance spectroscopy analysis. *Synthetic Metals*, 2009, **159**(21–22): 2162–2164.

Ong, S. H. Molecular epidemiology of dengue viruses from complete genome sequences. Inaugural Dissertation, 2010. https://edoc.unibas.ch/1155/1/my_thesis_edoc.pdf.

Ong, S. H.; Yip, J. T.; Chen, Y. L.; Liu, W.; Harun, S.; Lystiyaningsih, E.; Heriyanto, B.; Beckett, C. G.; Mitchell, W. P.; Hibberd, M. L.; Suwandono, A.; Vasudevan, S. G.; and Schreiber, M. J. Periodic re-emergence of endemic strains with strong epidemic potential-a proposed explanation for the 2004 Indonesian dengue epidemic. *Infection, Genetics and Evolution*, 2008, **8**: 191–204.

Osatomi, K.; Fuke, I.; Tsuru, D.; Shiba, T.; Sakaki, Y.; and Sumiyoshi, H. Nucleotide sequence of dengue type 3 virus genomic RNA encoding viral structural proteins. *Virus Genes*, 1988, **2**: 99–108.

Osorio, J. E.; Huang, C. Y.; Kinney, R. M.; and Stinchcomb, D. T. Development of DENVax: A chimeric dengue-2 PDK-53-based tetravalent vaccine for protection against dengue fever. *Vaccine*, 2011, **29**(42): 7251–7260.

Parkash, O.; Yean, C.; and Shueb, R. Screen printed carbon electrode based electrochemical immunosensor for the detection of dengue NS1 antigen. *Diagnostics*, 2014, **4**(4): 165–180.

Patil, J. A.; Cherian, S.; Walimbe, A. M.; Patil, B. R.; Sathe, P. S.; Shah, P.; and Cecilia, D. Evolutionary dynamics of the American African genotype of dengue type 1 virus in India (1962–2005). *Infection, Genetics and Evolution*, 2011, **11**: 1443–1448.

Perera, R. and Kuhn, R. J. Structural proteomics of dengue virus. *Current Opinion in Microbiology*, 2008, **11**: 369–377.

Perumal, V. and Hashim, U. Advances in biosensors: Principle, architecture and applications. *Journal of Applied Biomedicine*, 2014, **12**(1): 1–15.

Race, M. W.; Willians, M. C.; and Agostini, C. F. M. Dengue in the Caribbean: Virus isolation in a mosquito (*Aedes pseudoscutellaris*) cell line. *Transactions of Royal Society of Tropical Medicine and Hygiene*, 1979, **73**: 18–22.

Rai, V.; Hapuarachchi, H. C.; Ng, L. C.; Soh, S. H.; Leo, Y. S.; and Toh, C. S. Ultrasensitive cDNA detection of dengue virus RNA using electrochemical nanoporous membrane-based biosensor. *PloS One*, 2012, **7**(8): e42346.

Randolph, V. B.; Winkler, G.; and Stollar, V. Acidotrophic amines inhibit proteolytic processing of Flavivirus prM protein. *Virology*, 1990, **174**(2): 450–458.

Rice, C. M.; Lenches, E. M.; Eddy, E. R.; Shin, S. J.; Sheets, R. L.; and Strauss, J. H. Nucleotide sequence of yellow fever virus: Implications for flavivirus gene expression and evolution. *Science*, 1985, **229**: 726–733.

Rice, C. M.; Strauss, E. G.; and Strauss, J. H. Structure of the flavivirus genome. In *Togaviruses and Flaviviruses* (S. Schlesinger and M. Schlesinger, eds.), Plenum Publishing Corp, New York, 1986, pp. 279–327.

Rico-Hesse, R. Molecular evolution and distribution of dengue viruses type 1 and 2 in nature. *Virology*, 1990, **174**: 479–493.

Rigau-Pérez, J. G. and Gubler, D. J. Is there an inapparent dengue explosion? *Lancet*, 1999, **353**(9158): 1101.

Roche, R. R.; Alvarez, M.; Guzman, M. G.; Morier, L.; and Kouri, G. Comparison of rapid centrifugation assay with conventional tissue culture method for isolation of dengue 2 virus in C6/36-HT cells. *Journal of Clinical Microbiology*, 2000, **38**(9): 3508–3510.

Rollier, C. S.; Reyes-Sandoval, A.; Cottingham, M. G.; Ewer, K.; and Hill, A. Viral vectors as vaccine platforms: Deployment in sight. *Current Opinion Immunology*, 2011, **23**(3): 377–382.

Rosen, L. and Gubler, D. J. The use of mosquitoes to detect and propagate dengue viruses. *American Journal of Tropical Medicine and Hygiene*, 1974, **21**: 1153–1160.

Rosen, L. The Emperor's new clothes revisited, or reflections on the pathogenesis of dengue hemorrhagic fever. *American Journal of Tropical Medicine and Hygiene*, 1977, **26**: 337–343.

Rosen, L.; Shroyer, D. A.; Tesh, R. B.; Frier, J. E.; and Lien, J. C. Transovarial transmission of dengue virus by mosquitoes: *Aedes albopictus* and *Aedes aegypti*. *American Journal of Tropical Medicine and Hygiene*, 1983, **32**: 1108–1119.

Rush, B. An account of the bilious remitting fever, as it appeared in Philadelphia in the summer and autumn of the year 1780. In *Medical Inquiries and Observations*, Pritchard & Hall, Philadelphia, 1789, p. 104.

Russell, P. K. and Nisalak, A. Dengue virus identification by plaque reduction neutralization test. *Journal of Immunology*, 1967, **99**: 291–296.

Russell, P. K.; Brandt, W. E.; and Dalrymple, J. M. Chemical and antigenic structure of flaviviruses. In *The Togaviruses* (R. W. Schlesinger, ed.). Academic Press, Inc., New York, 1980, pp. 503–529.

Russell, P. K.; Nisalak, A.; Sukhavachana, P.; and Vivona, A. A plaque reduction test for dengue virus neutralizing antibodies. *Journal of Immunology*, 1967, **99**: 285–290.

Sabchareon, A.; Wallace, D.; Sirivichayakul, C.; Limkittikul, K.; Chanthavanich, P.; Jiwarivavei, V.; Dulvachai, W.; Pengsaa, K.; Wartel, T. A.; Moureau, A.; Saville, M.; Bouckenooghe, A.; Viviani, S.; Torniporth, N. G.; and Lang, J. Protective efficacy of the recombinant, live attenuated, CYD tetravalent dengue vaccine in Thai schoolchildren: A randomised, controlled phase 2b trial. *Lancet*, 2012, **380**(9853): 1559–1567.

Sabin, A. B. and Schlesinger, R. W. Production of immunity to dengue with virus modified by propagation in mice. *Science*, 1945, **101**: 640–642.

Sabin, A. B. Dengue. In *Viral and Rickettsial Infections of Man* (T. Rivers and F. Horsfall, eds.), J. B. Lippincott Co., Philadelphia, 1959, pp. 361–373.

Sabin, A. B. Research on dengue during World War II. *American Journal of Tropical Medicine and Hygiene*, 1952, **1**: 30–50.

Sabin, A. B. The dengue group of viruses and its family relationships. *Bacteriological Reviews*, 1950, **14**: 225–232.

Sahni, A. K.; Grover, N.; Sharma, A.; Khan, I. D.; and Kishore, J. Reverse transcription loop-mediated isothermal amplification (RT-LAMP) for diagnosis of dengue. *Medical Journal Armed Forces India*, 2013, **69**: 246–253.

Salas-Benito, J. S. and Angel, R. M. Identification of two surface proteins from C6/36 cell lines that bind dengue type virus. *Journal of Virology*, 1997, **71**(10): 7246–7252.

Saluzzo, J. F.; Cornet, M.; Adam, C.; Eyraud, M.; and Digoutte, J. P. Dengue 2 in eastern Senegal: Serologic survey in simian and human populations. 1974–85. *Bulletin de la Société de Pathologie Exotique*, 1986, **79**: 313–322.

Samsa, M. M.; Mondotte, J. A.; Caramelo, J. J.; and Gamarnik, A. V. Uncoupling *cis*-acting RNA elements from coding sequences revealed a requirement of the N-terminal region of dengue virus capsid protein in virus particle formation. *Journal of Virology*, 2012, **86**: 1046–1058.

Scaturro, P.; Cortese, M.; Chatel-Chaix, L.; Fischl, W.; and Bartenschlager, R. Dengue virus non-structural protein 1 modulates infectious particle production via interaction with the structural proteins. *PLoS Pathogens*, 2015, **11**(11): 1–32.

Schlesinger, R. W. *Dengue Viruses*. Springer-Verlag, New York, 1977.

Schreiber, M. J.; Holmes, E. C.; Ong, S. H.; Soh, H. S.; Liu, W.; Tanner, L.; Aw, P. P.; Tan, H. C.; Ng, L. C.; Leo, Y. S.; Low, J. G.; Ong, A.; Ooi, E. E.; Vasudevan, S. G.; and Hibberd, M. L. Genomic epidemiology of a dengue virus epidemic in urban Singapore. *Journal of Virology*, 2009, **83**(9): 4163–4173.

Schreiber, M. J.; Ong, S. H.; Holland, R. C. G.; Hibberd, M. L.; Vasudevan, S. G.; Mitchell, W. P.; and Holmes, E. C. DengueInfo: A web portal to dengue information resources. *Infection Genetics and Evolution*, 2007, **7**: 540–541.

Semenza, J. C.; Sudre, B.; Miniota, J.; Rossi, M.; Hu, W.; Kossowsky, D.; Suk, J. E.; van Bortel, W.; and Khan, K. International dispersal of dengue through air travel: Importation risk for Europe. *PLoS Neglected Tropical Disease*, 2014, 8:

Shapiro, D.; Brandt, W. E.; and Russell, P. K. Change involving a viral membrane glycoprotein during morphogenesis of group B arboviruses. *Virology*, 1972, **50**: 906–911.

Sharma, K.; Angel, B.; Singh, H.; Purohit, A.; and Joshi, V. Entomological studies for surveillance and prevention of dengue in arid and semi-arid districts of Rajasthan, India. *Journal of Vector Borne Diseases*, 2008, **45**: 140–149.

Sharma, S.; Dash, P. K.; Agarwal, S.; Shukla, J.; Parida, M. M.; and Rao, P. V. Comparative complete genome analysis of dengue virus type 3 circulating in India between 2003 and 2008. *Journal of General Virology*, 2011, **92**: 1595–1600.

Shin, D.; Richards, S. L.; Alto, B. W.; Bettinardi, D. J.; and Smartt, C. T. Genome sequence analysis of Dengue virus 1 isolated in Key West, Florida. *PLoS One*, 2013, **8**: 1–9.

Siler, J. F.; Hall, M. W.; and Hitchens, A. P. Dengue: Its history, epidemiology, mechanism of transmission, etiology, clinical manifestations, immunity and prevention. *The Philippine Journal of Science*, 1926, **29**: 1–304.

Simasathien, S.; Thomas, S. J.; Watanaveeradej, V.; Nisalak, A.; Barberousse, C.; Innis, B. L.; Sun, W.; Putnak, J. R.; Eckels, K. H.; Hutagalung, Y.; Gibbons, R. V.; Zhang, C.; De La Barrera, R.; Jarman, R. G.; Chawachalasai, W.; and Mammen, M. P. Jr. Safety and immunogenicity of a tetravalent live-attenuated dengue vaccine in flavivirus naive children. *American Journal of Tropical Medicine and Hygiene*, 2008, **78**(3): 426–433.

Simmons, J.; St John, J.; and Reynolds, F. Experimental studies of dengue. *The Philippine Journal of Science*, 1931, **44**: 1–252.

Sin, M. L.; Mach, K. E.; Wong, P. K.; and Liao, J. C. Advances and challenges in biosensor-based diagnosis of infectious diseases. *Expert Review of Molecular Diagnostics*, 2014, **14**(2): 225–244.

Singh, K. R. P. and Paul, S. D. Isolation of dengue viruses in *Aedes albopictus* cell cultures. *Bulletin W.H.O.*, 1969, **40**: 982–983.

Souza, E.; Nascimento, G.; Santana, N.; Ferreira, D.; Lima, M.; Natividade, E.; Martins, D.; and Lima-Filho, J. Label-free electrochemical detection of the specific oligonucleotide sequence of dengue virus type 1 on pencil graphite electrodes. *Sensors*, 2011, **11**(6): 5616–5629.

Speight, G. and Westaway, E. G. Positive identification of NS4a, the last of the hypothetical nonstructural proteins of flaviviruses. *Virology*, 1989, **170**: 299–301.

Speight, G.; Coia, G.; Parker, M. D.; and Westaway, E. G. Gene mapping and positive identification of the nonstructural proteins NS2a, NS2b, NS3, NS4b, and NS5 of the flavivirus Kunjin and their cleavage sites. *Journal of General Virology*, 1988, **69**: 23–34.

Sriurairatna, S. and Bhamarapravati, N. Replication of dengue virus in *Aedes albopictus*. *American Journal of Tropical Medicine and Hygiene*, 1977, **26**: 1199–1205.

St. John, A. L. Multiple theories of dengue immune pathogenesis. *PLOS Pathogens*, 2015. https://doi.org/10.1371/journal.ppat.1003783.g002

Stiasny, K.; Kössl, C.; Lepault, J.; Rey, F. A.; and Heinz, F. X. Characterization of a structural intermediate of flavivirus membrane fusion. *PLoS Pathogens*, 2007, **3**(2): 1–9.

Stollar, V. Studies on the nature of dengue viruses. IV. The structural proteins of type 2 dengue virus. *Virology*, 1969, **39**: 426–438.

Sukhavachana, P.; Nisalak, A.; and Halstead, S. B. Tissue culture technique for the study of dengue viruses. *Bulletin of the World Health Organization*, 1966, **35**: 65–66.

Suk-Yin, C.; Kautner, I.; and Sai-Kit, L. Detection and serotyping of dengue viruses by PCR: A simple, rapid method for the isolation of viral RNA from infected mosquito larvae. *South Asian Journal of Tropical Medicine and Public Health*, 1994, **25**: 258–261.

Sun, W.; Cunningham, D.; Wasserman, S. S.; Perry, J.; Putnak, J. R.; Eckels, K. H.; Vaughn, D. W.; Thomas, S. J.; Kanesa-Thasan, N.; Innis, B. L.; and Edelman, R. Phase 2 clinical trial of three formulations of tetravalent live-attenuated dengue vaccine in flavivirus-naive adults. *Human Vaccines and Immunotherapeutics*, 2009, **5**(1): 33–40.

Swee, H. O. Molecular epidemiology of dengue viruses from complete genome. PhD thesis. University of Basel, 2010.

Tanaka, M. Rapid identification of flavivirus using the polymerase chain reaction. *Journal of Virological Methods*, 1993, **41**: 311–322.

Tesh, R. B. A method for the isolation and identification of dengue viruses, using mosquito cell cultures. *American Journal of Tropical Medicine and Hygiene*, 1979, **28**: 1053–1059.

Thongcharoen, P.; Wasi, C.; and Puthavathana, P. Dengue viruses. In *Monograph on Dengue/Dengue Hemorrhagic Fever* (P. Thongcharoen, ed.), World Health Organization, New Delhi, India, 1993.

Trivedi, S. G. and Lloyd, C. M. Eosinophils in the pathogenesis of allergic airways disease. *Cell and Molecular Life Sciences*, 2007, **64**(10): 1269–1289.

Twiddy, S. S.; Farrar, J.; Chau, V. N.; Wills Gould, E. A.; Lloyd, G. T.; and Holmes, E. C. Phylogenetic relationships and differential selection pressures among genotypes of dengue-2 virus. *Virology*, 2002, **298**: 63–72.

Usawattanakul, W.; Jittmittraphap, A.; Tapchaisri, P.; Siripanichgon, K.; Buchachart, K.; Hong-ngarm, A.; Thongtarado, P.; and Endy, T. P. Detection of dengue viral RNA in patients' sera by nucleic acid sequence-based amplification (NASBA) and polymerase chain reaction (PCR). *Dengue Bulletin*, 2002, 131–139.

Varma, M. G. R.; Pudney, M.; and Leake, C. J. Cell lines from larvae of *Aedes* (*Stegomyia*) *malayensis* Colless and *Aedes* (S.) *pseudoscutellaris* (theobald) and their infection with some arboviruses. *Transactions of Royal Society of Tropical Medicine and Hyg*iene, 1974, **68**: 374–382.

Velasco-Salas, Z. I.; Sierra, G. M.; Guzman, D. M.; Zambrano, J.; Vivas, D.; Comach, G.; Wilschut, J. C.; and Tami, A. Dengue seroprevalence and risk factors for past and recent viral transmission in Venezuela: A comprehensive community-based study *American Journal of Tropical Medicine and Hygiene*, 2014, **91**(5): 1039–1048.

Vordam, V. and Kuno, G. Laboratory diagnosis of dengue virus infections. In *Dengue and Dengue Hemorrhagic Fever* (D. J. Guber and G. Kuno, eds.), CAB International, London, United Kingdom, 1997, pp. 313–334.

Wang, S. H.; Syu, W. J.; and Hu, S. T. Identification of the homotypic interaction domain of the core protein of dengue virus type 2. *Journal of General Virology*, 2004, **85**: 2307–2314.

Watanaveeradej, V.; Simasathien, S.; Nisalak, A.; Endy, T. P.; Jarman, R. G.; Innis, B. L.; Thomas, S. J.; Gibbons, R. V.; Hengprasert, S.; Samokoses, R.; Kerdpanich, A.; Vaughn, D. W.; Putnak, J. R.; Eckels, K. H.; Barrera, R. L.; and Mammen, M. P. Safety and immunogenicity of a tetravalent live-attenuated dengue vaccine in flavivirus-naive infants. *American Journal of Tropical Medicine and Hygiene*, 2011, **85**(2): 341–351.

Wengler, G. and Wengler, G. Cell-associated West Nile flavivirus is covered by E and preM protein heterodimers which are destroyed and reorganized by proteolytic cleavage during virus release. *Journal of Virology*, 1989, **63**: 2521–2526.

Wengler, G.; Wengler, G.; and Gross, H. J. Studies on virus specific nucleic acids synthesized in vertebrate and mosquito cells infected with flavivirus. *Virology*, 1978, **89**: 423–437.

WHO (World Health Organization). *Dengue and Dengue Hemorrhagic Fever*. World Health Organization, Geneva, revised April 2002, Fact sheet No. 117.

WHO (World Health Organization). *Dengue Guidelines for Diagnosis: Treatment, Prevention and Control*. WHO, TDR. New Edition, 2009, 1–160. https://apps.who.int/iris/bitstream/handle/10665/44188/9789241547871_eng.pdf;jsessionid=5182452BF19D797342F693FB3F4A2604?sequence=1

Wolfe, N. D.; Kilbourn, A. M.; Karesh, W. B.; Rahman, H. A.; Bosi, E. J.; Cropp, B. C.; Andau, M.; Spielman, A.; and Gubler, D. J. Sylvatic transmission of arboviruses among Bornean Orangutans. *American Journal of Tropical Medicine and Hygiene*, 2001, **64**(5, 6): 310–316.

Wu, P. C.; Lay, J. G.; Guo, H. R.; Lin, C. Y.; Lung, S. C.; and Su, H. J. Higher temperature and urbanization affect the spatial patterns of dengue fever transmission in subtropical Taiwan. *Science of Total Environment*, 2009, **407**: 2224–2233.

Wu, S. J.; Grouard-Vogel, G.; Sun, W.; Mascola, J. R.; Brachtel, E.; Putvatana, R.; Louder, M. K.; Filgueira, L.; Marovich, M. A.; Wong, H. K.; Blauvelt, A.; Murphy, G. S.; Robb, M. L.; Innes, B. L.; Birx, D. L.; Hayes, C. G.; and Frankel, S. S. Human skin Langerhans cells are targets of dengue virus infection. *Nature Medicine*, 2000, **6**: 816–820.

Wu, T. Z.; Su, C. C.; Chen, L. K.; Yang, H. H.; Tai, D. F.; and Peng, K. C. Piezoelectric immunochip for the detection of dengue fever in viremia phase. *Biosensors and Bioelectronics*, 2005, **21**(5): 689–695.

Young, P. R.; Hilditch, P. A.; Bletchly, C.; and Halloran, W. An antigen capture enzyme-linked immunosorbent assay reveals high levels of the dengue virus protein NS1 in the sera of infected patients. *Journal of Clinical Microbiology*, 2000, **38**: 1053–1057.

Yuill, T. M.; Sukhavachana, P.; Nisalak, A.; and Russell, P. K. Dengue virus recovery by direct and delayed plaques in LLCMK2 cells. *American Journal of Tropical Medicine and Hygiene*, 1968, **17**: 441–448.

Zaytseva, N. V.; Montagna, R. A.; and Baeumner, A. J. Microfluidic biosensor for the serotype-specific detection of dengue virus RNA. *Analytical Chemistry*, 2005, **77**(23): 7520–7527.

Zhang, W.; Chipman, P. R.; Corver, J.; Johnson, P. R.; Zhang, Y.; Mukhopadhyay, S.; Baker, T. S.; Strauss, J. H., Rossmann, M. G.; and Kuhn, R. J. Visualization of membrane protein domains by cryo-electron microscopy of dengue virus. *Nature Structural and Molecular Biology*, 2003, **10**: 907–912.

Zhao, B. T.; Mackow, E.; Buckler-White, A.; Markoff, L.; Chanock, R. M.; Lai, C. J.; and Makino, M. Cloning full-length dengue type 4 viral DNA sequences: Analysis of genes coding for structural proteins. *Virology*, 1986, **155**: 77–88.

Zhao, Y.; Soh, T. S.; Zheng, J.; Chan, K. W. K.; Phoo, W. W.; Lee, C. C.; et al. A crystal structure of the dengue virus NS5 protein reveals a novel inter-domain interface essential for protein flexibility and virus replication. *PLoS Pathogens*, 2015, **11**(3): e1004682.

Chapter 5

Chikungunya Fever: Emergence and Reality

Neelam Yadav,[a] Bennet Angel,[b] Jagriti Narang,[c] Surender Singh Yadav,[d] and Vinod Joshi[b]

[a]*Centre for Biotechnology, Maharshi Dayanand University, Rohtak, India*

[b]*Amity Institute of Virology and Immunology, Amity University, Sector-125, Noida, India*

[c]*Department of Biotechnology, Jamia Hamdard University, New Delhi, India*

[d]*Department of Botany, Maharshi Dayanand University, Rohtak, India*

vinodjoshidmrc@gmail.com, bennetangel@gmail.com

This chapter aims to decipher all aspects of chikungunya virus (CHIKV), including epidemiology, virus morphology, transmission route, replication, pathogenicity, treatment and diagnostics, and risk factors. Chikungunya epidemics have been reported in various countries, chiefly in Asian and African countries and also in Europe and America. CHIKV is of the genus Alphavirus, which comes under the Togaviridae family. Transmission of CHIKV is mediated by two mosquito vectors: *Aedes aegypti* and *Aedes albopictus*. There are two

Small Bite, Big Threat: Deadly Infections Transmitted by Aedes *Mosquitoes*
Edited by Jagriti Narang and Manika Khanuja

ISBN 978-981-4800-86-0 (Hardcover), 978-1-003-00329-8 (eBook)
www.jennystanford.com

distinct transmission cycles for the transmission of CHIKV: enzootic and urban. The infection of CHIKV shows various symptoms, which may be asymptomatic in the early stage but acute and chronic in later phases. Several detection methods have been discussed with their pros and cons.

5.1 Epidemiology

Chikungunya fever is an infectious fatal disease caused by CHIKV. Chikungunya epidemics have been reported in various countries across the world chiefly in Asian and African countries, and also in Europe and America. The disease first appeared in Makonde Plateau in 1952, and since then sporadic occurrences have been reported (www.chikungunyavirusnet.com). Later it was documented in the 1960s in Bangkok. In India, the first outbreak of chikungunya was documented in Calcutta, Maharashtra, and Vellore in 1973. A drastic epidemic of CHIKV in the western rim of Indian Ocean was reported between March 2005 and February 2006. Limited incidences have also been documented in Gabon, Madagascar, Maldives, Mauritius, Mayotte, and Seychelles.

In 2007, Italy witnessed chikungunya infection, and this epidemic was followed by epidemics in Singapore, Malaysia, Thailand, and Australia in 2008 and then in Thailand in 2009. The Caledonia chikungunya outbreak was documented in 2011 and 2013. America was also affected with the attack of this virus (Fig. 5.1).

St. Martin and St. Barthelemy along with three other places witnessed a number of cases during the beginning of 2014, and then in May, 31,000 people were suspected to be infected with chikungunya in 14 countries, including French Guiana. Brazil, Puerto Rico, Colombia, Venezuela, and Chiapas were next in line to witness the effects of the virus in 2014. Interestingly, in the same year, two cases of chikungunya infection were reported from persons who had not traveled outside the United States, and this added the phenomenon of transovarial transmission of the virus in nature. As of now, chikungunya has been identified in nearly 40 countries.

It has been reported that rapid globalization and tourism have increased the risk of virus spreading (Enserink, 2008; Epstein, 2007;

Jain et al., 2008; Lanciotti et al., 2006; Pistone et al., 2009; Simon et al., 2008).

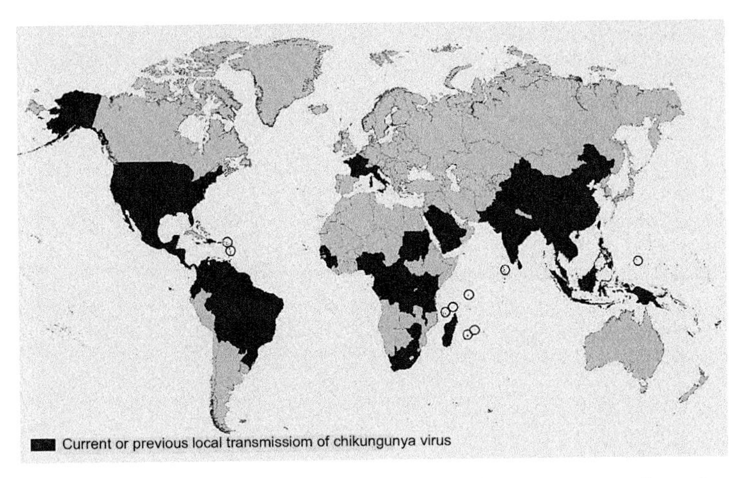

Figure 5.1 Countries and territories where chikungunya cases have been reported (without imported cases; as of March 10, 2015, CDC). Figure taken from https://www.cdc.gov/chikungunya/pdfs/chikungunyaworldmap_03-10-2015.pdf.

India witnessed many outbreaks in the past since 1963–1964. It first appeared in Calcutta (now Kolkata) with a severe disease manifestation leading to hemorrhage and death. This was followed by spread in Chennai infecting 300,000 people and later in Maharashtra (in 1973). Surprisingly, no reports of outbreaks appeared after the incidence in Maharashtra until 2005, after 32 years, when people started crowding in hospitals complaining of severe joint pain in certain parts of Andhra Pradesh and Karnataka. The investigation led by the World Health Organization (WHO) confirmed the cases to be typical of chikungunya infection. In 2006, a total of 1,109,033 suspected cases were reported from 129 districts in India (Lahariya and Pradhan, 2006; Padbidri and Gnaneswar, 1979; Sarkar et al., 1964; Shah et al., 1964). In 2007, 3.6 million people were infected with chikungunya in Kerala (Kumar et al., 2008; Srikant et al., 2010). A few cases were also reported in West Bengal in 2011 with approximately 36.89% sero-positivity (Fig. 5.2) (Chattopadhyay et al., 2016).

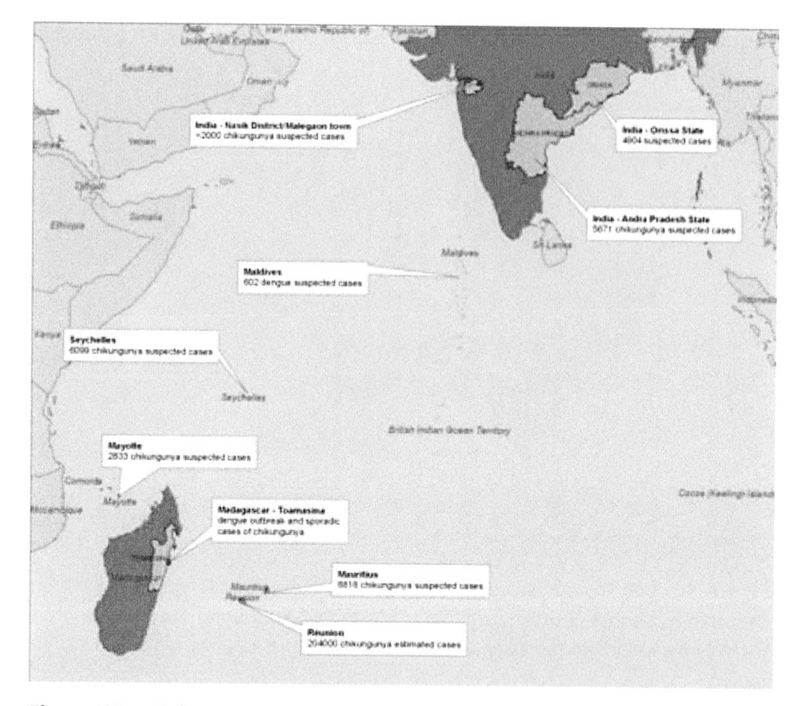

Figure 5.2 Chikungunya and dengue virus in the Indian Ocean, status as of March 17, 2006. Dark red indicates countries with occurrence of dengue and/or chikungunya. Light red indicates affected areas. Figure taken from https://www.who.int/csr/don/2006_03_17/en/.

5.2 Virus Morphology

The name chikungunya originates from the word "Makonde," which means "that which bends up." Makonde demonstrates the place from where the first incidence of chikungunya was documented. It is of the genus Alphavirus, which comes under the Togaviridae family. As per the Baltimore categorization of viruses, this virus falls in group IV ((+) ssRNA) viruses. It has an envelope and a nucleocapsid. The diameter is 60–70 nm (Higashi et al., 1967; Powers et al., 2001; Simizu et al., 1984; Strauss and Strauss, 1994). The size of genetic material includes about 11,805 nucleotides and codes for four nonstructural proteins—NS1, NS2, NS3, and NS4—and five structural proteins such as C, E3, E2, 6K, and E1. The 5′-end is capped with 7-methylguanosine, and the 3′-end of RNA molecule is poly-adenylated (Fig. 5.4). The entire viral sequences of CHIKV are available in the NCBI genome. Two sequences, strain Ross and S27 CHIKV African prototype of the 1952 epidemic, occurred in Tanzania (Khan et al., 2002), CHIKV (AY72732) isolated from *Aedes furcifer* during the 1983 outbreak in Senegal, and CHIKV of 2006 from Reunion Island (DQ4435.441), and six more sequences from Reunion and Seychelles Islands (Schuffenecker et al., 2006) have been reported. Partial sequences have also been submitted by many researchers (Powers et al., 2000; Yadav et al., 2003): L37661, L37661, AF490259, AF023283, AF192895, AF192907, and U94597. Yadav et al. (2003) have also published partial sequence of a Madras strain (M-713424). A total of 119 partial sequences are also available in the NCBI database from Reunion, Seychelles, Mauritius, Madagascar, and Mayotte Islands.

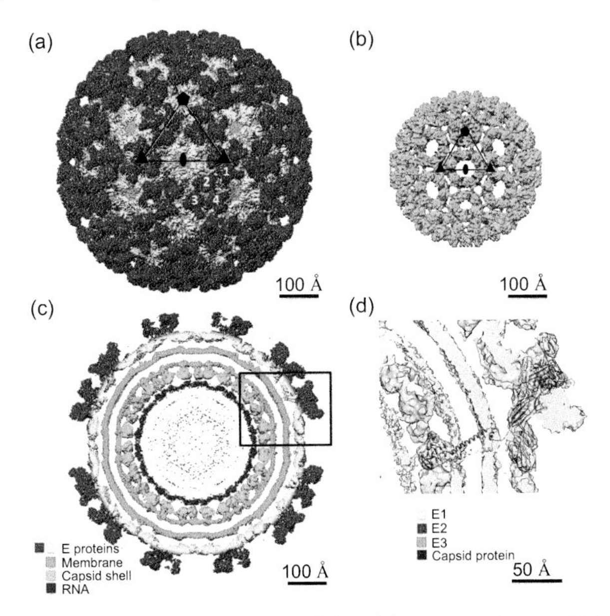

Figure 5.3 Structural representation of the chikungunya virus particle. Figure reprinted from Yap et al., 2017, Copyright National Academy of Sciences.

5.2.1 Structural Proteins

CHIKV is a mosquito-borne virus belonging to the genus Alphavirus in the Togaviridae family (Yap et al., 2017). These alphaviruses are composed of spherical envelop having diameter 700 Å and T = 4 quasi-icosahedral symmetry. The size of CHIKV genome is 12 kb positive-sense single-stranded RNA, which encodes four nonconfigural proteins (nsP1–4). These nonstructural proteins are essential for increasing the CHIKV population. There are about five structural proteins: capsid protein C, glycoproteins E1, E2, E3, and 6K (Strauss and Strauss, 1994). These configural proteins are produced in the form of long polyprotein, which break into C, E1, 6K, and p62 after post-translational modifications. The nucleocapsid contains a single copy of RNA genome complexed with 240 copies of Capsid protein (Melancon and Garoff, 1987). In addition, two other glycoproteins E1 and p62 interact to form heterodimers that subsequently trimerize into a viral spike in the endoplasmic reticulum (ER). The genes E2 and E3 are generated by the breakdown of p62 glycoprotein by cellular furin during its movement from the acidic environment of the Golgi and early endosomes to the neutral pH environment of the cell surface. The release of CHIV occurs from the cell membrane. Further configural modifications are made by the 6K protein. Alphaviruses and flaviviruses (Schmaljohn and McClain, 1996) have exhibited several resembling parameters such as their external glycoproteins are in icosahedral symmetry and wrapped in a lipid membrane, which further coats their RNA and associated with the capsid protein. The key difference between flaviviruses and alphaviruses lies in their maturation process. Flaviviruses in their initial association are "immature" non-pathogenic entities in the ER of the host cell, which are proteolytically converted into infectious viruses when they bud out from the host cell. Moreover, a characteristic icosahedral capsid coat is only visible in alphaviruses (Figs. 5.4 and 5.5).

When CHIKV infects a host cell, a conserved sequence on the N-terminal regions of the capsid proteins binds to the host cell's 60S ribosomal subunits and detaches the nucleocapsid and release the RNA from the nucleocapsid (Wengler et al., 1992). During assembly, the ribosome-binding site (RBS) is masked, but it is exposed at the end of the maturation process (Wengler, 1987). The fusion of CHIKV with the cell is mediated by E1 glycoprotein (Lescar et al., 2001). Glycoprotein E2 attaches to host receptors (Smith et al., 1990), while glycoprotein E3 helps in E1-p62 hetero-dimerization.

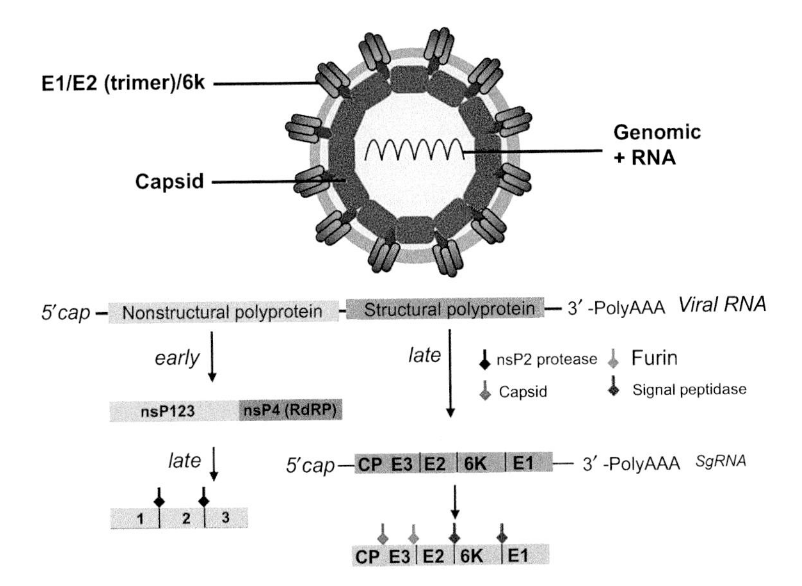

Figure 5.4 Structural organization of the chikungunya genome presenting the structural and nonstructural proteins. Figure taken from http://virologytidbits.blogspot.in/2014/08/chickungunya-virus-and-ndp52-deadly.html.

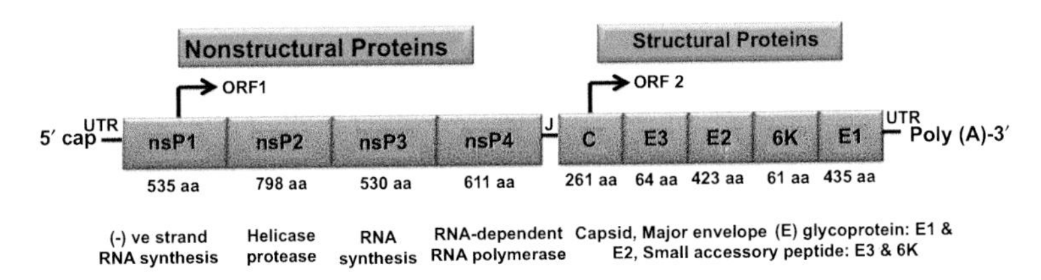

Figure 5.5 The chikungunya genome. Figure taken from https://f1000research.com/articles/6-1601/v1.

5.2.2 Nonstructural Proteins

5.2.2.1 NS1 protein

This is the first protein in line after the E protein. It is basically a glycoprotein with a molecular weight of 48,000 Da. This is a very important protein responsible for replication of viral genome and so is a marker of virus infection. It is also the major protein that interacts with the host immune system and is involved in evasion and pathogenesis, but its role in virus life cycle is not well established (Akey et al., 2014; Avirutnan et al., 2010; Avirutnan et al., 2011; Krishna et al., 2009; Young et al., 2000).

This protein is initially synthesized as a hydrophilic monomer but later changes to a hydrophobic dimer, which has two of its N-glycans modified on reaching the Golgi bodies to obtain a complex structure. The dimer has three domains: a β-roll domain, a Wing domain, and a β-ladder domain (see Fig. 4.11). Due to its varied role, it is now being used for experiment by researchers across the world for its utility as a target for antiviral therapy.

5.2.2.2 NS2 protein

The NS2 protein consists of two proteins, NS2a and NS2b, with molecular weights of approximately 20,000 and 14,500 Da, respectively. Both are hydrophobic in nature. While NS2a has a proteolytic role in processing the C-terminus of NS1 protein, the role of NS2b remains unknown (Chambers et al., 1989; Falgout et al., 1989; Speight et al., 1988).

5.2.2.3 NS3 protein

The dengue NS3 protein is a viral protease composed of six β-strands and arranged into two β-barrels formed of 1-180 residues. This is a hydrophilic protein with a molecular weight of 70,000 Da. Its function is like that of an RNA helicase and RTPase/NTPase (see Fig. 4.12). The active site involves His-51, Asp-75, and Ser-135, which is found between these two β-barrels and its activity depends on the presence of the NS2B cofactor. A six-stranded parallel β-sheet surrounded by four α-helices forms three subdomains: subdomain I to subdomain III (Henchal and Putnak, 1990; Perera and Kuhn, 2008).

5.2.2.4 NS4A protein

The dengue NS4A is a hydrophobic protein made of two proteins NS4a and NS4b with molecular weights of 16,000 and 27,000 Da, respectively (Chambers et al., 1989; Speight and Westaway, 1989; Speight et al., 1988). The protein is known to alter the cell membrane curvature (Miller et al., 2007) and induces autophagy. The NS4A acts as a scaffold for the virus replication complex and undergoes oligomerization (Lee et al., 2015). Mutations in NS4A that affect interaction with NS4B either inhibit or reduce virus replication, which shows the significance of NS4A. The interaction of NS4A with NS4B is essential for viral reproduction (Zou et al., 2015). It is also known to act as a cofactor along with the NS5 protein.

5.2.2.5 NS5 protein

The NS5 protein of dengue is a 900-residue peptide, which contains a methyltransferase domain at its N-terminal end (residues 1–296), whereas RNA-dependent RNA polymerase (RdRp) is located at its C-terminal end (residues 320–900) (Perera and Kuhn, 2008). It has a molecular weight of 105,000 Da. It is the most conserved protein of all the proteins, showing 67% sequence similarity between all the four dengue virus types. Figure 5.6 shows the structure of NS5 protein; MTase is the yellow-colored structure, RdRp fingers are green colored, palm is blue colored, and thumb is salmon colored. The linker helix (residues 263–267) between the two domains is in orange, while GTP and cofactor SAH are shown as sticks. Zinc ions are shown as spheres.

5.3 Transmission Route

The transmission of CHIKV is mediated by two mosquito vectors: *Aedes aegypti* and *Aedes albopictus*. *A. albopictus* has been found in 36 states of the United States and has been conferred as the metropolitan carrier for disease transmission in these areas (Lanciotti et al., 2007). Earlier *A. aegypti* was considered the key vector; however, due to mutation by replacing alanine with valine at the 226th position in the E1 gene, it led to flattering species change with *A. albopictus* replacing alanine (Santosh et al., 2008; Tsetsarkin et al., 2007; Vazeille et al., 2007). The female *Aedes* mosquitoes ingest blood as a meal from an infected person. The virus multiplies inside the mosquito, which makes the mosquito infective. Upright infection of CHIKV, that is, egg to larvae to adult and transmission in humans from mother to child, has also been reported. Ramful et al. (2007) have reported transplacental transmission of virus from mother to fetus, which led to the death of the growing fetus. Another research group reported similar condition in a study of 19 neonates (Gerardin et al., 2007). In addition to these, the virus circulating in humans and mosquitoes, a transmission cycle known as "zoonotic cycle" or "sylvatic cycle" also occurs (Fig. 5.6).

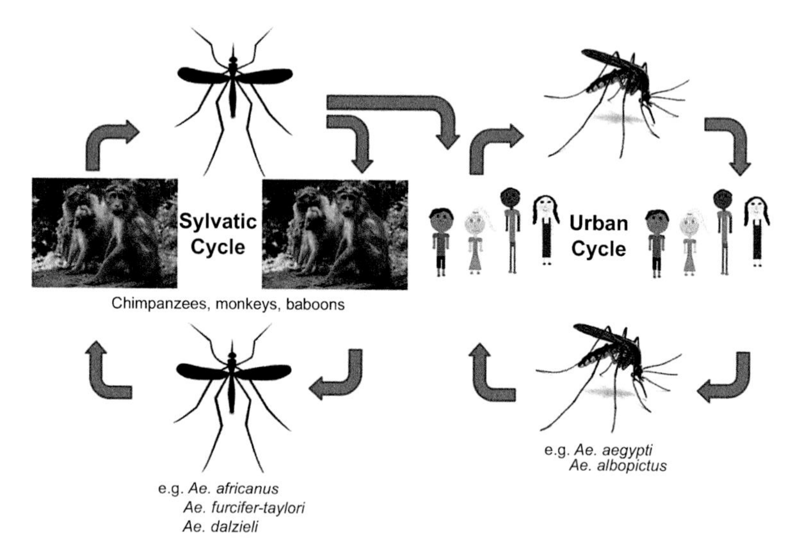

Figure 5.6 Transmission route of chikungunya virus. Reprinted from Thiboutot et al., 2010, under Creative Commons license.

5.3.1 Replication of Chikungunya Virus

There are two distinct cycles for the transmission of CHIKV: enzootic and urban. In African countries, CHIKV is transmitted via the enzootic cycle occurring in forest areas where arboreal mosquitoes, mainly *Aedes* spp., are present and act as vectors.

Nonhuman primates are the major reservoir and amplification hosts in the enzootic cycle based on their high rates of sero prevalence, documented infection and viremia in nature, and viremia levels in response to experimental infection (McIntosh et al., 1962).

The enzootic transmission cycle can infect people living in the vicinity of infected persons, and these mosquito vectors are involved in infection transmission in inter-human population. The outbreak of CHIKV has been reported in African urban areas where the more anthropophilic vectors such as *A. aegypti* and *A. albopictus* can transmit CHIKV to humans. The *A. aegypti* mosquitoes are ideal for epidemic transmission because adult females feed on humans as blood meals during a single gonotrophic cycle and deposit their ova in containers (larval sites), and relax inside houses where they get human as hosts (Weaver et al., 2012). *A. albopictus* mosquitoes act as both zoophilic and anthropophilic, are aggressive and silent, remain active the entire day, and have long a life span than other mosquitoes, approximately 56 days. *A. albopictus* lays vegetative eggs in timber and tires exported from Asia throughout the world (Reiter and Sprenger, 1987). The infectivity of different CHIKV strains varies for both *A. aegypti* and *A. albopictus*. Humans develop high-titer viremias that generally persist during the first 4 days after the onset of symptoms, with the peak estimated on the day of onset at approximately 109 viral RNA copies/ml14 and infectious titers sometimes exceeding 107PFU/ml (Leo et al., 2009). These titers generally exceed the oral infectious dose of 50% levels for both epidemic vector species, permitting efficient transmission among humans by mosquitoes (Tsetsarkin et al., 2007). For CHIKV, humans are the chief reservoir from where they transmit to other healthy individuals. Beside humans, monkeys, rodents, and birds are also found to be virus reservoir sustaining virus in the absence of human cases (Wolfe et al., 2001).

5.4 Pathogenicity

The incubation period for CHIKV fever is between 3 and 7 days (range, 2–12 days). Not all individuals infected with the virus develop symptoms. Sero surveys illustrate that about 3%–25% of infected persons with antibodies to CHIKV have asymptomatic infections (Queeyriaux et al., 2008). Symptoms of CHIKV infection include extreme high fever, that is, 102°F (39°C) for several days about 2 weeks and may be biphasic in nature (Deller and Russel, 1968). The onset of fever leads to devastating polyarthralgia. The CHIKV-infected person feels joint pains in wrists, elbows, fingers, knees, and ankles but can also affect more-proximal joints (Simen et al., 2007). The appearance of rashes after fever onset is typically maculopapular, involving the trunk and extremities but can also involve palms, soles, and the face. In addition, other dermatic symptoms are vesiculobullous lesions with desquamation, aphthous-like ulcers, and vasculitic lesions (Inamadar et al., 2008) (Fig. 5.7).

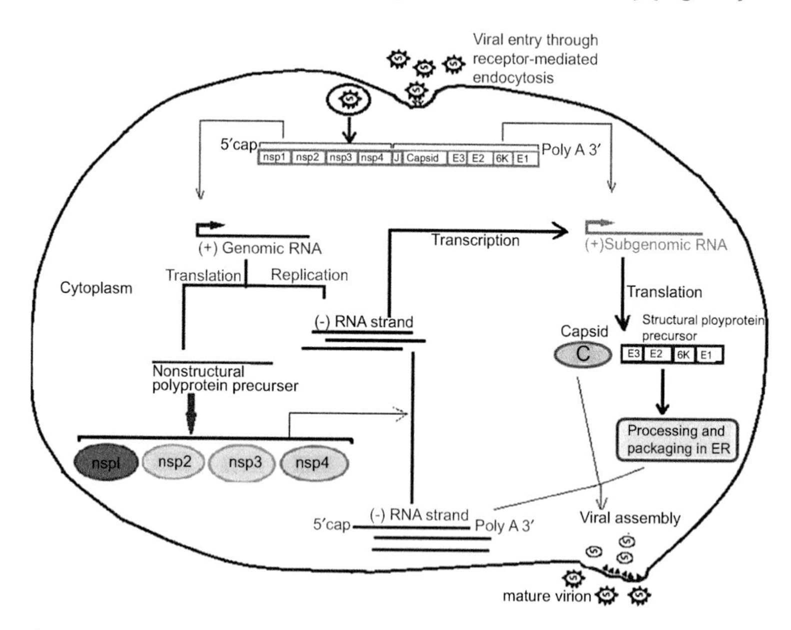

Figure 5.7 Chikungunya virus replication cycle. Figure reprinted from Deeba et al., 2016 , with permission from Oxford University Press.

Moreover, other acute symptoms include headache, fatigue, nausea, vomiting, and conjunctivitis; myalgia has also been seen. Blood deformities include leukopenia, thrombocytopenia, and hypocalcemia, and a mild to moderate increase in liver function test results (Simen et al., 2007) are seen with acute infection. Various other severe abnormalities are myocarditis, meningoencephalitis, and mild hemorrhage (Maiti et al., 1978). Besides these, neuroinvasive diseases have also been reported, such as Guillain-Barré syndrome, acute flaccid paralysis, and palsies. Other novel abnormalities include uveitis and retinitis (Lalitha et al., 2007), carpal or cubital tunnel syndrome, and Raynaud phenomenon after acute illness (Simon et al., 2007). CHIKV infections occurring during pregnancy do not transmit to the fetus (Gerardin et al., 2008). However, if the pregnant woman is infected with CHIKV during parturition, there are chances for transmission from mother to new borne, having a vertical transmission rate of 49%. CHIKV infection is disseminated through exposure of infected blood (Cordel et al., 2006).

Efficient replication results in the multiplication of CHIKV inside the host cells, that is, within the human as well as the mosquito cell where due to receptor-mediated endocytosis, the CHIKV enters the cell cytoplasm. The viral mRNA replicates by two rounds of translation. The genomic RNA undergoes partial translation, which leads to synthesis of four nonstructural proteins. The positive-sense RNA transforms into negative-sense/intermediate RNA. The structural proteins such as capsid and E2-6k-E1 of CHIKV are synthesized by translation of the sub-genomic mRNA. Assembly and package of the viral strands and structural proteins result in the formation of nascent virions, which is followed by their liberation from the cell cytoplasm.

5.5 Treatment and Diagnostics

The infection of CHIKV shows various symptoms, which may be asymptomatic in the early stage but acute and chronic in the later phase (Cunha and Trinta, 2017) (Fig. 5.8). The acute phase has been found to last for 21 days after which follows the chronic stage or sometimes the post-chronic phase with degenerative arthropathies and mooring joint stiffness. If the arthritis stage persists for more than 3 months, the chronic stage follows. No recommended treatment or vaccine is available against this disease; however, some drugs are prescribed. Various types of pains are associated with chikungunya infection, and the relieving measures have been recommended in the form of flowcharts (de Brito et al., 2016). The visual analogue scale (VAS) is employed for the measurement of pain level by the physician before counseling pain (Cunha and Trinta, 2017). The most commonly used antiviral drug is ribavirin (Ravichandran and Manian, 2008). In addition, dipyrone, paracetamol, and opioids in the form of analgesics have also been recommended as medicine, whereas NSAIDS and acetylsalicylic acid are not used due to their side effects. Likewise, anticonvulsant and anti-depressant drugs such as amitriptyline and gabapentin have been used in neuropathic pain.

Hydroxychloroquine has been prescribed in chronic phase infection in comparison to methotrexate. In some cases, sulfasalazine has also been used. These prescribed medicines are given according to a specific dose with necessary precautions that depend on the age of the individual and earlier medical history, and current physiological conditions should also be considered (de Brito et al., 2016). Bindarit has been found to be an effective anti-inflammatory drug in mouse models (Rulli et al., 2011). Homeopathic drugs such as Eupatorium perf, Pyroginum, Rhus-tox, Cedron, Influenzinum, China, and Arnica (www.chikungunyavirusnet.com) are also useful. Arbidol has been reported as an efficient viral-inhibiting drug (Delogu et al., 2011).

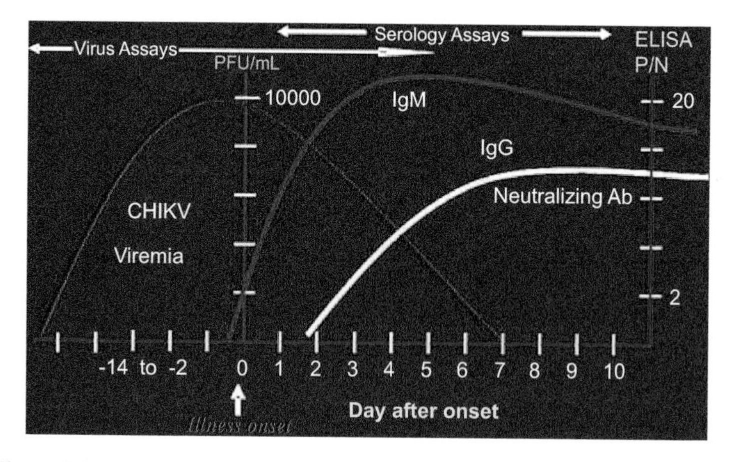

Figure 5.8 Time course of chikungunya virus viremia and immune response. Reprinted from Johnson et al., 2016, with permission from Oxford University Press.

5.5.1 Molecular Diagnostics

Scientific communities are monitoring the efficacy of vaccines at their different stages/phases. In the United States, phase II trial is going on for preparing a live, attenuated virus-based vaccine (www.chikungunyavirusnet.com). Efforts are also being made for developing formalin-inactivated vaccine, VLPs, DNA vaccines, and chimeric vaccines (Akahata et al., 2010; Mallilankaraman et al., 2011; Muthumani et al., 2008; Tiwari et al., 2009; Weaver et al., 2012). Studies based on siRNAs and shRNAs are also in the observing phase and found to be befitting (Dash et al., 2008; Lam et al., 2012).

The Centre for Disease Control (CDC) Atlanta has plotted a standard algorithm to understand the stage of infection. Serum samples of CHIKV-infected persons of less than 6 days can be investigated by real-time (RT) PCR assay, which employs particular primers. In addition to this, the infected serum sample of more than 6 days can be tested by Fpr antibodies using Mac-ELISA, and results have been confirmed by plaque reduction neutralization test (PRNT) assays (Johnson et al., 2016; Beaty et al., 1995; Lanciotti et al., 2006, 2014, 2016; Martin et al., 2000, 2004; Panning et al., 2009). Immunochromatographic (IC) assays are also being used for detecting viral infection.

5.5.2 Laboratory Diagnosis of Chikungunya Virus

CHIKV infection can be diagnosed based on clinical, epidemiological, and laboratory methods. Reverse transcriptase-polymerase chain reaction (RT-PCR) and serological tests such as enzyme-linked immunosorbent assay (ELISA) are the traditional ways of isolation of CHIKV. In the serum sample, acute CHIKV can be isolated within 8 days.

Various RT-PCR assays are available for the investigation of CHIKV RNA. Real-time PCR assays are more sensitive and have minimum risk of contamination. The Pan American Health Organization (PAHO) has advised the application of CHIKV RT-PCR protocols from the Centers for Disease Control and Prevention and the Institut Pasteur because of high sensitivity (Lanciotti et al., 2006; Panning et al., 2008).

5.5.3 Advanced Approaches

Singhal et al. (2018) have developed a genosensor based on 2D MoS_2 nanosheets for the detection of CHIKV. The genosensor exhibited a linear range of 0.1 nm to 0.1 μM, and the MoS_2-based genosensor was highly specific, sensitive, and efficient for detecting CHIKV. Lopez-Jimena et al. (2018) have developed a uni-tube one-step real-time reverse transcription loop-mediated isothermal amplification (RT-LAMP) assay for quick investigation of CHIKV, which is based on targeting the conserved 6K-E1. The assay has been tested with sera obtained from a CHIKV outbreak in Senegal in 2015.

5.6 Risk Factors

A severely CHIKV-infected individual showed encephalopathy, encephalitis, myocarditis, liver infection, and non-functioning of various organs. These ailments are lethal, and infants are also more prone to infection related with neurologic signs. The pace of infectivity gets more increased when mothers infected with CHIKV deliver their new borne babies. The person also feels pain in the joints (arthritis), hemorrhage, skin rashes, and sometimes tissue edema (Galán-Huerta et al., 2015).

5.7 Conclusion

CHIKV is a mosquito-borne virus and is classified under the family Togavirdae and genus *Alphavirus*. CHIKV is transmitted to humans through two mosquitoes, namely, *Aedes aegypti* and *Aedes albopictus*. Very high body temperature, polyarthritis, headache, rashes, and joint stiffness are some of the symptoms that are commonly seen in individuals infected with the virus. There are two transmission cycles for CHIKV, enzootic and urban. Conventionally, CHIKV has been detected by clinical, epidemiological, and laboratory techniques, RT-PCR, and ELISA. However, factors, such as lack of sensitivity and specificity of results, time-consuming diagnostic procedures, expensive instruments, and shortage of skilled professionals, have restricted the use of these techniques. However, these limitations of conventional techniques can be overcome by using advanced techniques, such as biosensing, that are simple, cost-effective, quick, highly sensitive, and specific. Drugs like ribavirin, dipyrone, and paracetamol have been recommended but no effective vaccine have been developed so far for the treatment of CHIKV-infected patients. Therefore, future research should be focused on developing more effective techniques for diagnosis and treatment of this disease.

References

Akahata, W.; Yang, Z. Y.; Andersen, H.; Sun, S.; Holdaway, H. A.; Kong, W. P.; Lewis, M. G.; Higgs, S.; Rossmann, M. G.; Rao, S.; and Nabel, G. J. A virus-like particle vaccine for epidemic chikungunya virus protects nonhuman primates against infection. *Nat. Med.*, 2010, **16**(3): 334–338.

Akey, D. L.; Brown, W. C.; Dutta, S.; Konwerski, J.; Jose, J.; Jurkiw, T. J.; DelProposto, J.; Ogata, C. M.; Skiniotis, G.; Kuhn, R. J.; and Smith, J. L. Flavivirus NS1 structures reveal surfaces for associations with membranes and the immune system. *Science*, 2014, **343**(6173): 881–885.

Avirutnan, P.; Fuchs, A.; Hauhart, R. E.; Somnuke, P.; Youn, S.; Diamond, M. S.; and Atkinson, J. P. Antagonism of the complement component C4 by flavivirus nonstructural protein NS1. *The Journal of Experimental Medicine*, 2010, **207**: 793–806.

Avirutnan, P.; Hauhart, R. E.; Somnuke, P.; Blom, A. M.; Diamond, M. S.; and Atkinson, J. P. Binding of flavivirus nonstructural protein NS1 to C4b binding protein modulates complement activation. *Journal of Immunology*, 2011, **187**: 424–433.

Bassetto, M.; De Burghgraeve, T.; Delang, L.; Massarotti, A.; Coluccia, A.; Zonta, N.; Gatti, V.; Colombano, G.; Sorba, G.; Silvestri, R.; Tron, G. C.; Neyts, J.; Leyssen, P.; and Brancale, A. Computer-aided identification, design and synthesis of a novel series of compounds with selective antiviral activity against chikungunya virus. *Antiviral Res.*, 2013, **98**(1): 12–18.

Beaty, B. J.; Calisher, C. H.; and Shope, R. E. Arboviruses. In: *Diagnostic Procedures for Viral, Rickettsial, and Chlamydial Infections* (Lennette, E. L. D. and Lennette, E., eds.). Washington, DC: American Public Health Association, 1995. pp. 189–212.

Chambers, T. J.; McCourt, D. W.; and Rice, C. Yellow fever virus proteins NS2a, NS2b, and NS4b: Identification and partial N-terminal amino acid sequence analysis. *Virology*, 1989, **169**: 100–109.

Chattopadhyay, S.; Mukherjee, R.; Nandi, A.; and Bhattacharya, N. Chikungunya virus infection in West Bengal, India. *Indian J. Med. Microbiol.*, 2016, **34**(2): 213–215.

Cordel, H. Investigation Group. Chikungunya outbreak on Réunion: Update. *Euro Surveill.*, 2006, **11**: E060302.3.

Cunha, R. V. and Trinta, K. S. Chikungunya virus: Clinical aspects and treatment—A review. *MemInstOswaldo Cruz, Rio de Janeiro*, 2017, **112**(8): 523–531.

Dash, P. K.; Tiwari, M.; Santhosh, S. R.; Parida, M.; and Lakshmana Rao, P. V. RNA interference mediated inhibition of chikungunya virus replication in mammalian cells. *Biochem. Biophys. Res. Commun.*, 2008, **376**(4): 718–722.

De Brito, C. A. A.; Sohsten, A. K. A. V.; de SáLeitão, C. C.; de Brito, R. C. C. M.; Valadares, L. D. D. A.; da Fonte, C. A. M.; de Mesquita, Z. B.; Cunha, R. V.; Luz, K.; Leão, K. M. C.; de Brito, C. M.; and Frutuoso, L. C. V. Pharmacologic management of pain in patients with chikungunya: A guideline. *Rev. Soc. Bras. Med. Trop.*, 2016, **49**(6): 668–679.

Deeba, F.; Islam, A.; Kazim, S. N.; Naqvi, I. H.; Broor, S.; Ahmed, A.; and Parveen, S. Chikungunya virus: recent advances in epidemiology, host pathogen interaction and vaccine strategies. *Pathog. Dis.*, **74**(3): 2016, ftv119.

Deller Jr., J. J. and Russell, P. K. Chikungunya disease. *Am. J. Trop. Med. Hyg.*, 1968, **17**(1): 107–111.

Delogu, I.; Pastorino, B.; Baronti, C.; Nougairède, A.; Bonnet, E.; and de Lamballerie, X. In vitro antiviral activity of arbidol against chikungunya virus and characteristics of a selected resistant mutant. *Antiviral Res.*, 2011, **90**(3): 99–107.

Enserink, M. Entomology. A mosquito goes global. *Science*, 2008, **320**: 864–866.

Enserink, M. Infectious diseases. Massive outbreak draws fresh attention to little-known virus. *Science*, 2006, **311**: 1085.

Epstein, P. R. Chikungunya fever resurgence and global warming. *Am. J. Trop. Med. Hyg.*, 2007, **76**: 403–404.

Falgout, B.; Chanock, R.; and Lai, C. J. Proper processing of dengue virus nonstructural glycoprotein NS1 requires the N-terminal hydrophobic signal sequence and the downstream nonstructural protein, NS2a. *Journal of Virology*, 1989, **63:** 1852–1860.

Galán-Huerta, K. A.; Rivas-Estilla, A. M.; Fernández-Salas, I.; Farfan-Ale, J. A.; and Ramos-Jiménez, J. Chikungunya virus: A general overview. *Medicina Universitaria*, 2015, **17**(68): 175–183.

Gerardin, P.; Barau, G.; Michault, A.; Bintner, M.; Randrianaivo, H.; Choker, G.; Lenglet, Y.; Touret, Y.; Bouveret, A.; Grivard, P.; Le Roux, K.; Blanc, S.; Schuffenecker, I.; Couderc, T.; Arenzana-Seisdedos, F.; Lecuit, M.; and Robillard, P. Y. Multidisciplinary prospective study of mother-to-child chikungunya virus infections on the Island of La Reunion. *PLoS Med.*, 2008, **5**: e60.

Henchal, E. A. and Putnak, R. J. The dengue viruses. *Clinical Microbiology Reviews*, 1990, **3**(4): 376–396.

Higashi, N.; Matsumoto, A.; Tabata, K.; and Nagatomo, Y. Electron microscope study of development of chikungunya virus in green monkey kidney stable (VERO) cells. *Virology*, 1967, **33**(1): 55–69.

http://www.chikungunyavirusnet.com/epidemiology.html

Inamadar, A. C.; Palit, A.; Sampagavi, V. V.; Raghunath, S.; and Deshmukh, N. S. Cutaneous manifestations of chikungunya fever: observations made during a recent outbreak in south India. *Int. J. Dermatol.*, 2008, **47**(2): 154–159.

Jain, M.; Rai, S.; and Chakravarti, A. Chikungunya: A review. *Trop. Doct.*, 2008, **38**: 70–72.

Johnson, W. B.; Russell, B. J.; and Goodman, C. H. Laboratory diagnosis of chikungunya virus infections and commercial sources for diagnostic assays. *J. Infect. Dis.*, 2016, **214**(5): S471–S474.

Kaur, P.; Thiruchelvan, M.; Lee, R. C.; Chen, H.; Chen, K. C.; Ng, M. L.; and Chu, J.J. Inhibition of chikungunya virus replication by harringtonine: A novel antiviral that suppresses viral protein expression. *Antimicrob. Agents Chemother.*, 2013, **57**(1): 155–167.

Khan, A. H.; Morita, K.; Parquet Md Mdel, C.; Hasebe, F.; Mathenge, E. G.; and Igarashi, A. Complete nucleotide sequence of chikungunya virus and evidence for an internal polyadenylation site. *J. Gen. Virol.*, 2002, **83**(12): 3075–3084.

Krishna, V. D.; Rangappa, M.; and Satchidanandam, V. Virus-specific cytolytic antibodies to nonstructural protein 1 of Japanese encephalitis virus effect reduction of virus output from infected cells. *J. Virol.*, 2009, **83**: 4766–4777.

Kumar, N. P.; Joseph, R.; Kamaraj, T.; and Jambulingam, P. A226V mutation in virus during the 2007 chikungunya outbreak in Kerala, India. *J. Gen. Virol.*, 2008, **89**(8): 1945–1948.

Lahariya, C. and Pradhan, S. K. Emergence of chikungunya virus in Indian subcontinent after 32 years: A review. *J. Vect. Borne Dis.*, 2006, **43**: 151–160.

Lalita, P.; Rathinam, S.; Banushree, K.; Maheshkumar, S.; Vijaykumar, R.; and Sathe, P. S. Ocular involvement associated with an epidemic outbreak of Chikungunya virus infection. *Am. J. Ophthalmol.*, 2007, **144**: 552–556.

Lam, S.; Chen, K. C.; Ng, M. M.; and Chu, J. J. Expression of plasmid-based shRNA against the E1 and nsP1 genes effectively silenced chikungunya virus replication. *PLoS One*, **2012**, **7**(10): e46396.

Lanciotti, R. S. and Lambert, A. J. Phylogenetic analysis of chikungunya virus strains circulating in the Western Hemisphere. *Am. J. Trop. Med. Hyg.*, 2016, **94**: 800–803.

Lanciotti, R. S. and Valadere, A. M. Transcontinental movement of Asian genotype chikungunya virus. *Emerg. Infect. Dis.*, 2014, **20**: 1400–1402.

Lanciotti, R. S.; Kosoy, O. L.; Laven, J. J.; Panella, A. J.; Velez, J. O.; Lambert, A. J.; and Campbell, G. L. Chikungunya virus in US travelers returning from India. *Emerg. Infect. Dis.*, 2006, **13**: 764–767.

Lee, C. M.; Xie, X.; Zou, J.; Li, S. H.; Lee, M. Y.; Dong, H.; Qin, C. F.; Kang, C.; and Shi, P. Y. Determinants of dengue virus ns4a protein oligomerization. *J. Virol.*, 2015, **89**(12): 6171–6183.

Leo, Y. S.; Chow, A. L.; Tan, L. K.; Lye, D. C.; Lin, L.; and Ng, L. C. Chikungunya outbreak, Singapore, 2008. *Emerg. Infect. Dis.*, 2009, **15**: 836–837.

Lescar, J.; Roussel, A.; Wien, M. W.; Navaza, J.; Fuller, S. D.; Wengler, G.; Wengler, G.; and Rey, F. A. The fusion glycoprotein shell of Semliki forest virus: An icosahedral assembly primed for fusogenic activation at endosomal pH. *Cell*, 2001, **105**: 137–148.

Lopez-Jimena, B.; Wehner, S.; Harold, G.; Bakheit, M.; Frischmann, S.; Bekaert, M.; Faye, O.; Sall, A. A.; and Weidmann, M. Development of a single-tube one-step RT-LAMP assay to detect the chikungunya virus genome. *PLoS Negl. Trop. Dis.*, 2018, **12**(5): e0006448.

Luo, D.; Xu, T.; Hunke, C.; Grüber, G.; Vasudevan, S. G.; and Lescar, C. Crystal structure of the NS3 protease-helicase from dengue virus. *J. Virol.*, 2008, **82**(1): 173–183.

Maiti, C. R.; Mukherjee, A. K.; Bose, B.; and Saha, G. L. Myopericarditis following chikungunya virus infection. *J. Indian Med. Assoc.*, 1978, **70**: 256–258.

Mallilankaraman, K.; Shedlock, D. J.; Bao, H.; Kawalekar, O. U.; Fagone, P.; Ramanathan, A. A.; Ferraro, B.; Stabenow, J.; Vijayachari, P.; Sundaram, S. G.; Muruganandam, N.; Sarangan, G.; Srikanth, P.; Khan, A. S.; Lewis, M. G.; Kim, J. J.; Sardesai, N. Y.; Muthumani, K.; and Weiner, D. B. A DNA vaccine against chikungunya virus is protective in mice and induces neutralizing antibodies in mice and nonhuman primates. *PLoS Negl. Trop. Dis.*, 2011, **5**(1): e928.

Martin, D. A.; Muth, D. A.; Brown, T.; Johnson, A. J.; Karabatsos, N.; and Roehrig, J. T. Standardization of immunoglobulin M capture enzyme-linked immunosorbent assays for routine diagnosis of arboviral infections. *J. Clin. Microbiol.*, 2000, **38**(5): 1823–1826.

Martin, D. A.; Noga, A.; Kosoy, O.; Johnson, A. J.; Petersen, L. R.; and Lanciotti, R. S. Evaluation of a diagnostic algorithm using immunoglobulin M enzyme-linked immunosorbent assay to differentiate human West Nile Virus and St. Louis Encephalitis virus infections during the 2002 West Nile Virus epidemic in the United States. *Clin. Diagn. Lab Immunol.*, 2004, **11**(6): 1130–1133.

McIntosh, B. M.; Paterson, H. E.; McGillivray, G.; and Desousa, J. Further studies on the chikungunya outbreak in southern Rhodesia in 1962. I. Mosquitoes, wild primates and birds in relation to the epidemic. *Ann. Trop. Med. Parasitol.*, 1964, **58**: 45–51.

Melancon, P. and Garoff, H. Processing of the Semliki forest virus structural polyprotein: Role of the capsid protease. *J. Virol.*, 1987, **61**: 1301–1309.

Miller, S.; Kastner, S.; Krijnse-Locker, J.; Bühler, S.; and Bartenschlager, R. The non-structural protein 4A of dengue virus is an integral membrane protein inducing membrane alterations in a 2K-regulated manner. *J. Biol. Chem.*, 2007, **282**(12): 8873–8882.

Muthumani, K.; Lankaraman, K. M.; Laddy, D. J.; Sundaram, S. G.; Chung, C. W.; Sako, E.; Wu, L.; Khan, A.; Sardesai. N.; Kim, J. J.; Vijayachari, P.; and Weiner, D. B. Immunogenicity of novel consensus-based DNA vaccines against chikungunya virus. *Vaccine*, 2008, **26**(40): 5128–5134.

Okabayashi, T.; Sasaki, T.; Masrinoul, P.; Chantawat, N.; Yoksan, S.; Nitatpattana, N.; Chusri, S.; Vargas, R. E. M.; Grandadam, M.; Brey, P. T.; Soegijanto, S.; Mulyantno, K. C.; Churrotin, S.; Kotaki, T.; Faye, O.; Sow, A.; Sall, A. A.; Puiprom, O.; Chaichana, P.; Kurosu, T.; Kato, S.; Kosaka, M.; Ramasoota, P.; and Ikutaf, K. Detection of chikungunya virus antigen by a novel rapid immunochromatographic test. *J. Clin. Microbiol.*, 2015, **53**(2): 382–388.

Padbidri, V. S. and Gnaneswar, T. T. Epidemiological investigations of chikungunya epidemic at Barsi, Maharashtra state, India. *J. Hyg. Epidemiol. Microbiol. Immunol.*, 1979, **23**(4): 445–451.

Panning, M.; Grywna, K.; van Esbroeck, M.; Emmerich, P., and Drosten, C. Chikungunya fever in travelers returning to Europe from the Indian Ocean region, 2006. *Emerg. Infect. Dis.*, 2008, **14**: 416–422.

Panning, M.; Hess, M.; Fischer, W.; Grywna, K.; Pfeffer, M.; and Drosten, C. Performance of the RealStar chikungunya virus real-time reverse transcription-PCR kit. *J. Clin. Microbiol.*, 2009, **47**(9): 3014–3016.

Perera, R. and Kuhn, R. J. Structural proteomics of dengue virus. *Curr. Opin. Microbiol.*, 2008, **11**: 369–377.

Pistone, T.; Ezzedine, K.; Schuffenecker, I.; Receveur, M. C.; and Malvy, D. An imported case of chikungunya fever from Madagascar: Use of the sentinel traveller for detecting emerging arboviral infections in tropical and European countries. *Travel Med. Infect. Dis.*, 2009, **7**: 52–54.

Pohjala, L.; Utt, A.; Varjak, M.; Lulla, A.; Merits, A.; Ahola, T.; and Tammela, P. Inhibitors of alphavirus entry and replication identified with a stable chikungunya replicon cell line and virus-based assays. *PLoS One*, 2011, **6**(12): e28923.

Powers, A. M.; Brault, A. C.; Shirako, Y.; Strauss, E. G.; Kang, W.; Strauss, J. H.; and Weaver, S. C. Evolutionary relationships and systematics of the alphaviruses. *J. Virol.*, 2001, **75**(21): 10118–10131.

Powers, A. M.; Brault, A. C.; Tesh, R. B.; and Weaver, S. C. Re-emergence of chikungunya and O'nyong-nyong viruses: Evidence for distinct geographical lineages and distant evolutionary relationships. *J. Gen. Virol.*, 2000, **81**(Pt 2): 471–479.

Queyriaux, B.; Simon, F.; Grandadam, M.; et al. Clinical burden of chikungunya virus infection. *Lancet Infect. Dis.*, 2008, **8**: 2–3.

Ramful, D.; Carbonnier, M.; Pasquet, M.; Bouhmani, B.; Ghazouani, J.; Noormahomed, T.; Beullier, G.; Attali, T.; Samperiz, S.; Fourmaintraux, A.; and Alessandri, J. L. Mother-to-child transmission of chikungunya virus infection. *Pediatr. Infect. Dis. J.*, 2007, **26**: 811–815.

Ravichandran, R. and Manian, M. Ribavirin therapy for chikungunya arthritis. *J. Infect. Dev. Ctries.*, 2008, **2**(2): 140–142.

Reiter, P. and Sprenger, D. The used tire trade: A mechanism for the world-wide dispersal of container breeding mosquitoes. *J. Am. Mosq. Control Assoc.*, 1987, **3**: 494–501.

Rulli, N. E.; Rolph, M. S.; Srikiatkhachorn, A.; Anantapreecha, S.; Guglielmotti, A.; and Mahalingam, S. Protection from arthritis and myositis in a mouse model of acute chikungunya virus disease by bindarit, an inhibitor of monocyte chemotactic protein-1 synthesis. *J. Infect. Dis.*, 2011, **204**(7): 1026–1030.

Santhosh, S. R.; Dash, P. K.; Parida, M. M.; Khan M.; Tiwari M.; and Lakshmana Rao, P. V. Comparative full genome analysis revealed E1: A226V shift in 2007 Indian chikungunya virus isolates. *Virus Res.*, 2008, **135**: 36–41.

Sarkar, J. K.; Chatterjee, S. N.; and Chakravarty, S. K. Haemorrhagic fever in Calcutta: Some epidemiological observations. *Indian J. Med. Res.*, 1964, **52**(7): 651–659.

Scaturro, P.; Cortese, M.; Chatel-Chaix, L.; Fischl, W.; and Bartenschlager, R. Dengue virus non-structural protein 1 modulates infectious particle production via interaction with the structural proteins. *PLoS Pathogens*, 2015, **11**(11): 1–32.

Schmaljohn, A. L. and McClain, D. Alphaviruses (Togaviridae) and Flaviviruses (Flaviviridae). In *Medical Microbiology*, 4th edn. (Baron, S., ed.) Galveston, TX: University of Texas Medical Branch at Galveston, 1996.

Schuffenecker, I.; Iteman, I.; Michault, A.; Murri, S.; Frangeul, L.; Vaney, M. C.; Lavenir, R.; Pardigon, N.; Reynes, J. M.; Pettinelli, F.; Biscornet, L.; Diancourt, L.; Michel, S.; Duquerroy, S.; Guigon, G.; Frenkiel, M. P.; Bréhin, A. C.; Cubito, N.; Desprès, P.; Kunst, F.; Rey, F. A.; Zeller, H.; and Brisse, S. Genome microevolution of chikungunya viruses causing the Indian Ocean outbreak. *PLoS Med.*, 2006, **3**(7): e263.

Shah, K. V.; Gibbs, C. J. Jr.; and Banerjee, G. Virological investigation of the epidemic of haemorrhagic fever in Calcutta: Isolation of three strains of chikungunya virus. *Indian J. Med. Res.*, 1964, **52**: 676–683.

Simizu, B.; Yamamoto, K.; Hashimoto, K.; and Ogata, T. Structural proteins of chikungunya virus. *J. Virol.*, 1984, **51**(1): 254–258.

Simon, F.; Parola, P.; Grandadam, M.; Fourcade, S.; Oliver, M.; Brouqui, P.; Hance, P.; Kraemer, P.; Ali Mohamed, A.; de Lamballerie, X.; Charrel, R.; and Tolou, H. Chikungunya infection: an emerging rheumatism among travelers returned from Indian Ocean islands. Report of 47 cases. *Medicine (Baltimore)*, 2007, **86**(3): 123–137.

Simon, F.; Paule, P.; and Oliver, M. Chikungunya virus-induced myopericarditis: Toward an increase of dilated cardiomyopathy in countries with epidemics? *Am. J. Trop. Med. Hyg.*, 2008, **78**: 212–213.

Singhal, C.; Khanuja, M.; Chaudhary, N.; et al. Detection of chikungunya virus DNA using two-dimensional MoS2 nanosheets based disposable biosensor. *Sci. Rep.*, 2018, **8**: 7734.

Smith, T. J.; Cheng, R. H.; Olson, N. H.; Peterson, P.; Chase, E.; Kuhn, R. J.; and Baker, T. S. Putative receptor binding sites on alphaviruses as visualized by cryoelectron microscopy. *Proc. Natl. Acad. Sci. USA*, 1995, **92**: 10648–10652.

Speight, G. and Westaway, E. G. Positive identification of NS4a, the last of the hypothetical nonstructural proteins of flaviviruses. *Virology*, 1989, **170**: 299–301.

Speight, G.; Coia, G.; Parker, M. D.; and Westaway, E. G. Gene mapping and positive identification of the nonstructural proteins NS2a, NS2b, NS3, NS4b, and NS5 of the flavivirus Kunjin and their cleavage sites. *J. General Virol.*, 1988, **69**: 23–34.

Srikanth, P.; Sarangan, G.; Mallilankaraman, K.; Nayar, S. A.; Barani, R.; Mattew, T.; Selvaraj, G. F.; Sheriff, K. A.; Palani, G.; and Muthumani, K. Molecular characterization of chikungunya virus during an outbreak in South India. *Indian J. Med. Microbiol.*, 2010, **28**: 299–302.

Strauss, J. H. and Strauss, E. G. The alphaviruses: Gene expression, replication, and evolution. *Microbiol. Rev.*, 1994, **58**(3): 491–562.

Thiboutot, M. M.; Kannan, S.; Kawalekar, O. U.; Shedlock, D. J.; Khan, A. S.; Sarangan, G.; Srikanth, P.; Weiner, D. B.; and Muthumani, K. Chikungunya: A potentially emerging epidemic? *PLoS Negl. Trop. Dis.*, 2010, **4**(4): e623. https://doi.org/10.1371/journal.pntd.0000623.

Thiruvengadam, K. V.; Kalyanasundaram, V.; and Rajgopal, J. Clinical and pathological studies on chikungunya fever in Madras City. *Indian J. Med. Res.*, 1965, **53**: 729–744.

Tiwari, M.; Parida, M.; Santhosh, S. R.; Khan, M.; Dash, P. K.; and Rao, P. V. Assessment of immunogenic potential of Vero adapted formalin inactivated vaccine derived from novel ECSA genotype of chikungunya virus. *Vaccine*, 2009, **27**(18): 2513–2522.

Tsetsarkin, K. A.; Vanlandingham, D. L.; McGee, C. E.; and Higgs, S. A single mutation in chikungunya virus affects vector specificity and epidemic potential. *PLoS Pathog.*, 2007, **3**: e201.

Vazeille, M.; Moutailler, S.; Coudrier, D.; Rousseaux, C.; Khun, H.; Huerre, M.; Thiria, J.; Dehecq, J. S.; Fontenille, D.; Schuffenecker, I.; Despres, P.; and Failloux, A. B. Two chikungunya isolates from the outbreak of La Reunion (Indian Ocean) exhibit different patterns of infection in the mosquito, Aedesalbopictus. *PLoS One*, 2007, **2**: e1168.

Weaver, S. C.; Osorio, J. E.; Livengood, J. A.; Chen, R.; and Stinchcomb, D. T. Chikungunya virus and prospects for a vaccine. *Expert Rev. Vaccines*, 2012, **11**(9): 1087–1101.

Wengler, G. The mode of assembly of alphavirus cores implies a mechanism for the disassembly of the cores in the early stages of infection. Brief review. *Arch Virol*, 1987, **94**: 1–14.

Wengler, G.; Wengler, G.; Boege, U.; and Wahn, K. Establishment and analysis of a system which allows assembly and disassembly of alphavirus core-like particles under physiological conditions in vitro. *Virology*, 1984, **132**: 401–412.

Wengler, G.; Würkner, D.; and Wengler, G. Identification of a sequence element in the alphavirus core protein that mediates interaction of cores with ribosomes and the disassembly of cores. *Virology*, 1992, **191**: 880–888.

Wolfe, N. D.; Kilbourn, A. M.; Karesh, W. B.; Rahman, H. A.; Bosi, E. J.; Cropp, B. C.; Andau, M.; Spielman, A.; and Gubler, D. J. Sylvatic transmission of arboviruses among *Bornean orangutans*. *Am. J. Trop. Med. Hyg.*, 2001, **64**: 310–316.

Yadav, P.; Shouche, Y. S.; Munot, H. P.; Mishra, A. C.; and Mourya, D. T. Genotyping of chikungunya virus isolates from India during 1963-

2000 by reverse transcription-polymerase chain reaction. *Acta Virol.*, 2003, **47**(2): 125–127.

Yap, M. L.; Klose, T.; Urakami, A.; Hasan, S. S.; Akahata, W.; and Rossmann, M. G. Structural studies of chikungunya virus maturation. *Proc. Natl. Acad. Sci. USA*, 2017, **114**(52): 13703–13707.

Young, P. R.; Hilditch, P. A.; Bletchly, C.; and Halloran, W. An antigen capture enzyme-linked immunosorbent assay reveals high levels of the dengue virus protein NS1 in the sera of infected patients. *J. Clin. Microbiol.*, 2000, **38**: 1053–1057.

Zou, J.; Xie, X.; Wang, Q. Y.; Dong, H.; Lee, M. Y.; Kang, C.; Yuan, Z.; and Shi, P. Y. Characterization of dengue virus NS4A and NS4B protein interaction. *J. Virol.*, 2015, **89**(7): 3455–3470.

Chapter 6

Zika: An Ancient Virus Incipient into New Spaces

Bennet Angel,[a] Neelam Yadav,[b] Jagriti Narang,[c] Surender Singh Yadav,[d] Annette Angel,[e] and Vinod Joshi[a]

[a]*Amity Institute of Virology and Immunology, Amity University, Sector-125, Noida, India*
[b]*Centre for Biotechnology, Maharshi Dayanand University, Rohtak, India*
[c]*Department of Biotechnology, Jamia Hamdard University, New Delhi, India*
[d]*Department of Botany, Maharshi Dayanand University, Rohtak, India*
[e]*Division of Zoonosis, National Centre for Disease Control, 22 Sham Nath Marg, Civil Lines, Delhi, India*
vinodjoshidmrc@gmail.com, bennetangel@gmail.com

6.1 Epidemiology

In 1947, a group of researchers found that fever developed in a caged rhesus macaque in the Zika forest (Zika means "overgrown" in the Luganda language), near the East African Virus Research Institute

Small Bite, Big Threat: Deadly Infections Transmitted by Aedes *Mosquitoes*
Edited by Jagriti Narang and Manika Khanuja

ISBN 978-981-4800-86-0 (Hardcover), 978-1-003-00329-8 (eBook)
www.jennystanford.com

in Entebbe, Uganda. The researchers isolated from the serum of the monkey a transmissible agent that was first described as Zika virus (ZIKV) in 1952. It was subsequently isolated from a human in Nigeria in 1954. From its discovery until 2007, confirmed cases of ZIKV infection from Africa and South East Asia were rare. In 2007, however, a major epidemic occurred in Yap Island, Micronesia. More recently, epidemics have occurred in Polynesia, Easter Island, the Cook Islands, and New Caledonia (Dick et al., 1952; Kindhauser et al., 2016; Macnamara, 1954).

In April 2007, the first outbreak of Zika fever was reported outside of Africa and Asia, on the island of Yap in the Federated States of Micronesia (Fig. 6.1). Symptoms of this fever were rashes, conjunctivitis, and arthralgia, and in the beginning, it was thought to be dengue. The chikungunya and Ross River viruses were also suspected. Serum samples of infected patients in the acute phase were confirmed by the presence of the RNA of ZIKV. In 2015, ZIKV outbreaks were documented in Africa, Asia, and Brazil (Chang et al., 2016; Olson et al., 1981).

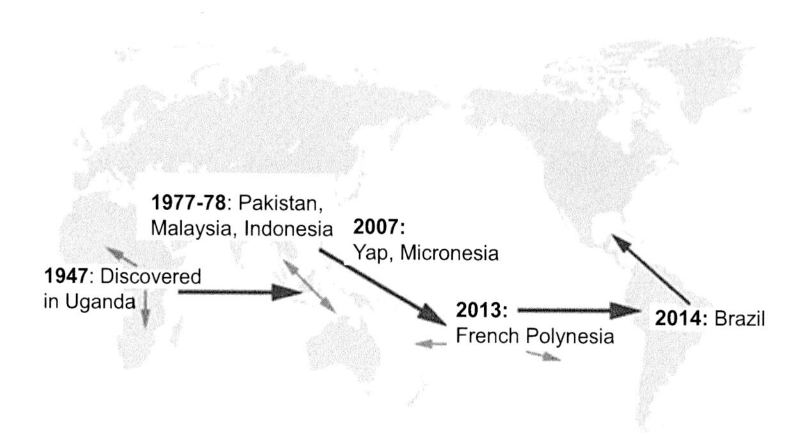

Figure 6.1 Spread of Zika virus from Uganda in Africa. Figure taken from http://www.research.lancs.ac.uk/portal/en/publications/zika-virus(716d8fe5-79d1-4843-aba8-0ae56451b5dd)/export.html.

In January 2016, the CDC issued a level 2 alert for people traveling to regions where ZIKV transmission was ongoing. The agency also advised that pregnant women should consult with doctor before traveling. According to a CDC report, Brazilian health authorities had documented more than 3500 microcephaly cases between October 2015 and January 2016 (Bearcroft, 1956; Olson et al., 1981; Paixão et al., 2016; Simpson, 1964).

In humans, Zika fever first appeared in Nigeria in 1954. Some of the outbreaks were documented in tropical Africa and in South East Asia. The major epidemic with 185 ZIKV cases affected Yap Island of the Federal States of Micronesia. About 108 affected cases were validated by either PCR or serological tests, while 72 cases were suspected. Persons infected with ZIKV showed symptoms such as rashes, fever, arthralgia, and conjunctivitis, but no casualty was reported. In the Yap epidemic, *Aedes hensilli* was the chief vector for the transmission of ZIKV, and this was the beginning phase of ZIKV infection in the exterior of Africa and Asia (Fig. 6.1). In French Polynesia, the second major outbreak was documented in 2013 (Macnamara, 1954).

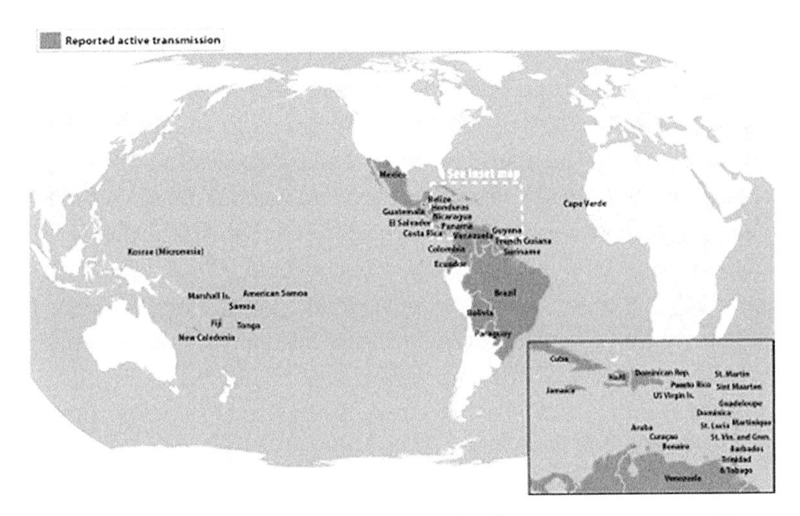

Figure 6.2 Global distribution of countries that have past or current evidence of ZIKV transmission (as of 18 April 2016). Figure taken from https://www.ecdc.europa.eu/en/publications-data/zika-virus-transmission-worldwide.

Information regarding the countries and areas with active Zika virus transmission has been availed. Moreover, the Pan America Health Organization (PAHO) had alarmed Brazil about the ZIKV infection in May 2015. As per the Brazilian Health Ministry, no official documentation on people infected with ZIKV was present in November 2015. The information on ZIKV infection in South America was disseminated to South and Central America in November 2015. It was not advisable for people to travel to these ZIKV-infected countries. Due to the increased severity index of ZIKV and microcephaly, the CDC had issued a travel alarm on January 15, 2016 to postpone travel to counties such as Brazil, Colombia, E1 Salvador, French Guiana, Guatemala, Haiti Honduras, Mexico, Panama, Paraguay, Suriname, Venezuela, and the Commonwealth of Puerto Rico (Fig. 6.2) (Bearcroft, 1956; Macnamara, 1954; Olson et al., 1981; Paixão et al., 2016; Petersen et al., 2016; Simpson, 1964).

6.2 Virus Morphology

Zika virus is an insect-borne virus (arbovirus), belonging to the *Flavivirus* genus in the Flaviviridae family, which causes fever. It is somewhat similar to dengue, yellow fever, West Nile fever, Japanese encephalitis, and viruses that are members of the family Flaviviridae. It is an enveloped, icosahedral, non-segmented, single-stranded, positive-sense RNA genetic material. The diameter of viral entities is 40 nm composed of external envelop and interior dense core. The size of ZIKV genome is 10,617 nucleotides that synthesize three configural proteins such as capsid, membrane, and envelope-154 glycosylation motif, which is responsible for causing virulency. Besides these structural proteins, viral genome also synthesizes several nonstructural proteins such as NS1, NS2A, NS2B, NS3, NS4A, NS4B, and NS5 (Fig. 6.3) (Modis et al., 2003; Rey et al., 1995; Zhang, W. et al., 2003; Zhang, Y. et al., 2003, 2004).

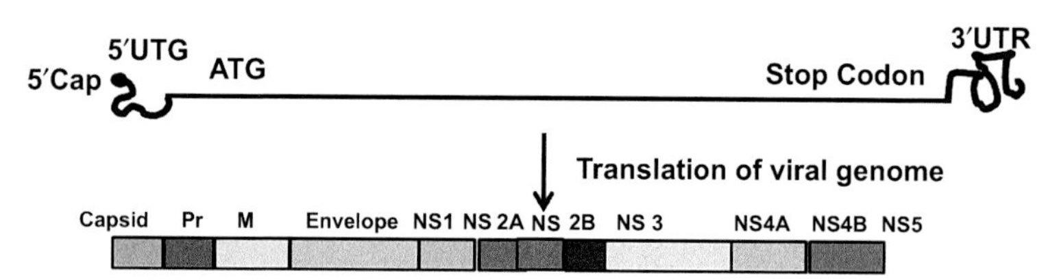

Figure 6.3 Zika virus genome structure. Figure taken from https://viralzone.expasy.org/6756?outline=all_by_species.

6.2.1 Structural Proteins

Structural proteins are mainly for interacting with the surrounding environment and are present on the outer membrane of ZIKV. Various functions of structural proteins are membrane fusion, recognition by the host immune system, and involvement in the binding of receptors (Modis et al., 2003; Rey et al., 1995; Zhang, Y et al., 2004). The outermost membrane that contains structural proteins is of icosahedral shape as confirmed through cryo-electron microscopy. The outermost shell is mainly composed of two proteins: (i) envelope protein (having 505 amino acids) and (ii) membrane protein (75 amino acids). The envelope protein is glycoprotein, and it forms association with the lipid membrane by transmembrane domains comprising three regions D1, DII, and DIII, and there is another region, that is, transmembrane, which helps in providing support to the whole framework. The D1, DII, and DIII regions are arranged in a manner that half of three E protein dimers results in the formation of a triangle, and ultimately 180 copies of the E protein dimer end produce the icosahedral shape of the outermost shell of ZIKV particles (Kostyuchenko et al., 2013, 2014). The tip of the DII domain contains a fusion loop that helps to interact with the DIII of the other monomer. Thus, it forms the dimer structure leaving a gap of hydrophobic residue of 98–109 m (Cruz-Oliveira et al., 2015). This fusion loop that joins the DII with the DIII of other dimer is highly conserved among all flavivirus family and also helps in interaction with the host cell membrane and viral membrane.

6.2.1.1 Membrane protein

Membrane proteins are present beneath the envelope protein and composed of loop and transmembrane domain. The loop is associated with the N terminus of protein, and transmembrane domains are associated with supporting the membrane protein in the lipid membrane. The transmembrane domains of envelope and membrane proteins are rooted in the lipid bilayer.

6.2.2 Nonstructural Proteins

There are seven nonstructural proteins present in the ZIKV genome: NS1, NS2A, NS2B, NS3, NS4A, NS4B, and NS5. Each protein has its unique function, and it is necessary for the replication of genetic material of virus, processing of proteins, and helping in protecting the virus from the host immune machinery. All are glycoproteins and expressed in the host cell machinery (Kostyuchenko et al., 2013, 2014). Among them, NS1 is mainly involved in evading the virus from the host cell immune system, and its role is also found to be prominent in fetal pathologies, that is, microcephaly. It is mainly expressed in the host cell and present on the surface of the ZIKV cell membrane to evade the human immune response (Kostyuchenko et al., 2014). NS2B and NS3 nonstructural proteins are involved in the processing of proteins and are involved in the protease function. NS5 is the main protein involved in the genome replication of virus and helps in maintaining the viral population. NS5 mainly consists of two domains, that is, C-terminal domain and N-terminal domain, which mainly act as enzymes for processing of proteins. C-terminal domains are mainly responsible for viral replication, and it has RNA-dependent RNA polymerase enzyme, which uses RNA as template. The C-terminal domain is mainly responsible for methyltransferase activity and controls the processing activity of viral genome (Cruz-Oliveira et al., 2015; Hamel et al., 2015).

6.3 Transmission Route

ZIKV can be transmitted by two routes: vector transmission and non-vector transmission. ZIKV showed vertical transmission as male *Aedes furcifer* mosquitoes also have ZIKV. These viruses were first found in *Aedes africanus*; however, other viruses were also isolated from this *Aedes* species (Hamel et al., 2015). This host preference was also seen in the case of this species as monkeys are more preferred host as compared to humans. The time between the vector infection and the ability to transmit the virus was found to be 15 days in the case of ZIKV. The time of incubation was found to be shorter compared to other viruses. The transmission efficiency in the case of ZIKV was also found to be good compared to yellow fever virus (Amara and Mercer, 2015; Mercer and Helenius, 2010). Other modes of transmission of ZIKV are found to be sexual intercourse, which was evident from many reported cases. Maternal fetal transmission was found to be most prominent in ZIKV (Liu et al., 2016) (Fig. 6.4).

However, other arboviruses can be transmitted to fetus through the maternal route as evident from many reported medical cases. Other routes of transmission are laboratory contamination and blood transfusion. The first case of ZIKV transmission by blood transfusion was reported in Brazil in December 2015 (Liu et al., 2016).

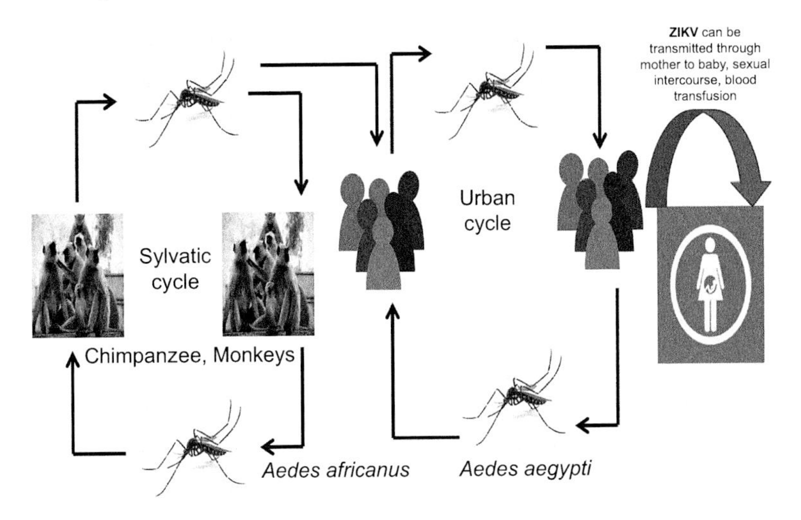

Figure 6.4 A pictorial representation of the transmission cycle of ZIKV infection. Figure taken from www.cdc.gov/zika/prevention/transmission-methods.html.

6.3.1 Replication of Zika Virus within Systems: The Extrinsic System

ZIKV is transmitted to humans by blood meals. First, a mosquito consumes an infected blood meal. The virus multiplies in the mosquito and reaches its salivary gland. When the mosquito bites, its salivary gland releases or injects ZIKV into the blood stream of a non-infected host. Human skin is permissive to ZIKV infection and allows the virus to replicate themselves (Amara and Mercer, 2015; Mercer and Helenius, 2010). After reaching the blood stream, the viruses are transported to hemolymph, where they multiply and show progressive growth, ultimately leading to viremia. After enough multiplication, the viruses are transported to other organs such as liver, kidney, and other visceral organs. Due to viremia, there is increased production of auto-phagosomes and anti-viral proteins (Amara and Mercer, 2015; Liu et al., 2016; Mercer and Helenius, 2010). T-cell mediated response was also elicited during viral infections. Most of the recent experimental in vitro studies focused on the cell types involved in transplacental transmission and neural damage, in order to get clues to ZIKV pathogenesis.

6.3.2 Replication of Zika Virus within Systems: The Intrinsic System

Once enough viremia develops, the virus infects the cells by attaching with the help of some receptors such as lectin. This receptor also mediates the attachment of other arbovirus such as dengue virus. However, other reported receptors of ZIKV are mucin domain, T-cell immunoglobin, etc. (Tabata et al., 2016). After the formation of endosomes, envelope protein dimers formed due to the fusion of DII domain with DIII dissociate and expose the hydrophobic residue, which ultimately results into rearrangement of the E monomers into trimmers (Harrison, 2015). The fusion loop of DII domain attaches to the endosome membrane, which ultimately leads to form association between viral lipid envelopes and the vesicle membrane. This association results into the release of virus RNA into the cytoplasm (Barba-Spaeth et al., 2016; Dai et al., 2016; Kostyuchenko et al., 2016; Sirohi et al., 2016). However, antibodies are produced due to the elicited immune response after infection. The produced antibodies can block the attachment of virus to the membrane by rearrangement in fusion proteins, but there should be enough amounts of antibodies produced and binding with the virus should also be at the epitope site (Kostyuchenko et al., 2016). When less amount of antibodies is produced or the virions bind to the antibodies at other sites, this antibody–virion complex still has the capacity to infect cells. Moreover, their attachment to cells having immunoglobin receptor gets further enhanced, and signals produced through these immunoglobin receptors cause the cell to be more susceptible to infection, and replication of the virus gets increased dramatically (Haddow et al., 2012).

ZIKV and dengue virus are closely related, and their vectors are also same, that is, *A. aegypti*. There is also possibility of cross-reactivity if a person was earlier infected with dengue virus (Barba-

Spaeth et al., 2016; Dai et al., 2016). Because of the association of dengue antibody and ZIKV due to cross-reactivity, there is antibody-dependent enhancement of infections. This association plays a major role in the pathogenesis of ZIKV. Immunoglobin receptors are expressed in the cells of the central nervous system and placenta.

6.3.3 Intracellular Replication of Zika Virus

Host protein machinery recognizes the viral genome and helps in synthesizing and sorting proteins for maintaining the replication of virus (Harrison, 2015). The main site for the synthesis of proteins is the membranes of endoplasmic reticulum. The proteins are sorted by both structural and nonstructural proteins as newly synthesized proteins are directed toward the lumen of ER membrane, while other proteins are present on the cytoplasmic side of the ER membrane. However, cleavage and processing of proteins are mainly done by the nonstructural proteins such as the NS3 protease domain in association with NS2B. However, cleavage is directed by the signal of host signalase (Sirohi et al., 2016). The nonstructural proteins with the dsRNA are directed toward membranous vesicle and become the site of RNA replication. Viral genomic RNA contains both 2° and 3° structures due to the untranslated regions at 3′ and 5′. Due to these 2° and 3° structures, the viral genome becomes circular and RNA is synthesized, and its modification and capping are done after viral replication in cytoplasm. However, there are other viral proteins that are transported to other locations for infection. But the transport of other Zika viral proteins is still not fully understood (Faye et al., 2014).

6.4 Pathogenicity

Immature virions present in the cytoplasm contain a lipid membrane having prM, and envelope proteins are present on the surface in icosahedral fashion (Kostyuchenko et al., 2016). Trimers of prM–envelope protein heterodimers are formed for the translation and processing of viral polyproteins. prM translocation is assisted by C protein, which contains a site for RNA interaction and shows interaction due to C-terminal trans-domain and also serves as signal sequence. Immature virions become mature when they move around the low-pH environment of trans-Golgi network (Dai et al., 2016). In the low-pH environment, prM can be cleaved by host protease. There is change in the conformation of E protein dimer, and the remaining portion is still associated with the virions (Barba-Spaeth et al., 2016). Upon increase in pH, the mature virion separates from the pr peptide as the virion is exported from the cell using host secretory machinery. The mature ZIKV envelope protein has only one glycosylation at Asn154, which makes it differ from the dengue virus envelope protein, which is glycosylated at two sites (Asn153 and Asn57) (Haddow et al., 2012). The molecular pattern present on infected cells can be easily recognized by pattern recognition receptors (RIG-1/MDA5 and TLR3) induced by the innate immune response. It ultimately leads to the production of interferons (IFN-I and IFN-III) (Faye et al., 2014). Transmembrane IFN inducible proteins are expressed, which inhibit the replication of ZIKV. IFITM3 and IFITM1 have also been shown to inhibit the replication of ZIKV (Barba-Spaeth et al., 2016; Dai et al., 2016; Kostyuchenko et al., 2016; Sirohi et al., 2016) (Figs. 6.5 and 6.6).

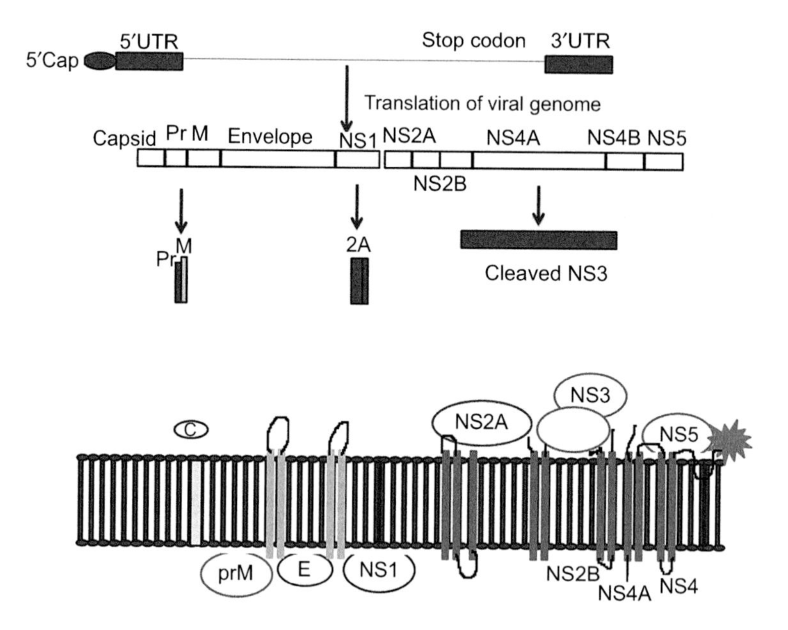

Figure 6.5 Diagrammatic representation of the virus replication for ZIKV pathogenicity. Figure adapted from Heinz and Stiasny, 2017.

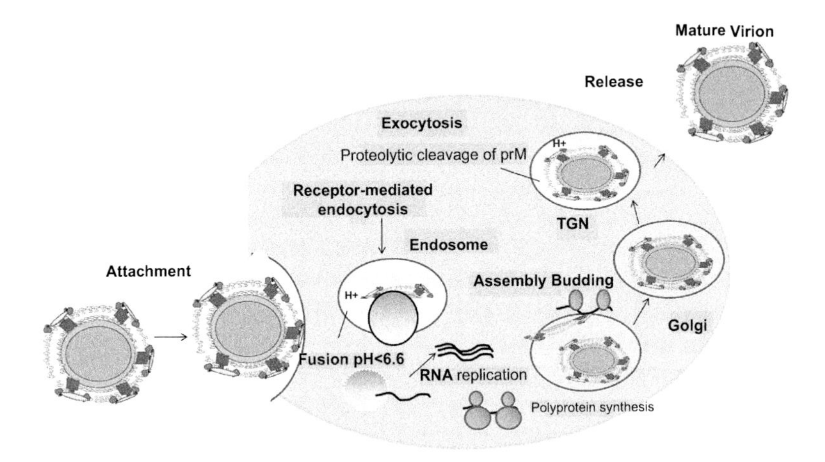

Figure 6.6 Pictorial expression of replication pathway of ZIKV. Figure adapted from Heinz and Stiasny, 2017.

6.5 Treatment and Diagnostics

6.5.1 Zika Vaccine Update

At present no vaccine or chemotherapy is available in the market. However, many institutions are working on the development of vaccines or some therapeutic agents against ZIKV. According to reports, the first trail in humans has also started (Cohen, 2016; Martines et al., 2016a). Various types of vaccines are being evaluated for their efficacy in mouse models, such as DNA vaccines, vaccines based on serotypes, envelope membrane protein expression vector, and inactivated ZIKV antigen (Abbink et al. 2016; Larocca et al., 2016). Various molecules that are being exploited for ZIKV vaccines are nucleic acids, expression vector, inactivated viral antigen, and recombinant vaccines (Guzman et al., 2013). The aforementioned vaccine types have the potential to produce antibodies against different serotypes of ZIKV. The major challenge in the case of flavivirus vaccines is antibody-dependent enhancement. Due to the close resemblance in the structure of flavivirus, there is a major problem of cross-reactivity, which increases the proliferation of flavivirus through the mechanism of antibody-dependent enhancement (Stettler et al., 2016). Due to vaccination, antibodies are produced upon infection with other flavivirus. These low-level antibodies interact with the virus but fail to neutralize the virus. However, they start targeting the immune complex, which prevents further deterioration of the virus and increase in infection rate (Stettler et al., 2016). An in vitro study also confirmed the antibody-dependent enhancement, as researchers have used plasma immune to dengue. Antibodies produced upon ZIKV infection showed cross-reactivity and led to ADE (Stettler et al., 2016). This is due to the structural homology between all flavivirus and its serotypes. Further studies also confirmed that preexisting vaccination might interfere with the subsequent vaccination of other flavivirus (Lanciotti et al., 2007). To overcome these limitations, monoclonal antibodies are used, which specifically neutralize ZIKV infections (Barba-Spaeth et al., 2016; Stettler et al., 2016). However, serum immune to other flavivirus shows poor neutralizing capacity (Swanstrom et al., 2016). These antibodies can be used in the case of immunocompromised patients and pregnant women (Luisa et al., 2016). No specific drugs are available for the treatment of any flavivirus. However, researches are underway to target specific flavivirus proteins, inhibitors of translational modifications, and disruptions of host proteins, which are necessary for viral replications (Makhluf et al., 2016). Recently, the FDA approved many drugs that were identified as inhibitors of ZIKV infections (Barrows et al., 2016). Various identified drugs include daptomycin, sertraline-HCl, pyrimethamine, bortezomib, and cyclosporine A (Barrows et al., 2016).

6.5.2 Disease Management

The World Health Organization (WHO) has laid down certain instructions, which should be advised to patients and followed by primary and secondary health centers (WHO, 2009 guidelines, www.who.int/emergencies/diseases/zika/india-november-2018/en/). A format for collecting information on Zika patients has been provided by the WHO to note down full details of the patient, which include family history of disease (WHO, 2009 guidelines, www.who.int/emergencies/diseases/zika/india-november-2018/en/).

Past medical details, locality, details of traveling to flavivirus-prone areas etc. should be captured. The patients mental state, hydration state, abdominal tenderness, and other parameters should be checked. The blood/serum sample should then be drawn to diagnose routine blood- and dengue-related tests. Plenty of fluids and ORS are recommended, and proper rest is advised. As per the WHO, no treatment is available for ZIKV infection or its associated diseases. Symptoms of ZIKV infection are usually mild. People with symptoms such as fever, rash, or arthralgia should get plenty of rest, drink fluids, and treat pain and fever with common medicines. If symptoms worsen, they should seek medical care and advice. Pregnant women living in areas with Zika transmission or who develop symptoms of ZIKV infection should seek medical attention for laboratory testing and other clinical care (WHO guidelines, 2009).

All the above studies show that consistent efforts are underway all across the world to combat the disease. Till then effective control/ prevention measures seem to be the only option for disease control. For disease control, it is very important to understand all the three components of the disease thoroughly: the human host, the vector, and the virus. The following sections evaluate the techniques that are currently employed to study and detect the virus multiplying in human hosts or, specifically, patients. The WHO, in its book, has provided proper guidelines for diagnostics, treatment, and control measures for dengue virus (WHO, 2009).

6.5.3 Laboratory Diagnosis of Zika Virus

6.5.3.1 Virus isolations

Traditionally ZIKV can be cultured, but it is not sensitive and requires more time. ZIKV is normally not diagnosed with culture system as it is very pathogenic and has lots of potential for infecting the biological samples. The isolation of virus is tedious and has the potential of infection. It requires biosafety lab III. It is not the widely used diagnostic technique for detection of ZIKV.

6.5.3.2 Immunological diagnostic assays

Immunological tests such ELISA and immunoblot assay are widely used techniques for the detection of ZIKV. Serum antibodies such as IgM and IgG are widely exploited for detecting ZIKV. These antibodies are expressed in cerebral spinal fluids (CSF). IgM is expressed in neonates (30/31) infected with microcephaly (Rabe et al., 2016). However cross-reactivity is a major problem in serological diagnosis of ZIKV as it is closely related to dengue virus. Envelope protein detection shows more cross-reactivity as compared to NS1 detection. The NS1 detection method is a more sensitive method because the ZIKV NS1 protein shows structural difference with its closely related flavivirus such as dengue (Cordeiro et al., 2016; Huzly et al., 2016). IgM is expressed in early stage infection, while IgG is expressed in a late stage of infection. In secondary infections or earlier infection with the arboviruses, IgG is also expressed in the early stage of infection, which can show cross-reactivity with other arboviruses. This can be excluded by using the NS1 detection method. Neutralization tests can also be performed for more specific detection of arboviruses. Neutralization can be carried out as standard plaque reduction test (Lanciotti et al., 2007). The test detects antibodies present in the infected serum sample by mixing it with the neat form of virus. It is allowed to react and then poured onto the Vero cell line (or mosquito cell line), which has been grown in a 24-well plate. In the early stage of infection, PCR can be well exploited, while in the late stage of infection, serological tests can be well exploited (Cordeiro et al., 2016; Huzly et al., 2016; Rabe et al., 2016).

6.5.3.3 Molecular diagnostic assays

The PCR method is most reliable, sensitive, and a specific method of detection. The genome of ZIKV is RNA; therefore, the RT-PCR technique is employed for the detection of all arbovirus (Barzon et al. 2015; Corman et al. 2016; Waggoner et al., 2016). Samples used for the extraction of viral RNA are urine, blood, amniotic fluid, cerebrospinal fluid, placenta, and saliva, but some samples require well-automated laboratories (Tappe et al., 2014). Whole blood can have viral RNA for a longer time. Urine samples can have viral RNA in the late stage of infection (Tappe et al., 2014). RT-PCR is more specific and sensitive compared to serological diagnosis, but the viral RNA is not present in the later stage of infection. The RT-PCR technique is very specific because assays have been developed in order to target the specific conserved regions of ZIKV (Barzon et al., 2015; Corman et al., 2016).

Next-generation sequencing is another viable tool for the detection of ZIKV. Sequencing is a very good technology as sequencing of the whole genome opens many doors of early and specific detection (Calvet et al., 2016). However, next-generation sequencing is not exploited for detection as it is still under investigation. ZIKV has a small genome, and it can be easily sequenced, which make the detection method more specific and sensitive (Faria et al., 2016). Sequencing of genome not only helps in detection but also provides many avenues for the therapeutic and epidemiological studies. Nowadays, sequencing method has also become more advanced and can be performed by simple and easy methods of sequencing (Lanciotti et al., 2008). A few web portals and consortiums have been created by scientists across the world so that the genome sequences and other relevant information be made available and convenient on a single platform (Calvet et al., 2016; Faria et al., 2016). These efforts will also strengthen and encourage eminent researchers to come under one roof and be a repository for sharing information and updates. Some of the portals are Dengu Info (Schreiber et al., 2007), Flavitrack (Misra and Schein, 2007), DENVirDB (Asnet et al., 2014), Flavivirus Toolkit, etc.

6.5.3.4 Nanotechnology-based detection methods: Biosensors

Advanced biosensors offer many advantageous features such as fast response time, specificity, sensitivity, fewer resource settings, and point of care applications (Kaushik et al., 2018). Therefore, biosensors can be considered a viable tool for disease management. They offer early detection as they reduce the test time, which also reduces the time gap between diagnostic and therapeutic. However, biosensors for Zika detection need to be investigated further. There are a few reported biosensors for the detection of ZIKV, which are as described in this section.

Kaushik et al. (2018) developed an electrochemical immunosensor for the detection of ZIKV. In this approach, the biological recognition element was taken as antibody, which was immobilized on an interdigitated Au electrode. This patterned electrode offers many advantages such as low reagent/sample volume. All electrodes are patterned on one platform and have the potential to interface with smartphones for data analysis. In this method, the Zika envelope antibody is immobilized as it has high binding capacity with the antigen, but the envelope protein can show more cross-reactivity with other flaviviruses compared to the NS1 antigen. Therefore, this can be a major limitation of the proposed work. However, the developed sensor was highly specific toward ZIKV as it showed no response toward dengue, chikungunya, and West Nile virus.

Afsahi et al. (2018) developed a biosensor on a graphene platform having a portable readout and can be connected to a monitor. In this sensor, a monoclonal antibody against ZIKV NS1 antigen is employed, which helps in the detection of ZIKV. The employment of ZIKV NS1 antigen offers more specificity, and it is validated by using other flavivirus for cross-reactivity. The developed sensors have high sensitivity, specificity, and portable readout, but they use commercialized electrodes, which can increase the cost of the device.

Pardee et al. (2016) developed a paper-based platform for the detection of ZIKV. It offers many advantageous features as it is highly specific because of immobilized CRISPR-Cas9, which shows high affinity toward RNA genome. The detection limit was found to be in femtomolar. A paper-based platform can prove to be the best diagnostic platform for ZIKV detection, but in this approach, RNA needs to be isolated from the viral genome, which increases time.

Tancharoen et al. (2019) developed a surface printing–based electrochemical biosensor for the detection of ZIKV. The sensor relies on the principle of change in the current with change in the virus concentration. The detection limit of the sensor is very low, and it is also validated with the conventional method. The sensor is applied to real samples such as serum.

6.6 Risk Factors

Various risk factors are associated with ZIKV, such as mosquito transmission risk, pregnancy risk, sexual transmission risk, blood transfusion risk, and regional risk.

Mosquito Transmission Risk

ZIKV belongs to the family Flaviviridae, and its genome is closely related with other flaviviruses. The vector of all flaviviruses is *A. aegypti*, which is active during the day. During the breeding season (i.e., in summer and spring), the mosquito population is more, which increases the risk of infection. Once a mosquito bites, ZIKV is inoculated and can easily cross the skin barrier. It booms in hot areas and in tropical and sub-tropical regions, Central America, India, South East Asia, and Northern Australia. ZIKV poses more risk because in most cases, the virus is asymptomatic and symptoms are very mild. If the ZIKV persists in the human body for more than a week, it can lead to serious health implications such as Guillain-Barré syndrome. In this syndrome, the body system begins to kill its own nerve cells.

Sexual Transmission Risk

Sexual transmission of ZIKV has been recently discovered. Earlier it was thought to be only a mosquito-spread disease, but ZIKV transmission has also been seen in mosquito-less areas. A recent study confirmed that ZIKV can be transmitted through sexual intercourse. In the case, a man infected with ZIKV came home from another country and mated with his wife, who also became infected with ZIKV. There was no mosquito in that area, so it was concluded after some evidence that ZIKV was transmitted due to sexual intercourse. It was also found that ZIKV can persist in semen for a longer time compared to vaginal fluids. Therefore, it was concluded that male can transmit ZIKV to female, but the opposite is not possible due to less survival percentage in vaginal fluids.

Blood Transfusion Risk

ZIKV can be transmitted through blood transfusion. However, the path followed by the virus is not clear. There are several cases that confirmed ZIKV transmission due to blood transfusion. Therefore, recent guidelines on screening blood during blood donation mandate that blood samples should be removed if the samples are found to be infected with ZIKV. Nowadays, blood is tested for all diseases, including ZIKV, before transmission.

Pregnancy Risk

When a mosquito bites a pregnant woman and inoculates ZIKV, the virus has the potential to pass through the placenta to the neonates in the first trimester. When undifferentiated stem cells start differentiating into new organs, this ZIKV infection can create various malformations. It can cause abortion or stillbirth. The most dangerous risk is when the virus infects the baby's brain and causes irreversible brain damage, which cannot be treated, i.e., microcephaly. However, this risk is limited only to the first trimester and after that the risk can be alleviated.

Regional Risk

Various areas that experienced ZIKV outbreaks in the recent years are Brazil, Venezuela, Puerto, Colombia, and the United States. There are various cities in the United States affected by ZIKV. Therefore, these areas can be considered areas that are at high risk of ZIKV infection. If a person visits these places, he or she needs to follow preventive guidelines.

6.7 Conclusion

ZIKV is an insect-borne virus which is transmitted by means of both vector as well as non-vector. Person infected with ZIKV have shown symptoms within 15 days. A number of conventional methods such as ELISA and RT-PCR have been used for the diagnosis of ZIKV. However these methods have been found less efficient as a result researchers have used modern diagnostic analytical tools such as biosensors for the fast, economic and efficient detection of ZIKV. Presently there has not been recommendable chemotherapy for the treatment of ZIKV. There are number of vaccines are under trails on mice for the prevention of ZIKV infection. Hence, researchers should primarily focused to design superior nano-based approaches which can provide capable solution for the fast, cost-effective, mobile and efficient diagnosis of ZIKV as compared to conventional laboratory serological and molecular approaches.

References

Abbink, P.; Larocca, R. A.; De La Barrera, R. A. et al. Protective efficacy of multiple vaccine platforms against Zika virus challenge in rhesus monkeys. *Science*, 2016, **353**: 1129–1132.

Afsahi, S.; Lerner, M. B.; Goldstein, J. M. et al. Novel graphene-based biosensor for early detection of Zika virus infection. *Biosens. Bioelectron.*, 2018, **100**: 85–88.

Amara, A. and Mercer, J. Viral apoptotic mimicry. *Nat. Rev. Microbiol.*, 2015, **13**: 461–469.

Asnet, M. J.; Rubia, A. G. P.; Ramya, G. et al. DENVirDB: A web portal of dengue virus sequence information on Asian isolates. *J. Vector Borne Dis.*, 2014, **51**: 82–85.

Barba-Spaeth, G.; Dejnirattisai, W.; Rouvinski, A. et al. Structural basis of potent Zika-dengue virus antibody cross-neutralization. *Nature*, 2016, **536**: 48–53.

Barrows, N. J.; Campos, R. K.; Powell, S. T. et al. A screen of FDA-approved drugs for inhibitors of Zika virus infection. *Cell Host Microbe*, 2016, **20**: 259–270.

Barzon, L.; Pacenti, M.; Ulbert, S. et al. Latest developments and challenges in the diagnosis of human West Nile virus infection. *Expert Rev. Anti-Infect. Ther.*, 2015, **13**: 327–342.

Barzon, L.; Trevisan, M.; Sinigaglia, A. et al. Zika virus: From pathogenesis to disease control. *FEMS Microbiol. Lett.*, 2016, **363**, fnw202.

Bearcroft, W. G. Zika virus infection experimentally induced in a human volunteer. *Trans. R. Soc. Trop. Med. Hyg.*, 1956, **50**: 442–448.

Boskey, E. Causes and risk factors of the Zika virus. Infectious Diseases. Last updated October 22, 2019. www.verywellhealth.com/zika-virus-causes-risk-factors-4083059.

Calvet, G.; Aguiar, R. S.; Melo, A. S. et al. Detection and sequencing of Zika virus from amniotic fluid of fetuses with microcephaly in Brazil: A case study. *Lancet Infect. Dis.*, 2016, **16**: 653–660.

Chang, C.; Ortiz, K.; Ansari, A. et al. The Zika outbreak of the 21st century. *J. Autoimmun.*, 2016, **68**: 1–13.

Cohen, J. Zika vaccine has a good shot. *Science*, 2016; **353**: 529–530.

Cordeiro, M. T.; Pena, L. J.; Brito, C. A. et al. Positive IgM for Zika virus in the cerebrospinal fluid of 30 neonates with microcephaly in Brazil. *Lancet*, 2016, **387**: 1811–1812.

Corman, V. M.; Rasche, A.; Baronti, C. et al. Clinical comparison, standardization and optimization of Zika virus molecular detection. *Bull. World Health Organ.*, 2016, DOI: 10.2471/BLT.16.175950.

Cruz-Oliveira, C.; Freire, J. M.; Conceicao, T. M. et al. Receptors and routes of dengue virus entry into the host cells. *FEMS Microbiol. Rev.*, 2015, **39**: 155–170.

Dai, L.; Song, J.; Lu, X. et al. Structures of the Zika virus envelope protein and its complex with a flavivirus broadly protective antibody. *Cell Host Microbe.*, 2016, **19**: 696–704.

Dick, G. W.; Kitchen, S. F.; and Haddow, A. J. Zika virus: (I) isolations and serological specificity. *Trans. R. Soc. Trop. Med. Hyg.*, 1952, **46**: 509–520.

Duffy, M. R.; Chen, T. H.; Hancock, W. T. et al. Zika virus outbreak on Yap Island, Federated States of Micronesia. *N. Engl. J. Med.*, 2009, **360**: 2536–2543.

European Centre for Disease Prevention and Control (ECDC). ECDC Proposed Case Definition for Surveillance of Zika Virus Infection. Solna, Sweden: ECDC, 2016a. http://ecdc.europa.eu/en/healthtopics/zika virus infection/patient-case-management/ Pages/case-definition.aspx#sthash.T4Ref5nF.dpuf (August 13, 2016, date last accessed).

Faria, N. R.; Azevedo Rdo, S.; Kraemer, M. U. et al. Zika virus in the Americas: Early epidemiological and genetic findings. *Science*, 2016, **352**: 345–349.

Faye, O.; Freire, C. C. M.; Iamarino, A. et al. Molecular evolution of Zika virus during its emergence in the 20(th) century. *PLoS Negl. Trop. Dis.*, 2014, **8**: e2636.

Guzman, M. G.; Alvarez, M.; and Halstead, S. B. Secondary infection as a risk factor for dengue hemorrhagic fever/dengue shock syndrome: A historical perspective and role of antibody-dependent enhancement of infection. *Arch. Virol.*, 2013, **158**: 1445–1459.

Haddow, A. D.; Schuh, A. J.; Yasuda, C. Y. et al. Genetic characterization of Zika virus strains: Geographic expansion of the Asian lineage. *PLoS Negl. Trop. Dis.*, 2012, **6**: e1477.

Hamel, R.; Dejarnac, O.; Wichit, S. et al. Biology of Zika virus infection in human skin cells. *J. Virol.*, 2015, **89**: 8880–8896.

Harrison, S. C. Viral membrane fusion. *Virology*, 2015, **479–480**: 498–507.

Heinz, F. X. and Stiasny, K. The antigenic structure of Zika virus and its relation to other flaviviruses: Implications for infection and immunoprophylaxis. *Microbiol. Mol. Biol. Rev.*, 2017, **81**: e00055-16.

https://viralzone.expasy.org/6756?outline=all_by_species

https://www.who.int/emergencies/diseases/zika/india-november-2018/en/

https://www.who.int/news-room/fact-sheets/detail/zika-virus

Huzly, D.; Hanselmann, I.; Schmidt-Chanasit, J. et al. High specificity of a novel Zika virus ELISA in European patients after exposure to different flaviviruses. *Euro. Surveill.*, 2016, **21**: 30203.

Kaushik, A.; Yndart, A.; Kumar, S. et al. A sensitive electrochemical immunosensor for label-free detection of Zika-virus protein. *Sci. Rep.*, 2018, **8**: 9700.

Kindhauser, M. K.; Allen, T.; Frank, V. et al. Zika: The origin and spread of a mosquito-borne virus. *Bull. World Health Organ.*, 2016, **94**(9): 675–686.

Kostyuchenko, V. A.; Chew, P. L.; Ng, T. S. et al. Near-atomic resolution cryo-electron microscopic structure of dengue serotype 4 virus. *J. Virol.*, 2014, **88**: 477–482.

Kostyuchenko, V. A.; Lim, E. X. Y.; Zhang, S. et al. Structure of the thermally stable Zika virus. *Nature*, 2016, **533**: 425–428.

Kostyuchenko, V. A.; Zhang, Q.; Tan, J. L. et al. Immature and mature dengue serotype 1 virus structures provide insight into the maturation process. *J. Virol.*, 2013, **87**: 7700–7707.

Lanciotti, R. S.; Kosoy, O. L.; Laven, J. J. et al. Genetic and serologic properties of Zika virus associated with an epidemic, Yap State, Micronesia, 2007. *Emerg. Infect. Dis.*, 2008, **14**: 1232–1239.

Larocca, R. A.; Abbink, P.; Peron, J. P. et al. Vaccine protection against Zika virus from Brazil. *Nature*, 2016, **536**: 474–478.

Liu, S.; DeLalio, L. J.; Isakson, B. E. et al. AXL-mediated productive infection of human endothelial cells by Zika virus. *Circ. Res.*, 2016, **119**: 1183–1189.

Macnamara, F. N. Zika virus: A report on three cases of human infection during an epidemic of jaundice in Nigeria. *Trans. R. Soc. Trop. Med. Hyg.*, 1954, **48**: 139–145.

Makhluf, H.; Kim, K.; and Shresta, S. Novel strategies for discovering inhibitors of Dengue and Zika fever. *Expert Opin. Drug. Discov.*, 2016, **11**: 921–923.

Martines, R. B.; Bhatnagar, J.; de Oliveira Ramos, A. M. et al. Pathology of congenital Zika syndrome in Brazil: A case series. *Lancet*, 2016, **388**: 898–904.

Mercer, J. and Helenius, A. Apoptotic mimicry: Phosphatidylserine mediated macropinocytosis of vaccinia virus. *Ann. N. Y. Acad. Sci.*, 2010, **1209**: 49–55.

Misra, M and Schein, C. H. Flavitrack: An annotated database of flavivirus sequences. *Bioinformatics*, 2007, **23**(19): 2645–2647.

Modis, Y.; Ogata, S.; Clements, D. et al. A ligand-binding pocket in the dengue virus envelope glycoprotein. *Proc. Natl. Acad. Sci. USA*, 2003, **100**: 6986–6991.

Morrison, C. DNA vaccines against Zika virus speed into clinical trials. *Nat. Rev. Drug Discov.*, 2016, **15**: 521–522.

Olson, J. G.; Ksiazek, T. G.; Suhandiman, T. et al. Zika virus, a cause of fever in Central Java, Indonesia. *Trans. R. Soc. Trop. Med. Hyg.*, 1981, **75**: 389–393.

Paixão, E. S.; Barreto, F.; Teixeira, M. G. et al. History, epidemiology, and clinical manifestations of Zika: A systematic review. *Am. J. Public Health*, 2016, **106**(4): 606–612.

Pardee, K., Green, A. A., Takahashi, M. K. et al. Rapid, low-cost detection of Zika virus using programmable biomolecular components. *Cell*, 2016, **165**: 1255–1266.

Petersen, L. R.; Jamieson, D. J.; Powers, A. M. et al. Zika virus. *N. Engl. J. Med.*, 2016, **374**: 1552–1563.

Rabe, I. B.; Staples, J. E.; Villanueva, J. et al. Interim guidance for interpretation of Zika virus antibody test results. *MMWR Morb. Mortal Wkly. Rep.*, 2016, **65**: 543–546.

Rey, F. A.; Heinz, F. X.; Mandl, C. et al. The envelope glycoprotein from tick-borne encephalitis virus at 2 A resolution. *Nature*, 1995, **375**: 291–298.

Schreiber, M. J.; Ong, S. H.; Holland, R. C. G. et al. DengueInfo: A web portal to dengue information resources. *Infect. Genet. Evol.*, 2007, **7**: 540–541.

Simpson, D. I. Zika virus infection in man. *Trans. R. Soc. Trop. Med. Hyg.*, 1964, **58**: 335–338.

Sirohi, D.; Chen, Z.; Sun, L. et al. The 3.8 Å resolution cryo-EM structure of Zika virus. *Science*, 2016, **352**: 467–470.

Stettler, K.; Beltramello, M.; Espinosa, D. A. et al. Specificity, cross-reactivity and function of antibodies elicited by Zika virus infection. *Science*, 2016, **353**: 823–826.

Swanstrom, J. A.; Plante, J. A.; Plante, K. S. et al. Dengue virus envelope dimer epitope monoclonal antibodies isolated from dengue patients are protective against Zika virus. *mBio*, 2016, **7**: e01123–16.

Tabata, T.; Petitt, M.; Puerta-Guardo, H. et al. Zika virus targets different primary human placental cells, suggesting two routes for vertical transmission. *Cell Host Microbe.*, 2016, **20**: 155–166.

Tancharoen, C.; Sukjee, W.; Thepparit, C. et al. Electrochemical biosensor based on surface imprinting for Zika virus detection in serum. *ACS Sens.*, 2019, **4**: 69–75.

Tappe, D.; Rissland, J.; Gabriel, M. et al. First case of laboratory confirmed Zika virus infection imported into Europe, November 2013. *Euro. Surveill.*, 2014, **19**: 20685.

Waggoner, J. J. and Pinsky, B. A. Zika virus: Diagnostics for an emerging pandemic threat. *J. Clin. Microbiol.*, 2016, **54**: 860–867.

Waggoner, J. J.; Gresh, L.; Mohamed-Hadley, A. et al. Single reaction multiplex reverse transcription PCR for detection of Zika, chikungunya, and dengue viruses. *Emerg. Infect. Dis.*, 2016, **22**: 1295–1297.

Zhang, W.; Chipman, P. R.; Corver, J. et al. Visualization of membrane protein domains by cryo-electron microscopy of dengue virus. *Nat. Struct. Biol.*, 2003, **10**: 907–912.

Zhang, X.; Ge, P.; Yu, X. et al. Cryo-EM structure of the mature dengue virus at 3.5-A resolution. *Nat. Struct. Mol. Biol.*, 2013, **20**: 105–110.

Zhang, Y.; Corver, J.; Chipman, P. R. et al. Structures of immature flavivirus particles. *EMBO J.*, 2003, **22**: 2604–2613.

Zhang, Y.; Zhang, W.; Ogata, S. et al. Conformational changes of the flavivirus E glycoprotein. *Structure*, 2004, **12**: 1607–1618.

Chapter 7

Yellow Fever: Emergence and Reality

Neelam Yadav,[a] Bennet Angel,[b] Jagriti Narang,[c] Surender Singh Yadav,[d] Vinod Joshi,[b] and Annette Angel[e]

[a]*Centre for Biotechnology, Maharshi Dayanand University, Rohtak, India*
[b]*Amity Institute of Virology and Immunology, Amity University, Sector-125, Noida, India*
[c]*Department of Biotechnology, Jamia Hamdard University, New Delhi, India*
[d]*Department of Botany, Maharshi Dayanand University, Rohtak, India*
[e]*Division of Zoonosis, National Centre for Disease Control, 22 Sham Nath Marg, Civil Lines, Delhi, India*
vinodjoshidmrc@gmail.com, bennetangel@gmail.com

Yellow fever (YF) is a mosquito-borne flaviviral hemorrhagic fever. This fever may cause hepatitis, kidney malfunctioning, hemorrhage, and multiorgan failure. Attack of yellow fever virus (YFV) has caused several outbreaks in the world. Conventionally, YFV is detected by using clinical approaches, but better results could not be obtained. Therefore, development of some advanced nano-based technologies is the need of hour. For treating this fever, a live, attenuated vaccine (YF 17D) has been used. However, this YF 17D vaccine has shown encephalitis in children below 9 months and persons above 60

Small Bite, Big Threat: Deadly Infections Transmitted by Aedes *Mosquitoes*
Edited by Jagriti Narang and Manika Khanuja

ISBN 978-981-4800-86-0 (Hardcover), 978-1-003-00329-8 (eBook)
www.jennystanford.com

years. Therefore, due to these setbacks of conventional YF vaccine, it is necessary to develop some novel techniques as well as drugs that can be beneficial for humankind. Some antiviral compounds such as ribavirin have also been used for the treatment of YFV infection. Future research can be focused on cytokine-mediated therapy, production of novel antibodies and interferons, and related immunocomponents, which should be effective when introduced before or within a narrow time window after infection. Herein we describe the epidemiology of YFV, virus morphology, historical aspects, pathogenicity, transmission, detection methods, and treatment associated with YF.

7.1 Epidemiology

YF is an acute viral hemorrhagic disease. It is transmitted from infected mosquitoes. The word "yellow" refers to the jaundice that affects some patients. In the past, YF was reported in countries such as Africa, the Americas, Europe, and the Caribbean. YF evolved in Africa around 3000 years ago (Figs. 7.1 and 7.2). From West Africa, YF was imported into the western hemisphere. The first evidence of YF was documented in an outbreak of the disease in Yucatán and Guadeloupe. Epidemics of YF were reported in eastern United States, including New York, in 1668, while epidemics were reported in Boston and Charleston in 1691 and 1699, respectively. In one of the first epidemics of Cadiz, Spain, approximately 2200 deaths were documented. After this, the next epidemic was in French and British seaports. A widespread outbreak was documented in tropical and subtropical regions of the Americas, as well as the West Indies, Central America, and the United States.

Until the mid-1800s, it was believed that YF was transmitted by infected individuals or items. In 1848 and 1891, American physician Josiah Clark Nott and Cuban physician Carlos Finlay stated that mosquitoes are responsible for transmission of infection as vectors. In 1900, The Reed Yellow Fever Commission proved that YF infection is transmitted to humans by the *Aedes aegypti* mosquito. In 1905, an epidemic was reported in New Orleans, the United States.

In Panama and Havana, Cuba, after it became known that YF was spread by *A. aegypti* mosquitoes from virus to humans, various hygienic campaigns were organized at a large scale. Consequently, YF disease was abolished in these regions. In 1906, abolition of YF led to the completion of the Panama Canal. Two vaccines against YFV were developed in 1930: 17D and the French neurotropic vaccines. In Nigeria, approximately 120,000 cases with 24,000 mortalities were reported. From a public health perspective, the main concern was inefficient vaccination, despite its accessibility and being inexpensive. The French neurotropic vaccine increased the encephalitis cases, and hence this vaccine was banned.

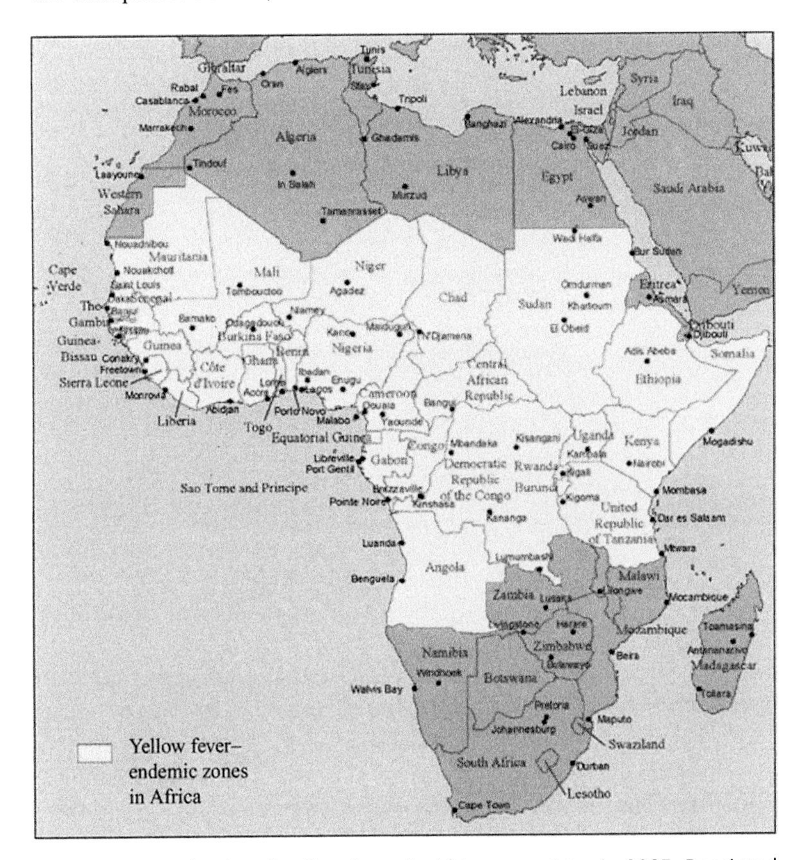

Figure 7.1 Epidemics of yellow fever in African countries in 2005. Reprinted from Barnett, 2007, by permission of Oxford University Press.

Figure 7.2 Epidemics of yellow fever reported in Central and South America countries in 2005. Reprinted from Barnett, 2007, by permission of Oxford University Press.

In 1982, the 17D vaccine was used globally for immunization of YFV. In several South American and African countries, YF vaccine was recommended for childhood vaccinations. This approach declined cases vulnerable to YFV; still people who were not exposed to vaccination were at risk. According to the World Health Organization (WHO), several hundred infected persons with YFV have been reported from the endemic countries particularly in South America and Africa annually. Recently, two virus sub-strains are being used that provide defense system against YFV: (i) 17DD, mainly used for immunization in Brazil, and (ii) 17D-204 in all other vaccines, such as Sanofi Pasteur vaccine and YF-VAX®, used in the United States. The WHO reports that several infected cases of YF are not diagnosed because of their asymptomatic nature and inefficient surveillance. Figure 7.3 shows the WHO report of vaccination in different countries. The International Health Regulations (IHR) of the WHO allow the entry requirements for YF vaccination for the protection of people, thereby preventing transmission of YF within their borders. Presently, many government agencies require authorized evidence of vaccination against the disease before the entry of travelers to their country boundaries (CDC, 2002; https://www.cdc.gov/yellowfever/vaccine/index.html).

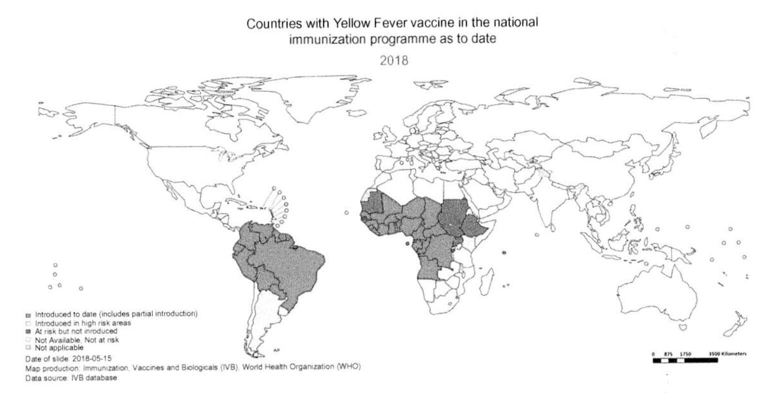

Figure 7.3 The WHO report for vaccination in different countries against YFV. Figure taken from www.who.int/immunization/monitoring_surveillance/burden/vpd/surveillance_type/passive/yellow_fever/en/.

In 1973, about 100 deaths were reported from YF epidemic in Philadelphia. After a few days, the number of deaths increased from 100 to 5000 people. YF or American plague disease starts with fever and ache in muscles. Subsequent victims become jaundiced; therefore, its name is "yellow" fever. YF patients are more susceptible to liver and kidney failure. Like dengue virus, YFV is also transmitted though mosquitoes. In the United States, the first YF epidemic was reported in the late 1690s. Nearly 100 years later, in the summer of 1793, immigrants from a YF epidemic in the Caribbean fled to Philadelphia. Within weeks, people throughout the city were experiencing symptoms. At present, even after vaccination for YF, approximately 20,000 people die annually (http://www.history.com/this-day-in-history/yellow-fever-breaks-out-in-philadelphiawww.history.com/this-day-in-history/yellow-fever-breaks-out-in-philadelphia).

A YF outbreak was reported in 34 African countries and 13 Central and South American countries. In 2013, investigation based on African data sources revealed 84,000–170,000 severe cases of YF, which accounted for the death of 29,000–60,000 people.

From 17th to 19th centuries, YF outbreaks disrupted economies and devastation of human lives (http://www.who.int/mediacentre/factsheets/fs100/en/).

7.2 Virus Morphology

The size of YFV varies between 40 and 50 nm. It is an enveloped RNA containing virus and belongs to the family *Flaviviridae* and the group of hemorrhagic fevers (Fontenille et al., 1997). The RNA is positive-sense, single-stranded RNA and contains 11,000 nucleotides with a single open reading frame encoding a polyprotein. The host proteases break the viral polyprotein into three structural proteins—C, prM, and E—and seven nonstructural proteins: NS1, NS2A, NS2B, NS3, NS4A, NS4B, and NS5 (Gouy et al., 2010). The 3′ UTR region of YFV is essential for the inactivation of the host 5′-3′ exonuclease XRN1. The PKS3 pseudoknot structure is found in the UTR region, which acts as a molecular signal for the inactivation of exonuclease, which is only a viral requirement for the production of sub-genomic flavivirus RNA (sfRNA). Partial digestion of sfRNAs of the viral genome leads to potent pathogenicity (Pisano et al., 1997).

The YFV infects cells such as monocytes, macrophages, and dendritic cells. These viruses attach to the surface of cells by specific receptors and get internalized to form endosomal vesicle. The environment inside the endosome is acidic, which stimulates the fusion of endosomal membrane with the virus envelope. The viral capsid penetrates the cytosol and releases its own genetic material. The binding of viral receptors and fusion with membrane are mediated by protein E, which alters their configuration in the acidic environment and followed by the arrangement of the 90 homo dimers to 60 homo trimers (Gouy et al., 2010).

As the YFV enters the host cell, the viral genetic material is replicated in the rough endoplasmic reticulum (RER). Initially, immature forms of YFV particles are generated inside the RER. The M-protein of these immature viral particles is not cleaved to form mature, and hence they are known as precursor M (prM) protein. The prM protein makes a complex with protein E. The processing of immature particles into mature form occurs by the furin protein of host in Golgi apparatus, which cleaves prM to M. As a result, E protein is released from the complex and finally converted into mature, infectious virion (Gouy et al., 2010). The genetic map of YFV is depicted in Fig. 7.4, while Fig. 7.5 shows the electron microscopy of YFV.

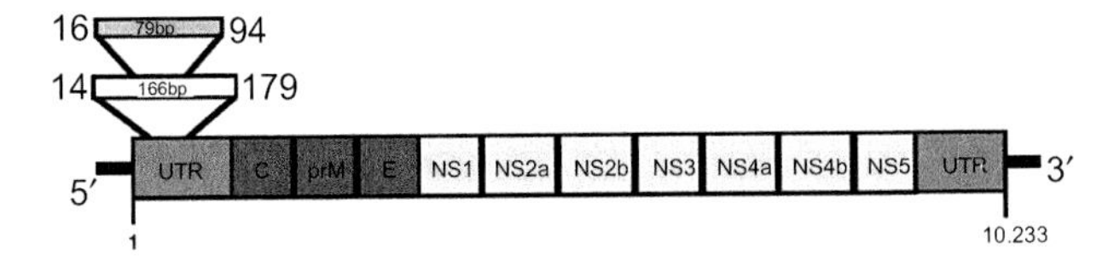

Figure 7.4 Genetic map of yellow fever virus. Figure reprinted from Méndez et al., 2013.

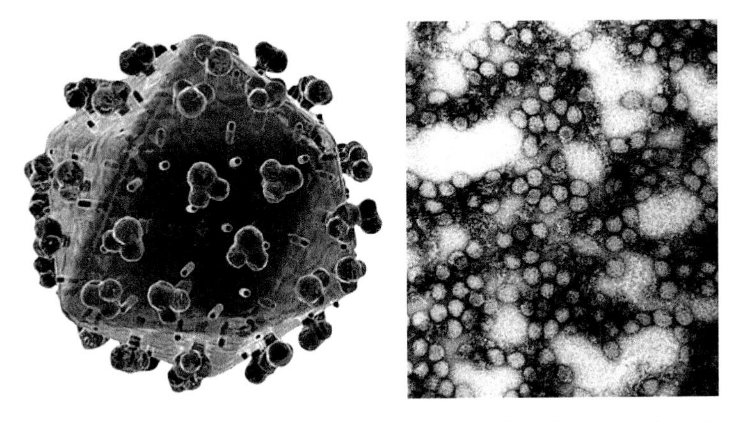

Figure 7.5 (a) Computer-generated image of the yellow fever virus based on electron microscopy. Figure taken from Rayur.com. (b) Photograph of yellow fever virus. Figure taken from https://www.sciencephoto.com/dennis-kunkel-microscopy-collection.

7.3 Historical Overview

As a disease, YF made the authorities and people significantly cautious when the first outbreak was reported in Philadelphia in 1793, with 9% casualty. In Mississippi, YFV infection resulted in the death of 20,000 people in 1878. The most recent epidemic was reported in New Orleans, the United States, in 1905 with 452 deaths (Mutebi and Barrett, 2002). This quick decline in epidemics was due to the significant efforts of the US Army Medical Corps Yellow Fever Commission directed by Walter Reed, Virginian physician, at the time of the Spanish–Cuban–American War. Reed applied the work of Cuban scientist Carlos Finlay and proved the role of mosquitoes, mainly *A. aegypti*, in the transmission of YFV. Hence, he succeeded in developing hygiene techniques for decreasing the mosquito abundance (Barrett and Monath, 2003). YFV-infected persons were identified by general flu-like symptoms, and about 200,000 cases with 30,000 mortalities were observed annually (Robertson et al., 1997).

7.4 Pathogenicity

YF is caused by prototype YFV that belongs to the genus *Flavivirus* and the family *Flaviviridae.* Symptoms of YFV infection may vary from mild to hemorrhagic, which leads to failure of vital organs. There is no therapy for the treatment of YF; however, a YF vaccine has been developed for the prevention of YFV(1-2). The WHO has recommended the effective and live attenuated YFV-17D, though in African countries approximately 300,000 annual mortality has been reported (Mutebi and Barrett, 2002). A dramatic increase in YF patients is due to reduced vaccination, urbanization, migration of the human population, and re-infestation of *A. aegypti* in the last 30 years (Barrett and Monath, 2003; Robertson et al., 1997). In 2005, the "YF initiative" was taken by the Global Alliance of Vaccine and Immunization (GAVI) to control the YF infection through combined preventive vaccination campaigns with routine immunization. The YFV transmits the infection to nonhuman primates by the sylvatic cycle where mosquitoes of the genera *Aedes* in Africa and *Haemagogus* in South America are acting as vector. In epidemic areas, humans are more at risk of YFV infection. Consequently, these infected humans spread the viral infection in cities as well. Therefore, an infected YF human population is responsible for the emergence of epidemics through *A. aegypti* as an urban vector (Barrett and Monath, 2003; Gould et al., 2008). One more pathway is vertical transmission of YFV in drought from infected female mosquitoes to their eggs. In such conditions, the stability of YFV gets increased and followed by their reactivation at the time when the progeny comes out in favorable conditions (Barrett and Monath, 2003; Fontenille et al., 1997).

The prevalence of YF relies on the transmission by mosquito vector. Hence, it is confined to African and South American tropical countries. However, YF has not been reported in Asia; yet the vector for transmission of YF, that is, *A. aegypti*, is present in these regions (Barrett and Monath, 2003; Barrett and Higgs, 2007; Gould et al., 2008). Nina et al. (2013) and Stokes et al. (1928) discovered that rhesus monkeys were prone to YFV infection. Bearcroft (1957) demonstrated the pathology of YFV in several organs of *Macaca mulatta*. Theiler (1936) showed the similarity of YF pathology between monkey and human. Tigertt et al. (1960) comprehensively studied the pathology of YF in the liver of *M. mulatta*. Lloyd et al. (1936) demonstrated that the Asibi strain of YFV lost its virulence during continuous passage in cell culture. Hardy (1963) and Hearn et al. (1966) reported viral virulence during serial passage in animal cell cultures. This was because of differences in the rates of virus attenuation upon passage of many cell lines, which may reflect the nutritional or metabolic differences of culturing cells. John et al. (1971) studied the altered virulence pattern of YFV by using cell lines and comparative pathology of passaged and unpassaged virus.

7.5 Transmission/Reservoirs

YFV is transmitted through mosquito vectors such as *Aedes* and *Haemagogus* spp. and prevalent in several primate communities. Transmission of YFV involves three different cycles, which include the sylvatic (jungle) cycle where mosquitoes transmit virus via monkey intermediates to other mosquitoes with rare infection to humans (Fig. 7.6). The second is the intermediate (Savannah) cycle in which humans and monkeys exist as intermediates with negligible epidemics, and the third is the urban cycle, which is well known for epidemics. Through the urban cycle, YFV-infected humans spread the virus to new mosquito population and non-immune people (Robertson et al., 1996).

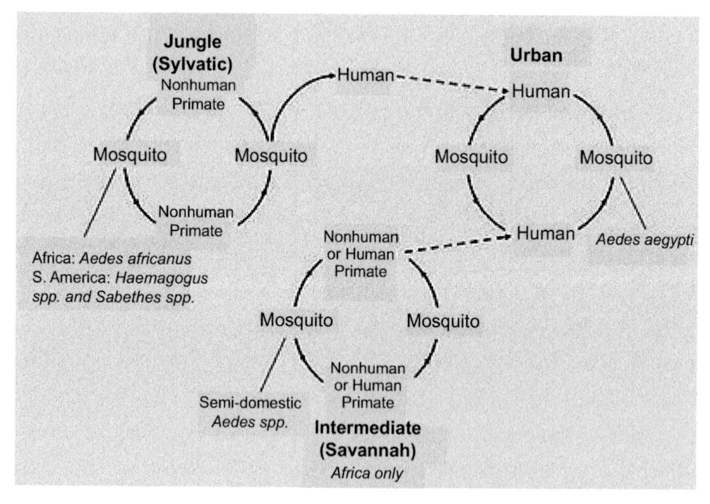

Figure 7.6 Diagram of the various transmission cycles of yellow fever virus. Figure taken from https://www.cdc.gov/yellowfever/transmission/index.html.

7.5.1 Incubation/Colonization

The entry of YFV occurs through mosquito bites where the virus remains dormant for 6 to 8 days and is responsible for severe and rapid viral infection to host cells such as dendritic cells. This initiates acute YFV infection in the affected cells (Monath, 2001). The aforementioned virus particles enter the host by the process of receptor-mediated endocytosis. Now the virus synthesizes the proteins required for virus replication in the endoplasmic reticulum of the host by using its mRNA (Gould and Solomon, 2008).

7.6 Diagnosis and Treatment of YFV Infection

7.6.1 Clinical Diagnosis of YF

Doctors diagnose YF by seeing the symptoms of the patient and conducting blood tests, which help in the detection of other related signs such as dengue fever, leptospirosis, malaria, poisoning, typhoid, hepatitis, and hemorrhagic fevers. From the blood test, doctors identify either the infection of YFV or production of antibodies in the patient's body. Moreover, blood test can also detect the declined level of white blood cells, or leucopenia (https://www.medicalnewstoday.com/articles/174372.phpwww.medicalnewstoday.com/articles/174372.php). Furthermore, other symptoms observed in endemic areas include fever, bradycardia, and appearance of jaundice. Many tests are carried out for the confirmation of YFV infection: complete blood count, urinalysis, testing of liver functions, coagulation tests, viral blood culture, and serologic tests. Other symptoms such as leukopenia with a relative neutropenia, thrombocytopenia and more time required for clot formation and increasing the prothrombin formation are frequent. The level of bilirubin and aminotransferase increases for several months (Gardner and Ryman, 2010).

Clinical diagnosis of YF in the field, particularly diagnosis of isolated cases, remains difficult for several reasons. Case-by-case differences in severity, and in the clusters of symptoms observed, make this a difficult disease to recognize, and a mild disease often escapes diagnosis.

7.6.2 Laboratory Diagnosis of YF

Diagnosis of YF is essential, but for this highly skilled lab personnel are required for operating laboratory equipment and reagents. Gardner and Ryman (2010) reported that laboratory criteria for diagnosis are any one of the following (i) presence of either IgM which is specific for YFV or elevated level of IgG in the sera of severe infection and recovery sera, respectively, when the patient does not undergo vaccination; (ii) YFV detection; (iii) positive postmortem liver histopathology; (iv) suspicion of antigen specific for YFV in tissues using immunohistochemistry. Suspected cases are characterized by very high fever after 2 weeks of appearance of jaundice symptoms.

7.6.3 Treatment of YFV Infection

Treatment for YF infection is not well known, but helpful care is essential. A person with severe infection is put in an intensive care unit (ICU) and ventilation where vasoactive medicines, fluid resuscitation, and ventilator support are provided. Medicines are given for hemorrhage, kidney and liver non-functioning, and secondary infections are treated. Doctors also use salicylates in the case of increased risk of bleeding. YFV does not transmit the infection to other person, yet virus-infected patients should be isolated by mosquito netting. As such no pertinent antiviral drug is used. Various therapeutic compounds have been prescribed in vitro, such as ribavirin and interferon-α. However, these ribavirin compounds have been found to be conflicting in model animals such as monkeys (Huggins et al., 1991). In addition, this treatment of infected monkeys by interferon-γ has shown delayed onset of YFV symptoms but no effect on their survival (Arroyo et al., 1988).

7.6.4 Vaccination for YFV

7.6.4.1 Development of live attenuated YF vaccines

There are two YF vaccines which was developed in 1930 and their attenuation methods are different. Both vaccines, i.e., French neurotropic vaccine (FNV) and 17 KD stain lost the ability to cause fever but showed antigenic property. In Senegal, the FNV vaccines was developed in 1927 by 128 serial passages of wild type French viscerotropic virus (FVV) in mouse brain (Staples and Monath, 2008). Approximately 40 million doses of FNV were administered in a vaccination campaign by scarification of skin along with the smallpox vaccine in French-speaking countries of West Africa in the 1940s and early 1950s. In these countries, YFV cases declined drastically, though vaccination with FNV has increased the prevalence of encephalitis in children. Therefore, in 1961, FNV vaccination in children below age 10 was banned, and its manufacturing was discontinued in 1980 (Gardner et al., 2015).

The 17D vaccine was synthesized by Theiler and Smith in 1937. It was developed by attenuation of Asibi YFV in 176 serial passages in murine and chick embryo tissue cultures. In 1945, the problems of over- or under-attenuation were solved. The 17D vaccine has been considered a safe and efficient live attenuated viral vaccine (Barrett, 1997; Monath, 2005).

7.6.4.2 Immune response generated during YFV vaccination

After immunization with the live attenuated 17D vaccine, the levels of circulating viral RNA and infectious virus (<100 pfu/mL) declined in blood (Reinhardt et al., 1998). The live attenuated 17D vaccines use toll-like receptors that accelerate the activation and maturation of human DCs (Barba-Spaeth et al., 2005; Querec et al., 2006). Proinflammatory cytokines interleukin-1β 93 and tumor necrosis factor–α are also released (Hacker et al., 1998; Reinhardt et al., 1998). There was activation of humoral immunity by activating CD_4^+ T-lymphocytes and B-lymphocytes. Production of YFV-specific IgM plays a significant role in host defense (Bonnevie-Nielsen et al., 1995; Hacker et al., 2001).

7.6.4.3 Adverse effects of YFV vaccines

The 17D vaccine is the most efficient vaccine till date. However, vaccination with 17D produces some side effects such as pain or redness at the injection site, headache, malaise, and myalgia. There are two categories of severe side effects related to the aforementioned vaccination: (i) vaccine-associated neurotropic disease (YEL-AND) (Merlo et al., 1993; Schoub et al., 1990) and (ii) vaccine-associated viscerotropic disease (YEL-AVD) (Belsher et al., 2007; CDC, 2002).

7.7 Modern Biology

There are about seven genotypes of YFV (Mutebi et al., 2001; von Lindern et al., 2006). Among these, two genotypes have been reported in South America, while five in Africa. Previous evolutionary studies of YFV do not include the entire viral genome sequence (Chang et al., 1995; Lepiniec et al., 1994; Sall et al., 2010; Wang et al., 1996). There is limited availability of the entire genome sequencing of YFV, mainly the wild strains isolated from Africa (sylvatic). Hence, to understand the molecular mechanism of evolution, it is necessary to sequence the whole genome of YFV, which will lead to the evolution of novel genotypes of YFV. The nucleotide sequences of distinct genotypes of YFV vary between 25% and 30%. This shows the high level of similarity between the different genotypes of YFV (Deubel et al., 1985; Lepiniec et al., 1994). In addition, the infection of YF has not been consistently reported in the epidemic regions. The largest number of YFV outbreaks has been reported in the West African countries, while the lowest in East African countries. This was due to the difference in the nucleotide sequences of YFV found in these African countries. It has been found that the genetic variability is more in West African genotype I of YFV than the East African genotype II.

Moreover, the 3′-nontranslated region (3′-NTR) differs in repeating sequences among distinct genotypes of YFV. These redundant sequences of 3′-NTR are crucial for causing virulency and amplification of YFV. In addition to viral genetic information, the extent of virulency of YFV also relies on several other parameters such as adaptation conditions of host and vector, host behavior, weather, and environmental conditions. These interactions are very typical and create issues for understanding the mechanism of YFV outbreak as well as management at right time. The epigenetic study of YFV is focused on the sylvatic strains of YFV in West Africa. This study provides important information related to evolution: (i) In Senegal, six distinct lineages of YFV have been identified, which show more diversity in this region; (ii) the forest area of Kedougou is the chief reservoir of YFV lineages; (iii) vertical transmission of YFV in environment. Therefore, we can investigate the heterogeneity in distinct lineages found in Senegal and West African countries. Besides, we can also study the relationship with insect and human hepatic cells. From the nucleotide sequence obtained from the GenBank database, we can perform the phylogenetic investigation of the whole genome of different genotypes of YFV.

7.8 Conventional Methods for Detection of YFV

YF is caused by an arthropod-borne virus and is one of the most threatened diseases. There are various reasons for the spread of

YFV such as rapidly growing population, migration, infection by vector transmission, relocation, transport and immigration of infected people (Méndez et al., 2013). Therefore, it is necessary to diagnose YF accurately and timely, particularly in the epidemic countries. Serological methods include presence of IgM antibody captured by enzyme-linked immunosorbent assay (MAC-ELISA), hemagglutination inhibition, complement fixation, and neutralization (Méndez et al., 2003; Monath, 2001). But these serological methods have certain limitations, which include delayed detection of IgM antibodies. It takes about 5–6 days; therefore, severity of infection increases manifolds and chances of cross-reactivity with other flaviviruses increase. In view of these limitations, molecular diagnostic methods have come in the investigation of YFV. These methods detect YFV based on the viral genome. The extent of severity of YF infection can be determined by immunohistochemistry using monoclonal antibodies (Méndez et al., 2003; Monath, 2001; Ricaurte et al., 1993). The PCR-based methods detect the YFV within 7–10 days. The sensitivity of this method is poor as it is not capable of detecting a few copies of viral particles in the sample. Secondly, the presence of toxic salts in the serum is not detected, which retards the growth of cells (Deubel et al., 1997). Therefore, it necessitates the specific and quick detection of YF in its early stage of infection. Consequently, people are more focused toward hygienic strategies and campaigning. Hence, in order to improve the existing PCR approach for YF diagnosis, researchers have started to investigate YFV by using RT-PCR followed by a nested PCR. The RT-PCR-based detection of YFV has increased the sensitivity and helps in fast detection.

Conversely, direct detection methods for YFV provide a transferable, facile, and robust technique, which is convenient in limited resource settings as well as in diagnosis in the field as well during epidemic (Escadafal et al., 2014). Thus, novel molecular techniques based on isothermal amplification have been developed for the detection of YFV. These techniques include real-time reverse transcription loop-mediated isothermal amplification (RT-LAMP) (Kwallah et al., 2013) and helicase-dependent amplification assays (HDA) (Domingo et al., 2012), reverse transcriptase recombinase polymerase amplification (RPA) assay for YFV detection (Escadafal, et al., 2014).

7.9 Risk Factors

Areas with no routine vaccination are more prone to YFV infection. Furthermore, vaccination provides 99% defense against YFV infection to children above 9 months and adults up to 60 years. Children under 9 months are not vaccinated for YFV; in such conditions, the risk of YF is inevitable. Besides, babies breastfed by infected mothers, persons allergic to eggs and with a weak immune system, HIV-infected people, as well as patients undergoing chemotherapy and radiotherapy are prone to YFV infection. Although typical cases of YF infection can be identified, sometimes YFV infection does not include in the differential diagnosis of patients showing symptoms such as headache, nausea, backache, and fever during the early stages of the infection. Sometimes symptoms of YFV infection are confused with the symptoms of dengue fever, Lassa fever, Ebola fever, malaria, typhoid, and hepatitis.

7.10 Conclusion

YF is caused by an arthropod-borne virus belonging to the family Flaviviridae. The severe infection of YFV has emerged into a large number of epidemics worldwide. No effective drug has been synthesized for the prevention and control of YFV infection. The WHO has recommended the routine vaccination of 17D vaccine for children above 10 months and adults up to 60 years. However, the use of 17D vaccines increases the risk of encephalitis among the age group between 10th month to 60 years. Recently, there is no advanced technology for the diagnosis and treatment of YF. Therefore, future research should be focused on developing some prominent drugs, which should be economic, highly effective, and reliable; show rapid response against YFV; and can overcome the side effects of existing therapeutics.

References

Arroyo, J. I.; Apperson, S. A.; Cropp, C. B.; Marafino, J. B.; Monath, T. P.; Tesh, R. B.; and Garcia-Blanco, M. A. Effect of human gamma interferon on yellow fever virus infection. *Am. J. Trop. Med. Hyg.*, 1988, **38**(3): 647–650.

Barba-Spaeth, G.; Longman, R. S.; Albert, M. L.; and Rice, C. M. Live attenuated yellow fever 17D infects human DCs and allows for presentation of endogenous and recombinant T cell epitopes. *J. Exp. Med.*, 2005, **202**(9): 1179–1184.

Barnett, E. D. Yellow fever: Epidemiology and prevention. *Clin. Infect. Dis.* 2007, **44**(6): 850–856.

Barrett, A. D. and Higgs, S. Yellow fever: A disease that has yet to be conquered. *Annu. Rev. Entomol.*, 2007, **52**: 209–229.

Barrett, A. D. and Monath, T. P. Epidemiology and ecology of yellow fever virus. *Adv. Virus Res.*, 2003, **61**: 291–315.

Barrett, A. D. Yellow fever vaccines. *Biologicals*, 1997, **25**(1): 17–25.

Bearcroft, W. G. C. The histopathology of the liver of yellow fever-infected rhesus monkeys. *J. Pathol. Bacteriol.*, 1957, **74**: 295–303.

Belsher, J. L.; Gay, P.; Brinton, M.; Della Valla, J.; Ridenour, R.; Lanciotti, R.; and Zhu, T. Fatal multiorgan failure due to yellow fever vaccine-associated viscerotropic disease. *Vaccine*, 2007, **25**(50): 8480–8485.

Bonnevie-Nielsen, V.; Heron, I.; Monath, T. P.; and Calisher, C. H. Lymphocytic 2′, 5′-oligoadenylate synthetase activity increases prior to the appearance of neutralizing antibodies and immunoglobulin M and immunoglobulin G antibodies after primary and secondary immunization with yellow fever vaccine. *Clin. Diagn. Lab. Immunol.*, 1995, **2**(3): 302–306.

Centers for Disease Control and Prevention (CDC). Adverse events associated with 17D-derived yellow fever vaccination—United States, 2001-2002. *Morb. Mort. Wkly. Rep.*, 2002, **51**(44): 989–993.

Chang, G. J.; Cropp, B. C.; Kinney, R. M.; Trent, D. W.; and Gubler, D. J. Nucleotide sequence variation of the envelope protein gene identifies two distinct genotypes of yellow fever virus. *J. Virol.*, 1995, **69**: 5773–5780.

Converse, J. L.; Kovatch, R. M.; Pulliam, J. D.; Nagle, S. C.; and Snyder, E. M. Virulence and pathogenesis of yellow fever virus serially passaged in cell culture. *Appl. Environ. Microbiol.*, 1971, **21**(6): 1053–1057.

Deubel, V; Pailliez, J. P.; Cornet, M.; Schlesinger, J. J.; Diop, M.; Diop, A.; Digoutte, J. P.; and Girard, M. Homogeneity among Senegalese strains of yellow fever virus. *Am. J. Trop. Med. Hyg.*, 1985, **34**: 976–983.

Deubel, V.; Huerre, M.; Cathomas, G.; Drouet, M. T.; Wuscher, N.; Le Guenno, B., and Widmer, A. F. Molecular detection and characterization of yellow fever virus in blood and liver specimens of a non-vaccinated fatal human case. *J. Med. Virol.*, 1997, **53**(3): 212–217.

Domingo, C.; Patel, P.; Yillah, J.; Weidmann, M.; Mendez, J. A.; Nakouné, E. R.; and Niedrig, M. Advanced yellow fever virus genome detection in point-of-care facilities and reference laboratories. *J. Clin. Microbiol.*, 2012, **50**: 4054–4060.

Escadafal, C.; Faye, O.; Faye, O.; Weidmann, M.; Strohmeier, O.; von Stetten, F.; and Patel, P. Rapid molecular assays for the detection of yellow fever virus in low-resource settings. *PLoS Negl. Trop. Dis.*, 2014, **8**(3): e2730.

Fontenille, D; Diallo, M; Mondo, M; Ndiaye, M; and Thonnon, J. First evidence of natural vertical transmission of yellow fever virus in *Aedes aegypti*, its epidemic vector. *Trans. R. Soc. Trop. Med. Hyg.*, 1997, **91**: 533–535.

Gardner, C. L. and Ryman, K. D. Yellow fever: A reemerging threat. *Clin. Lab. Med.*, 2010, **30**(1): 237–260.

Gould, E. A.; de Lamballerie, X.; Zanotto, P. M.; and Holmes, E. C. Origins, evolution, and vector/host co-adaptations within the genus Flavivirus. *Adv. Virus Res.*, 2003, **59**: 277–314.

Gould, E. A. and Solomon, T. Pathogenic flaviviruses. *Lancet*, 2007, **371**: 500–509.

Gouy, M.; Guindon, S.; and Gascuel, O. SeaView version 4: A multiplatform graphical user interface for sequence alignment and phylogenetic tree building. *Mol. Biol. Evol.*, 2010, **27**: 221–224.

Hacker, U. T.; Erhardt, S.; Tschöp, K.; Jelinek, T.; and Endres, S. Influence of the IL-1Ra gene polymorphism on in vivo synthesis of IL-1Ra and IL-1β after live yellow fever vaccination. *Clin. Exp. Immunol.*, 2001, **125**(3): 465–469.

Hacker, U. T.; Jelinek, T.; Erhardt, S.; Eigler, A.; Hartmann, G.; Nothdurft, H. D.; and Endres, S. In vivo synthesis of tumor necrosis factor-α in healthy humans after live yellow fever vaccination. *J. Infect. Dis.*, 1998, **177**(3): 774–778.

Hardy, F. M. The growth of Asibi strain yellow fever virus in tissue cultures. II. Modification of virus and cells. *J. Infect. Dis.*, 1963, **113**: 9–14.

Hearn, H. J. Jr.; Chappell, W. A.; Demchak, P.; and Dominik, J. W. Attenuation of aerosolized yellow fever virus after passage in cell culture. *Bacteriol. Rev.*, 1966, **30**: 615–623.

https://www.cdc.gov/yellowfever/transmission/index.html.

https://www.cdc.gov/yellowfever/vaccine/index.html

Huggins, J. W.; Hsiang, C. M.; Cosgriff, T. M.; Guang, M. Y.; Smith, J. I.; Wu, Z. O.; and Oland, D. D. Prospective, double-blind, concurrent, placebo-controlled clinical trial of intravenous ribavirin therapy of hemorrhagic fever with renal syndrome. *J. Infect. Dis.*, 1991, **164**(6): 1119–1127.

Kwallah, A. O.; Inoue, S.; Muigai, A. W.; Kubo, T.; Sang, R., Morita, K; and Mwau, M. A real-time reverse transcription loop-mediated isothermal amplification assay for the rapid detection of yellow fever virus. *J. Virol. Methods*, 2013, **193**: 23–27.

Lepiniec, L.; Dalgarno, L.; Huong, V. T.; Monath, T. P.; Digoutte, J. P.; and Deubel, V. Geographic distribution and evolution of yellow fever viruses based on direct sequencing of genomic cDNA fragments. *J. Gen. Virol.*, 1994, **75**(2): 417–423.

Lloyd, W.; Theiler, M.; and Ricci, N. I. Modification of the virulence of yellow fever virus by cultivation in tissues in vitro. *Trans. R. Soc. Trop. Med. Hyg.*, 1936, **29**: 481–529.

Méndez, J. A.; Rodríguez, G.; del Pilar Bernal, M.; de Calvache, D.; and Boshell, J. Detección molecular del virus de la fiebre amarilla en muestras de suero de casos fatales humanos y en cerebros de ratón. *Biomédica*, 2003, **23**(2): 232–238.

Méndez, M. C.; Domingo, C.; Tenorio, A.; Pardo, L. C.; Rey, G. J.; and Méndez, J. A. Development of a reverse transcription polymerase chain reaction method for yellow fever virus detection. *Biomédica*, 2013, **33**: 190–196.

Merlo, C.; Steffen, R.; Landis, T.; Tsai, T.; and Karabatsos, N. Possible association of encephalitis and 17D yellow fever vaccination in a 29-year-old traveller. *Vaccine*, 1993, **11**(6): 691.

Monath, T. P. Treatment of yellow fever. *Antiviral Res.*, 2008, **78**: 116–124.

Monath, T. P. Yellow fever: An update. *Lancet Infect. Dis.*, 2001, **1**: 11–20.

Monath, T. P. Yellow fever vaccine. *Expert Rev. Vaccines*, 2005, **4**(4): 553–574.

Mutebi, J. P. and Barrett, A. D. The epidemiology of yellow fever in Africa. *Microbes Infect.*, 2002, **4**: 1459–1468.

Mutebi, J. P.; Wang, H.; Li, L.; Bryant, J. E.; and Barrett, A. D. Phylogenetic and evolutionary relationships among yellow fever virus isolates in Africa. *J. Virol.*, 2001, **75**: 6999–7008.

PBS. The Great Fever. *PBS*, 2006, Sept. 29. Web.

Pisano, M. R.; Nicoli, J.; and Tolou, H. Homogeneity of yellow fever virus strains isolated during an epidemic and a post-epidemic period in West Africa. *Virus Genes*, 1997, **14**: 225–234.

Querec, T.; Bennouna, S.; Alkan, S.; Laouar, Y.; Gorden, K.; Flavell, R.; and Pulendran, B. Yellow fever vaccine YF-17D activates multiple dendritic cell subsets via TLR2, 7, 8, and 9 to stimulate polyvalent immunity. *J. Exp. Med.*, 2006, **203**(2): 413–424.

Reinhardt, B.; Jaspert, R.; Niedrig, M.; Kostner, C.; and L'age-Stehr, J. Development of viremia and humoral and cellular parameters of immune activation after vaccination with yellow fever virus strain 17D: A model of human flavivirus infection. *J. Med. Virol.*, 1998, **56**(2): 159–167.

Ricaurte, O.; Sarmiento, L.; and Caldas, M. L. Evaluación de un método inmunohistoquímico para el diagnóstico de la fiebre amarilla. *Biomédica*, 1993, **13**(1): 15–19.

Robertson, S. E.; Hull, B. P.; Tomori, O.; Bele, O.; LeDuc, J. W.; and Esteves, K. Yellow fever: A decade of reemergence. *JAMA*, 1996, **276**: 1157–1162.

Sall, A. A.; Faye, O.; Diallo, M.; Firth, C; Kitchen, A.; and Holmes, E. C. Yellow fever virus exhibits slower evolutionary dynamics than dengue virus. *J. Virol.*, 2010, **84**: 765–772.

Schindler, R. and Hallauer, C. Prufung von Gelbfiebervirus-Variantem aus menschlichen Gewebekulturen in Affenuersuch. *Arch. Gesamte Virusforsch.*, 1963, **13**: 345–357.

Schoub, B. D.; Dommann, C. J.; Johnson, S.; Downie, C.; and Patel, P. L. Encephalitis in a 13-year-old boy following 17D yellow fever vaccine. *J. Infect.*, 1990, **21**(1): 105–106.

Smith, H. H. and Theiler, M. The adaptation of unmodified strains of yellow fever virus to cultivation in vitro. *J. Exp. Med.*, 1937, **65**(6): 801–808.

Staples, J. E. and Monath, T. P. Yellow fever: 100 years of discovery. *JAMA*, 2008, **300**(8): 960–962.

Stocka, N. K.; Larawaya, H.; Fayeb, O.; Diallo, M.; Niedrig, M.; and Sallb, A. A. Biological and phylogenetic characteristics of yellow fever virus lineages from West Africa. *J. Virol.*, 2013, **87**: 52895–52907.

Stokes, A.; Bauer, J. H.; and Hudson, N. P. Experimental transmission of yellow fever to laboratory animals. *Am. J. Trop. Med.*, 1928, **8**: 103–164.

Theiler, M. Yellow fever. In *Viral and Rickettsial Infections of Man* (Rivers, T. M. and Horsfall, F. L., eds.), J. P. Lippincott, Philadelphia, 1959, pp. 343–360.

Theiler, M. and Smith, H. H. The use of yellow fever virus modified by in vitro cultivation for human immunization. *J. Exp. Med.*, 1937, **65**(6): 787–800.

Tigertt, W. D.; Berge, T. O.; Gochenour, W. S.; Gleiser, C. A.; Eveland, W. C.; Bruegge, C. V.; and Smetana, H. F. Experimental yellow fever. *Trans. N. Y. Acad. Sci.*, 1960, **22**: 323–333.

University of Virginia. Yellow Fever Commission. Philip S. Hench Walter Reed Yellow Fever Collection. University of Virginia, The Claude Moore Health Sciences Library, 2004, Feb. 24. Web.

von Lindern, J. J.; Aroner, S.; Barrett, N. D.; Wicker, J. A.; Davis, C. T.; and Barrett, A. D. Genome analysis and phylogenetic relationships between east, central and west African isolates of Yellow fever virus. *J. Gen. Virol.*, 2006, **87**: 895–907.

Wang, E.; Weaver, S. C.; Shope, R. E.; Tesh, R. B.; Watts, D. M.; and Barrett, A. D. Genetic variation in yellow fever virus: Duplication in the 3′ noncoding region of strains from Africa. *Virology*, 1996, **225**: 274–281.

World Health Organization (WHO). Yellow Fever. WHO Media Centre, 2014, Mar. Web.

Chapter 8

West Nile Virus: The Silent Neuro-Invasive Terror

Vinod Joshi,[a] Annette Angel,[b] Bennet Angel,[a] Neelam Yadav,[c] Jagriti Narang,[d] and Surender Yadav[e]
[a]*Amity Institute of Virology and Immunology, Amity University, Sector-125, Noida, India*
[b]*Division of Zoonosis, National Centre for Disease Control, 22 Sham Nath Marg, Civil Lines, Delhi, India*
[c]*Centre for Biotechnology, Maharshi Dayanand University, Rohtak, India*
[d]*Department of Biotechnology, Jamia Hamdard University, New Delhi, India*
[e]*Department of Botany, Maharshi Dayanand University, Rohtak, India*
annetteangel_15@yahoo.co.in, vinodjoshidmrc@gmail.com, bennetangel@gmail.com

West Nile virus (WNV) is a mosquito-borne flavivirus. Recent epidemics of WNV have been reported in the United States and other parts of the world. In nature, the virus maintains the mosquito–bird–mosquito transmission cycle. The infection of WNV can be symptomatic and asymptomatic. Most of the asymptomatic cases of WNV are fatal if they are not properly diagnosed and treated. Scientists face a big challenge for the adequate diagnosis and prevention of WNV. Presently no effective vaccine or antiviral chemicals have been synthesized. Therefore, the present chapter

Small Bite, Big Threat: Deadly Infections Transmitted by Aedes *Mosquitoes*
Edited by Jagriti Narang and Manika Khanuja

ISBN 978-981-4800-86-0 (Hardcover), 978-1-003-00329-8 (eBook)
www.jennystanford.com

describes the epidemiology, virus morphology, transmission, and replication of virus, detection, and prevention methods of WNV.

8.1 Epidemiology

Another important virus of arbovirus origin is the WNV. The virus was first identified in Uganda in 1937 and since then has spread to almost all countries worldwide (Chancey et al., 2015; Eybpoosh et al., 2019; Hayes et al., 2005). Smithburn et al., (1940) identified this new virus during an endemic in Uganda. When the blood from an infected woman (37 years of age) was drawn and experimented for probable yellow fever viral infection by inoculating the serum intracerebrally in 10 mice. It was observed that the only one mice survived. Afterwards sub-inoculation was done but none of the mice survived. Later when the serum was inoculated into a Rhesus monkey, it was observed that the monkey developed encephalitis. The spread continued initially in the form of infrequent outbreaks in certain parts of the globe, such as India, Israel, Egypt, France, and South Africa. Then it increased to a higher magnitude affecting Bucharest, Romania, Russia, etc. and then spread to the United States by 1999. Molecular studies revealed that two lineages of WNV evolved: Lineage 1 and Lineage 2. Lineage 1 emerged in Central and Northern Africa, Europe, Australia, and the Americas in 1999, and Lineage 2 emerged in Southern Africa and Central Europe in 2005 (Fall et al., 2017; Rizzoli et al., 2015). In 1957, the virus was reported to cause meningoencephalitis in one of the patients in Israel. In the United States, seasonal outbreaks occur annually, while large outbreaks are known to occur throughout the country. More than 3 million people have been reported to be infected with the virus during 2010 (Peterson et al., 2003). Barrett reported that the economic burden due to the disease in the 15 years of its arrival, that is, from 1999 to 2014, has tremendously increased to 17,367 cases and 1654 deaths.

Meanwhile the enzootic form of the virus has also started to appear. Birds, cows, and some exotic captive birds in the Bronx Zoo in New York City incidentally died and were seen to have been suffered from meningoencephalitis and severe myocarditis. Initially the virus could not be isolated, but later molecular studies done at the Center for Disease Control (CDC), Atlanta, confirmed that it was similar to that of the virus isolated from the human brain tissue of infected persons in New York (Sampathkumar, 2003). In Morocco (1996, 2003), Italy (1998), Israel (2000), and Southern France (2000, 2003, 2004), the virus was found to be infecting horses (Schuffenecker et al., 2005; Zeller and Schuffenecker, 2004; Zeller et al., 2004). Birds were also reported to be infected with the WNV in New Brunswick and Nova Scotia (www.phacaspc.gc.ca/wnv-vwn).

From 1999 through 2001, there were 149 cases of WNV human illness reported in the United States and 18 deaths. In 2002, the reported cases increased dramatically, with 4156 laboratory-confirmed human cases and 284 deaths.

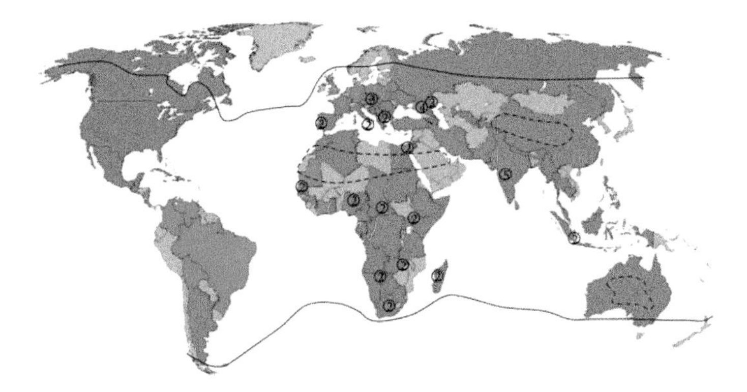

Figure 8.1 Global distribution of WNV by country: Red—human cases or human seropositivity; blue—nonhuman/mosquito cases or seropositivity; gray—no data or no positives reported. Black lines represent worldwide distribution of the main WNV mosquito vectors, excluding areas of extreme climate denoted by dashed lines. Circled numbers indicate the reported presence of WNV lineages other than Lineage 1 in that specific area. Reprinted from Chancey et al., 2015, under Creative Commons Attribution license.

India had also reported WNV infection in the past with the first record in 1952 (Banker, 1952; Bondre et al., 2007). The serologically confirmed cases were seen in the Vellore and Kolar districts in 1977, 1978, and 1981. Besides these cases, infection and encephalitis conditions had also been reported in Rajasthan (Udaipur), Maharashtra (Buldhana, Marathwada, Khandesh), Tamil Nadu, Karnataka, Andhra Pradesh, Maharashtra, Gujrat, Madhya Pradesh, and Odisha. In 2006, the virus was reported from Assam along with some cases of acute encephalitis syndrome (AES). An AES outbreak was also seen in 2011 in Kerala, with some confirmed cases of WNV.

The state of Kerala continues to be the place of regular infections (https://www.nhp.gov.in/disease/communicable-disease/west-nile-fever). In 2019, a 6-year-old boy from the Malappuram district of Kerala developed cold and fever-like symptoms accompanied with meningitis. He was diagnosed by the Christian Medical College, Vellore, and the National Institute of Virology (NIV), Pune, to be suffering from WNV infection. The boy did not survive and died after 2–3 days (www.hindustantimes.com/india-news/ncdc-team-from-delhi-in-kerala-after-boy-tests-positive-for-west-nile-virus/story-f6NVRDGSSg1v4s3Uf12gMJ.html).

8.2 Virus Morphology

The virus belongs to the family *Flaviviridae*. It is an enveloped virus with approximately 50 nm diameter. It consists of a single-stranded positive-sense RNA genome, with 10 genes: three structural and seven coding for nonstructural components. The total size of the genome is 11 kb, and there is no polyadenylation at the 3′ end. The 3′ end, thus, has a conserved CU_{OH} (Brinton et al., 1986; Rice et al., 1985; Wengler and Wengler, 1991). The 5′ end has a non-coding region of 90 nucleotides, while that toward the 3′ non-coding region has 337 to 649 nucleotides. The single polyprotein genome is co- and post-translationally cleaved into capsid (C), pre-M/membrane (prM/M), and envelope (E) proteins (which form the structural proteins) followed by NS1, NS2A, NS2B, NS3, NS4A, NS4B, and NS5 (as nonstructural proteins), as shown in Fig. 8.2. The cleavage is done with the help of host and viral proteases (Colpitts et al., 2012). The envelope protein E is glycosylated at the residue 154. The nucleocapsid protein is associated with the RNA genome and helps in viral assembly (Khromykh and Westaway, 1996; Markoff et al., 1997). During the viral assembly, the heterodimeric form of E and M proteins gets embedded within the bilipid layer and appears as protrusions of the viral surface. The membrane protein is known to protect the immature form of virus so that it does not undergo premature fusion before the virus buds off from the cell surface. This is achieved by blocking the fusion loop of E proteins, which is then cleaved off during the viral maturation process (Guirakhoo et al., 1991; Heinz et al., 1994; Stadler et al., 1997; Zhang et al., 2003; Zhang et al., 2007). The nonstructural proteins play a multifunctional role in viral synthesis and assembly. NS1 is involved in the replication process during the early stages. NS3 (serine protease) functions along with NS2B (RNA helicase) and NS4a (NTPase) and NS5 (RNA-dependent RNA polymerase (RdRp) in the C-terminal region and methyltransferase activity in the N-terminal region) to cleave the other nonstructural proteins from the main viral polyprotein. The nonstructural proteins are also involved in modulating cell signaling and immune response and are also known for inhibiting complement activation (Avirutnan et al., 2010; Avirutnan et al., 2011; Chung et al., 2006; Kyung et al., 2007; Laurent et al., 2010; Liu et al., 2004, Melian et al., 2013; Muñoz-Jordán and Fredericksen, 2010; Wen et al., 2004). A few researchers have also published the whole genome sequence of the WNV isolated in their country, for example in Cyprus (Richter et al., 2017), Slovakia, Central Europe (Drzewnioková et al., 2019), from infected horse in South Africa (Mentoor et al., 2016), from Crow in Belgium (Dridi et al., 2015), etc. Phylogenetic analysis and evolution studies of gene sequences from 2003 to 2012 have been conducted by Wedin from South Eastern United States (Wedin, 2013). Genomic sequences have also been reported from India. One complete sequence from a human isolate and 14 partial sequences isolated in the period 1955–82 have been reported by the NIV, Pune (Bondre et al., 2012). The isolates were from *Culex vishnui, Rousettus leschenaulti* (fruit bat), *Anopheles subpictus, Culex whitmorei, Culex bitaeniorhynchus*, and human host.

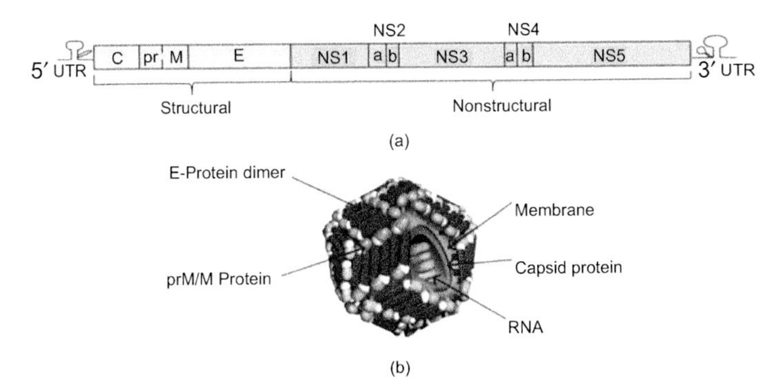

Figure 8.2 Structure of West Nile virus. Reprinted from Chancey et al., 2015, under Creative Commons Attribution license.

8.3 Transmission Route

The WNV is maintained in nature in the enzootic cycle, that is, the bird–arthropod cycle. The arthropod vectors responsible for transmitting WNV are varied. Up to 65 species and eight genera have been reported from the United States (CDC, Atlanta, 2009; Hamer et al., 2008; Reisen et al., 2006; Vitek et al., 2008). Mosquitoes, as we know, have different host-feeding patterns, and those which feed on birds and mammals form a bridge/connection for transmission. But such transmissions are known to transmit less viremia (CDC, Atlanta, 2009; Kilpatrick et al., 2005; Turell et al., 2002). The most commonly observed mosquito species is the *Culex* sp., with *Culex pipiens* dominating in the northeastern, north-central, and mid-Atlantic United States and *Culex quinquefasciatus* in the south and southwest, *Culex tarsalis* in the west, and *Culex vishnui* and *Culex bitaeniorhynchus* from India (Andreadis, 2012; Kilpatrick et al., 2005). Other species include *Culex stigmatosoma, Culex thriambus, Culex salinarius,* and *Culex nigripalpus* (Hamer et al., 2008; Reisen et al., 2006; Unlu et al., 2010; Vitek et al., 2008). The *Aedes* species, though reported by a few researchers as vectors, are not considered the primary vector for WNV transmission (Colpitts et al., 2012; Cupp et al., 2007; Farajollahi and Nelder, 2009; Holick et al., 2002; Sardelis et al., 2002; Tiawsirisup et al., 2005; Unlu et al., 2010; Vanlandingham et al., 2007). *Aedes triseriatus* has been reported to be found in the wild collected males (Unlu et al., 2010). The ornithophilous species from the order Passeriformes, especially crow of the Corvidae family, blue hays, and raven, are found to be capable of virus transmission (Diamond, 2009; Hubálek and Halouzka, 1999; Pfeffer and Dobler, 2010; Turell et al., 2005). It has also been observed that SPF chickens and Aigamo ducks may carry WNV (Dridi et al., 2013; Phipps et al., 2007; Senne et al., 2000; Shirafuji et al., 2009; Totani et al., 2011). Humans, on the other hand, serve as dead-end points of viral transmission due to less viremia (Bowen and Nemeth, 2007). This wide range of host preference along geographic barriers proves the virus to have high variation and global ecology (Fig. 8.3).

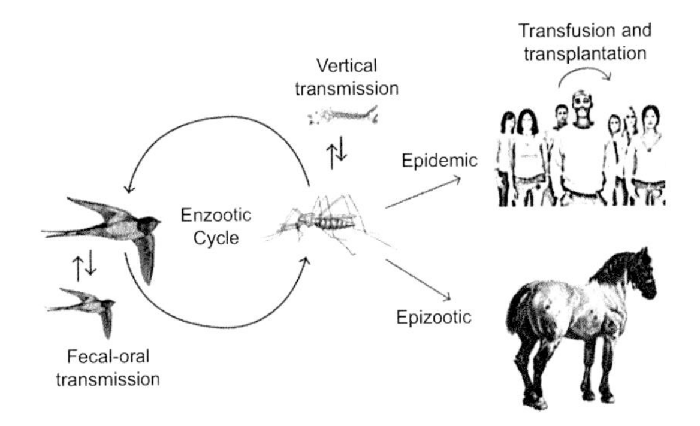

Figure 8.3 Enzootic cycle of West Nile virus. Reprinted from Chancey et al., 2015, under Creative Commons Attribution license.

8.4 Pathogenicity

The WNV is a neuro-invasive virus that causes asymptomatic infection in nearly 80% of the cases, while 20% of the cases are symptomatic, characterized by fever, flu-like symptoms, headaches, nausea, vomiting, swollen lymph glands, and sometimes skin rashes (Petersen and Marfin, 2002; Watson et al., 2004; www.who.int). Besides this ocular manifestation, meningoencephalitis, muscle weakness, cognitive impairment tremors, and poliomyelitis-like flaccid paralysis are also seen (Bakri and Kaiser, 2004; Petersen and Marfin, 2002; Samuel and Diamond, 2006; Sejvar et al., 2005). There is damage to the brain stem, and parts such as hippocampus region, cerebellum, anterior-horn neurons of the spinal cord, etc. are severely affected (Guarner et al., 2004; Kleinschmidt-DeMasters et al., 2004). The pathogenesis of brain invasion is still unclear; sometimes leukocyte infiltration and host inflammatory response have been reported (Samuel and Diamond, 2009; Samuel et al., 2007). Some of the risk factors include old age, immunosuppressed patients, and chronic conditions, including but not limited to hypertension, blood pressure, and chronic renal failure (Bode et al., 2006; Busch et al., 2006; Carson et al., 2012; Danis et al., 2011; Fratkin et al., 2004; Hayes and Leary, 2004; Jean et al., 2007; Kopel et al., 2011; Ladbury et al., 2013; Lindsey et al., 2009, 2010, 2012; Mostashari et al., 2001; Nash et al., 2001; Patnaik et al., 2006; Zou et al., 2010).

When an adult mosquito takes a blood meal, it inoculates the virus inside it. The virus then replicates within the midgut epithelium and finally reaches the salivary glands. The mosquito is now capable of infecting a new host during another blood meal. The virus that enters the human system (skin) is accompanied by many proteins of the mosquito's salivary gland to combat the host's hemostatic system. Within the skin, the Langerhans dendritic cells first encounter the virus (Byrne et al., 2001). Keratinocytes and neutrophils are also reported to first encounter the pathogen (Bai et al., 2010; Lim et al., 2011; Welte et al., 2009). A variety of factors are responsible for WNV transmission depending on the host infected. From there the virus enters the lymph nodes and later infects other tissues such as spleen and kidneys. Usually by the end of the first week of infection, the virus gets cleared from the system but may enter the central nervous system of immunocompromised patients. Persistent infection has also been seen in such cases where the primary infection was not detected until 60 days of infection (Brenner et al., 2005). The WNV is also known to be transmitted via blood transfusion, organ transplantation, and transplacental transmission via breast milk (CDC, 2002a, 2002b, 2002c; Hinckley et al., 2007; Iwamoto et al., 2003; Pealer et al., 2003).

Pathogenesis of WNV has been attempted by researchers using the mice model. Interferons (IFNs) are known to act during viral response. Of these, type I (IFN-α and IFN-β), type II (IFN-γ), and type III (IFN-λ) are important, and more specifically IFN-α/β are known to be produced by almost all cell types (Pestka et al., 2004; Platanias, 2005). Interferons also known to be a link between innate and adaptive immune responses. These are involved in the maturation of dendritic cells and activation of B and T cells (Asselin-Paturel, 2005; Le Bon et al., 2006). On the other hand, viral proteins such as NS2a, NS4b, and NS5 are known to counteract with Type I IFN through various mechanisms. Macrophages are also known to behave like an antigen-presenting cell (APC) and inhibit viral replication. This is done by the production of reactive oxygen species (ROS) (Kulkarni et al., 1991; Lin et al., 1997; Saxena et al., 2007). The mechanism by which the virus enters the brain is still not clear as stated in the earlier paragraph, yet two hypotheses have been put forward to understand the mechanism; one is entry via the blood–brain barrier (BBB) by infecting leucocytes and the other by directly affecting the brain endothelial cells (Fig. 8.4) (Pardridge, 1983).

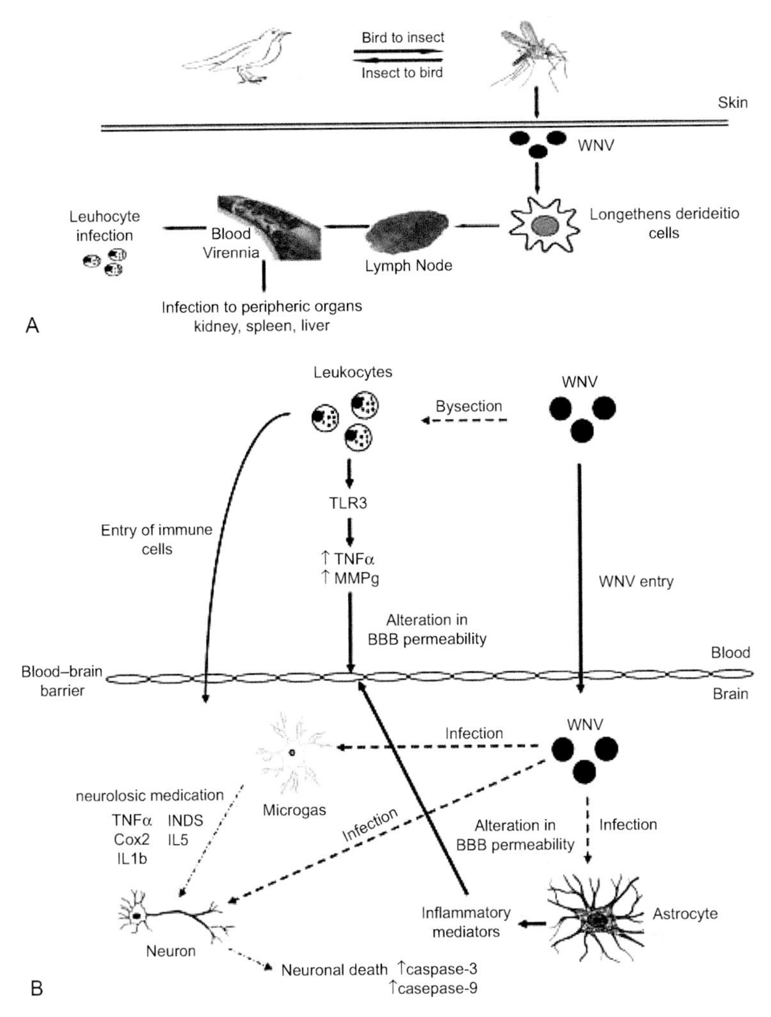

Figure 8.4 Pathogenesis of West Nile virus infection (A: initial pathogenesis during inoculation into skin. B: regulation of WNV after entry into brain). Reprinted from Donadieu et al., 2013, under Creative Commons Attribution license.

8.5 Detection Methods for WNV

8.5.1 Conventional Methods

Although conventional serological approaches have been used for WNV detection, these approaches have limitations such as more susceptibility for cross-reactivity during investigation; more time and high biosafety level required to avoid contamination (Prince et al., 2005). In view of these setbacks of serological methods, molecular methods came into existence. A number of rapid assays, such as enzyme immunoassays (EIA) and enzyme-linked immunosorbent assays (ELISA) (CDC), combination of both EIA and ELISA forms plaque reduction neutralization assays (PRNT) (Calisher et al., 1989; CDC, 2000), indirect immunofluorescent antibody staining (Koraka et al., 2002), complement fixation test, and microsphere immunoassay (Prince et al., 2005), have been described for the detection of WNV that are sensitive and resolve the challenges of intensive surveillance. Specimen used for the investigation of WNV by the assays includes tissues of birds and mammals, mosquitoes, as well as sera from human, equine, and birds (Turell et al., 2002). In addition to these rapid assays, there are various other approaches for WNV detection: nucleic-acid-based techniques, polymerase chain reaction (PCR) techniques (Shi et al., 2001), reverse transcription loop-mediated isothermal amplification (RT-LAMP) assay (Parida et al., 2004), cell culturing methods (Hunt et al., 2002), immunohistochemistry (Shieh et al., 2000; Swayne et al., 2000), and VecTest^{TM} dipsticks for detecting WNV antigens (Stone et al., 2004). Although these molecular methods are sensitive for the detection of WNV, these are expensive, complicated, and have delayed response time.

8.5.2 Nanotechnology-Based Methods for WNV Detection

Future research should be focused on developing updated nano-based technologies such as designing biosensing devices. Biosensing methods can overcome all the aforementioned limitations of conventional and molecular techniques because biosensor-based technologies are rapid, easy, and cost effective in detection than conventional and molecular methods (Cosnier et al., 2006). For instance, Cosnier et al. (2006) have constructed an electrochemical enzymatic polypyrrole integrated sensor for the detection of WNV. The WNV-specific IgG antibodies have been isolated by developing an amperometric immunosensor (Ionescu et al., 2007). Zhang et al. (2011) have detected WNV using surface-enhanced Raman scattering integrated with Raman active gold nanoparticles. Channon et al. (2018) have developed an electrochemical paper device for the detection of WNV.

8.6 Treatment

Till date there is no vaccine or chemotherapy available for treating WNV infection. Yet attempts to develop live attenuated, recombinant subunit forms, vectorized vaccine forms, and DNA vaccine forms are under way. A chimeric vaccine based on the prM and E parts of the virus was developed by incorporating it into the yellow fever 17D vaccine moiety (ChimeriVax-WN02), and it also reached phase II clinical trial but is now no longer available (Biedenbender, 2011). A monoclonal antibody E16 (MGAWNI) is under phase I trial (Beasley, 2011). Therapeutics are also on trails by different workers across the world, especially in the development of one that can cross the brain barriers. Ribavirin, which is used for treating many infectious diseases, has shown to have inhibitory activity during cell culture experiments (Anderson and Rahal, 2002; Day et al., 2005; Jordan et al., 2000), but hamsters have shown to have mortality when treated (Morrey et al., 2004). But during an outbreak in Israel in 2000, 37 patients were given ribavirin and a high mortality rate was observed (Chowers et al., 2001). Mycophenolic acid (MPA) was also observed to cause death in treated mice (Diamond, 2009). The use of siRNA (short interfering RNA) and IFN-α in mice models showed some protection and decrease in viremia, respectively, against the virus (Bai et al., 2005; Brooks and Phillpotts, 1999; Kumar et al., 2006; Morrey et al., 2004). But whether they would cross the membrane barrier is still not concluded. Antisense oligomers, which are known to inhibit viruses by binding to places of sequence specificity, have been tried for WNV (Kinney et al., 2005; Ma et al., 2000). This has been attempted by AVI Biopharma (www.clinicaltrials.gov/ct/show/NCT00091845). Fusion peptides have been studied to inhibit WNV infection (Hrobowski et al., 2005). Deoxynorjirimycin or castanospermine, iminosugar derivatives inhibit ER enzymes, which guides assembly and further secretion of viruses. These have also been studied during in vitro and in vivo experiments (Chang et al., 2009; Courageot et al., 2000; Gu et al., 2007; Schul et al., 2007; Whitby et al., 2005; Wu et al., 2002). High-throughput screens with small molecule libraries have been utilized by many workers to identify WNV inhibitors (Borowski et al., 2002; Goodell et al., 2006; Gu et al., 2006; Johnston et al., 2007; Nouiery et al., 2007; Puig-Basagoiti et al., 2006). In the same line, screening of 32,000 compounds have been done by a group for inhibition of WNV (Mueller et al., 2008).

As far as diagnosis of WNV infection is concerned, the presence of anti-WNV IgM is checked in the cerebrospinal fluid of the infected person (Colpitts et al., 2012). Plaque neutralization assay (PNA) is used to check the cross-reactivity of WNV infection with other flaviviruses such as Japanese encephalitis virus, St. Louis encephalitis virus, yellow fever virus, and dengue virus (Murray et al., 2011). The nucleic acid testing (NAT) kit is used to screen blood samples to ascertain WNV infection before giving them to required individuals (Bai et al., 2009; Busch et al., 2008; Rios et al., 2006; Zou et al., 2010).

8.7 Modern Biology

A few genetic determinants have been found to be associated with WNV infection. Genes such as OASL, CCR5, OAS1, IRF3, MX1, RFCi, SCN1A, and ANPEP have been found to be associated mostly with increased risk toward WNV infection when compared with healthy controls. And it is expected that in the time to come, more such correlations will be reported (Bigham et al., 2011; Loeb et al., 2011; Rios et al., 2010). Certain salivary proteins present in the saliva of *Aedes aegypti* and *Culex tarsalis* have been found to affect the WNV. This was found during experiments conducted on mouse models.

One important feature of our nervous system is that neurons either regenerate or do not regenerate at all. This may indirectly lead to elimination of virus without further multiplication as is the case in WNV infection. During this, the immune system also comes into action (Arjona et al., 2011; Cho and Diamond, 2012; Durrant et al., 2013; Samuel and Diamond, 2006). Interferons are known to play a very important role in virus clearance. It has been studied that T cell and macrophages travel to the brain during WNV infection (Glass et al., 2005, 2006; Shrestha and Diamond, 2004; Shrestha et al., 2006; Wang et al., 2003). On the other hand, CXCL10 (chemokine) is released by neuron cells. T-cell mediated immunity, though, proves to be an effective component during virus clearance but has also shown to cause irreversible damage to the patient (Wang et al., 2003). The regulatory T cells, that is, Treg cells help in controlling viral infection as observed by the presence of high level of Treg cells during asymptomatic infections (Lanteri et al., 2009).

To conclude, more studies are required to understand the viral mechanism of crossing the blood–brain barrier so that appropriate therapeutics may be formulated, and thus prevention of this serious form of neurotropic disease can be targeted. The only problem that exists as of today is to find and formulate drugs/inhibitors that can cross this blood–brain barrier (Diamond, 2009). It is also feared that it may take years to complete clinical trials due to a wide range of patients across borders and due to the various genetic variations the WNV exhibits (Donadieu et al., 2013).

8.8 Conclusion

Flavivirus WNV is transmitted when infected Culex mosquitoes bite. Outbreaks of WNV have been reported in the United States and other countries. The virus causes severe fever and several neuro-invasive diseases in humans. Therefore, it is necessary to mitigate the serious issues of proper management of WNV outbreaks. Various molecular techniques such as PCR-based techniques and biosensing techniques have been used for the detection of WNV, though nano-based approaches have not been developed. Hence, future research should be focused to develop more refined nano-based technologies and miniaturized techniques that can provide an effective platform for the investigation of WNV, which must be fast, economic, transportable, and efficient in comparison to serological and molecular techniques.

References

Anderson, J. F. and Rahal, J. J. Efficacy of interferon alpha-2b and ribavirin against West Nile virus in vitro. *Emerg. Infect. Dis.*, 2002, **8**: 107–108.

Andreadis, T. G. The contribution of *Culex pipiens* complex mosquitoes to transmission and persistence of West Nile virus in North America. *J. Am. Mosq. Control Assoc.*, 2012, **28**(4): 137–151.

Arjona, A.; Wang, P.; Montgomery, R. R.; and Fikrig, E. Innate immune control of West Nile virus infection. *Cell Microbiol.*, 2011, **13**: 1648–1658.

Asselin-Paturel, C.; Brizard, G.; Chemin, K.; Boonstra, A.; O'Garra, A.; Vicari, A.; and Trinchieri, G. Type I interferon dependence of plasmacytoid dendritic cell activation and migration. *J. Exp. Med.*, 2005, **201**: 1157–1167.

Avirutnan, P.; Fuchs, A.; Hauhart, R. E.; Somnuke, P.; Youn, S.; Diamond, M. S.; and Atkinson, J. P. Antagonism of the complement component C4 by flavivirus non-structural protein NS1. *J. Exp. Med.*, 2010, **207**(4): 793–806.

Avirutnan, P.; Hauhart, R. E.; Somnuke, P.; Blom, A. M.; Diamond, M. S.; and Atkinson, J. P. Binding of flavivirus nonstructural protein NS1 to C4b binding protein modulates complement activation. *J. Immunol.*, 2011, **187**(1): 424–433.

Bai, F.; Kong, K. F.; Dai, J.; Qian, F.; Zhang, L.; Brown, C. R.; Fikrig, E.; and Montgomery, R. R. A paradoxical role for neutrophils in the pathogenesis of West Nile virus. *J. Infect. Dis.*, 2010, **202:** 1804–1812.

Bai, F.; Town, T.; Qian, F.; Wang, P.; Kamanaka. M.; Connolly, T. M.; Gate, D.; Montgomery, R. R.; Flavell, R. A.; and Fikrig, E. IL-10 signaling blockade controls murine West Nile virus infection. *PLoS Pathog.*, 2009, e1000610. doi: 10.1371/journal.ppat.1000610.

Bai, F.; Wang, T.; Pal, U.; Bao, F.; Gould, L. H.; and Fikrig, E. Use of RNA interference to prevent lethal murine West Nile virus infection. *J. Infect. Dis.*, 2005, **191**: 1148–1154.

Bakri, S. J. and Kaiser, P. K. Ocular manifestations of West Nile virus. *Curr. Opin. Ophthalmol.*, 2004, **15**: 537–540.

Banker, D. D. Preliminary observations on antibody patterns against certain viruses among inhabitants of Bombay City. *Indian J. Med. Sci.*, 1952, **6**: 733–746.

Barrett, A. D. T. Economic burden of West Nile virus in the United States. *Am. J. Trop. Med. Hyg.*, 2014, **90**(3): 389–390.

Beasley, D. W. Vaccines and immunotherapeutics for the prevention and treatment of infections with West Nile virus. *Immunotherapy*, 2011, **3**: 269–285.

Bhatnagar, J.; Guarner, J.; Paddock, C. D.; Shieh, W. J.; Lanciotti, R. S.; Marfin, A. A.; and Zaki, S. R. Detection of West Nile virus in formalin-fixed, paraffin-embedded human tissues by RT-PCR: A useful adjunct to conventional tissue-based diagnostic methods. *J. Clin. Virol.*, 2007, **38**(2): 106–111.

Biedenbender, R.; Bevilacqua, J; Gregg, A. M.; Watson, M.; and Dayan, G. Phase II, randomized, double-blind, placebo-controlled, multicenter study to investigate the immunogenicity and safety of a West Nile virus vaccine in healthy adults. *J. Infect. Dis.*, 2011, **203**: 75–84.

Bigham, A. W.; Buckingham, K. J.; Husain, S.; Emond, M. J.; Bofferding, K. M.; Gildersleeve, H.; et al. Host genetic risk factors for West Nile virus infection and disease progression. *PloS One*, 2011, **6**(9): e24745.

Bode, V.; Sejvar, J. J.; Pape, W. J.; Campbell, G. L.; and Marfin, A. A. West Nile virus disease: A descriptive study of 228 patients hospitalized in a 4-county region of Colorado in 2003. *Clin. Infect. Dis.*, 2006, **42**(9): 1234–1240.

Bondre, V. P.; Jadi, R. S.; Mishra, A. C.; Yergolkar, P. N.; and Arankalle, V. A. West Nile virus isolates from India: Evidence for a distinct genetic lineage. *J. Gen. Virol.*, 2007, **88**: 875–884.

Borowski, P.; Lang, M.; Haag, A.; Schmitz, H.; Choe, J.; Chen, H. M.; Hosmane, R. S. Characterization of imidazo [4,5-d]pyridazine nucleosides as modulators of unwinding reaction mediated by West Nile virus nucleoside triphosphatase/helicase: Evidence for activity on the level of substrate and/or enzyme. *Antimicrob. Agents Chemother.*, 2002, **46**: 1231–1239.

Bowen, R. A. and Nemeth, N. M. Experimental infections with West Nile virus. *Curr. Opin. Infect. Dis.*, 2007, **20**: 293–297.

Brenner, W.; Storch, G.; Buller, R.; Vij, R.; Devine, S.; and DiPersio, J. West Nile virus encephalopathy in an allogeneic stem cell transplant recipient: Use of quantitative PCR for diagnosis and assessment of viral clearance. *Bone Marrow Transplant.*, 2005, **36**: 369–370.

Brinton, M. A.; Fernandez, A. V.; and Dispoto, J. H. The 3′-nucleotides of flavivirus genomic RNA form a conserved secondary structure. *Virology*, 1986, **153**(1): 113–121.

Brooks, T. J. and Phillpotts, R. J. Interferon-alpha protects mice against lethal infection with St Louis encephalitis virus delivered by the aerosol and subcutaneous routes. *Antiviral Res.*, 1999, **41**: 57–64.

Busch, M. P., Kleinman, S. H., Tobler, L. H., Kamel, H. T., Norris, P. J., Walsh, I., Matud, J. L., Prince, H. E., Lanciotti, R. S., Wright, D. J., Linnen, J. M., and Caglioti, S. Virus and antibody dynamics in acute West Nile virus infection. *J. Infect. Dis.*, 2008, **198**: 984–993.

Busch, M. P.; Wright, D. J.; Custer, B.; Tobler, L. H.; Stramer, S. L.; Kleinman, S. H.; Prince, H. E.; Bianco, C.; Foster, G.; Petersen, L. R.; Nemo, G.; and Glynn, S. A. West Nile virus infections projected from blood donor screening data, United States, 2003. *Emerg. Infect. Dis.*, 2006, **12**(3): 395–402.

Byrne, S. N.; Halliday, G. M.; Johnston, L. J.; and King, N. J. Interleukin-1 but not tumor necrosis factor is involved in West Nile virus-induced Langerhans cell migration from the skin in C57BL/6 mice. *J. Investig. Dermatol.*, 2001, **117**: 702–709.

Calisher, C. H.; Karabatsos, N.; Dalrymple, J. M.; Shope, R. E.; Porterfield, J. S.; Westaway, E. G.; and Brandt, W. E. Antigenic relationships between flaviviruses as determined by cross-neutralization tests with polyclonal antisera. *J. Gen. Virol.*, 1989, **70**(1): 37–43.

Carson, P. J.; Borchardt, S. M.; and Custer, B. Neuro invasive disease and West Nile virus infection, North Dakota, USA, 1999–2008. *Emerg. Infect. Dis.*, 2012, **18**(4): 684–686.

Ceausu, E.; Erscoiu, S.; Calistru, P.; Ispas, D.; Dorobat, O.; Homos, M.; Barbulescu, C.; Cojocaru, I.; Simion, C. V.; Cristea, C.; Oprea, C.; Dumitrescu, C.; Duiculescu, D.; Marcu, I.; Mociornita, C.; Stoicev, T.; Zolotusca, I.; Calomfirescu, C.; Rusu, R.; Hodrea, R.; Geamai, S.; and Paun, L. Clinical manifestations in the West Nile virus outbreak. *Rom. J. Virol.*, 1997, **48**: 3–11.

Centers for Disease Control and Prevention (CDC). Entomology. 2009, www.cdc.gov/ncidod/dvbid/westnile/mosquitoSpecies.htm.

Centers for Disease Control and Prevention (CDC). Guidelines for surveillance, prevention, and control of West Nile virus infection: United States. *Morb. Mortal. Wkly. Rep.*, 2000, **49**(02): 25–28.

Centers for Disease Control and Prevention (CDC). Intrauterine West Nile virus infection—New York, 2002. *Morb. Mortal. Wkly. Rep.*, 2002a, **51**(50): 1135–1136.

Centers for Disease Control and Prevention (CDC). Possible West Nile virus transmission to an infant through breastfeeding—Michigan. *Morb. Mortal. Wkly. Rep.*, 2002b, **51**(39): 877–878.

Centers for Disease Control and Prevention (CDC). Update: investigations of West Nile virus infections in recipients of organ transplantation and blood transfusion—Michigan, 2002c. *MMWR*, 2002, **51**: 879.

Chancey, C.; Grinev, A.; Volkova, E.; and Rios, M. The global ecology and epidemiology of West Nile virus. *Biomed. Res. Int.*, 2015, 376230. http://dx.doi.org/10.1155/2015/376230.

Chang, J; Wang, L.; Ma, D.; Qu, X.; Guo, H.; Xu, X.; Mason, P. M.; Bourne, N.; Moriarty, R.; Gu, B.; Guo, J. T.; and Block, T. M. Novel imino sugar derivatives demonstrate potent antiviral activity against flaviviruses. *Antimicrob. Agents Chemother.*, 2009, **53**: 1501–1508.

Channon, R. B.; Yang, Y.; Feibelman, K. M.; Geiss, B. J.; Dandy, D. S.; and Henry, C. S. Development of an electrochemical paper-based analytical device for trace detection of virus particles. *Anal. Chem.*, 2018, **90**(12): 7777–7783.

Cho, H. and Diamond, M. S. Immune responses to West Nile virus infection in the central nervous system. *Viruses*, 2012, **4**: 3812–3830.

Chowers, M. Y.; Lang, R.; Nassar, F.; Ben-David, D.; Giladi, M.; Rubinshtein, E.; Itzhaki, A.; Mishal, J.; Siegman-Igra, Y.; Kitzes, R.; Pick, N.; Landau, Z.; Wolf, D.; Bin, H.; Mendelson, E.; Pitlik, S. D.; and Weinberger, M. Clinical characteristics of the West Nile fever outbreak, Israel, 2000. *Emerg. Infect. Dis.*, 2001, **7**: 675–678.

Chung, K. M.; Liszewski, M. K.; Nybakken, G.; Davis, A. E., Townsend, R. R., Fremont, D. H., Atkinson, J. P., and Diamond, M. S. West Nile virus non-structural protein NS1 inhibits complement activation by binding the regulatory protein factor H. *Proc. Nat. Acad. Sci. USA*, 2006, **103**(50): 19111–19116.

Colpitts, T. M.; Conway, M. J.; Montgomery, R. R.; and Fikrig, E. West Nile virus: Biology, transmission, and human infection. *Clin. Microbiol. Rev.*, 2012, **25**(4): 635– 648.

Cosnier, S.; Ionescu, R. E.; Herrmann, S.; Bouffier, L.; Demeunynck, M.; and Marks, R. S. Electroenzymatic polypyrrole-intercalator sensor for the determination of West Nile virus cDNA. *Anal. Chem.*, 2006, **78**(19): 7054–7057.

Courageot, M. P.; Frenkiel, M. P.; Dos Santos, C. D.; Deubel, V.; and Despres, P. Alpha-glucosidase inhibitors reduce dengue virus production by affecting the initial steps of virion morphogenesis in the endoplasmic reticulum. *J. Virol.*, 2000, **74**: 564–572.

Cupp, E. W. West Nile virus infection in mosquitoes in the mid-south USA, 2002–2005. *J. Med. Entomol.*, 2007, **44**: 117–125.

Danis, K.; Papa, A.; Theocharopoulos, G.; Dougas, G.; Athanasiou, M.; Detsis, M.; Baka, A.; Lytras, T.; Mellou, K.; Bonovas, S.; and Panagiotopoulos, T. Outbreak of West Nile virus infection in Greece, 2010. *Emerg. Infect. Dis.*, 2011, **17**(10): 1868–1872.

Day, C. W.; Smee, D. F.; Julander, J. G.; Yamshchikov, V. F.; Sidwell, R. W.; and Morrey, J. D. Error-prone replication of West Nile virus caused by ribavirin. *Antiviral Res.*, 2005, **67**: 38–45.

Diamond, M. S. Progress on the development of therapeutics against West Nile virus. *Antiviral Res.*, 2009, **83**(3): 214–227. doi:10.1016/j.antiviral.2009.05.006.

Diamond, M. S. Virus and host determinants of West Nile virus pathogenesis. *PLoS Pathog.*, 2009, **5**(6): e1000452.

Donadieu, E.; Bahuon, C.; Lowenski, S.; Zientara, S.; Coulpier, M.; and Lecolline, S. Differential virulence and pathogenesis of West Nile viruses. *Viruses,* 2013, **5**(11): 2856–2880.

Drzewnioková, P.; Barzon, L.; Franchin, E.; Lavezzo, E.; Bakonyi, T.; Pistl, J.; and Csank, T. The complete genome sequence analysis of West Nile virus strains isolated in Slovakia (central Europe). *Arch. Virol.*, 2019, **164**(1): 273–277.

Dridi, M.; Rauw, F.; Muylkens, B.; Lecollinet, S.; van den Berg, T.; and Lambrecht, B. Setting up a SPF Chicken Model for the pathotyping of West Nile Virus (WNV) strains. *Transbound. Emerg. Dis.*, 2013, **60**(2): 51–62.

Dridi, M.; Van Den Berg, T.; Lecollinet, S.; and Lambrecht, B. Evaluation of the pathogenicity of West Nile virus (WNV) lineage 2 strains in a SPF chicken model of infection: NS3-249Pro mutation is neither sufficient nor necessary for conferring virulence. *Vet. Res.*, 2015, **46**: 130. doi:10.1186/s13567-015-0257-1.

Durrant, D. M.; Robinette, M. L.; and Klein, R. S. IL-1R1 is required for dendritic cell-mediated T cell reactivation within the CNS during West Nile virus encephalitis. *J. Exp. Med.*, 2013, **210**: 503–516.

Eybpoosh, S.; Fazlalipour, M.; Baniasadi, V.; Pouriayevali, M. H.; Sadeghi, F.; Ahmadi Vasmehjani, A.; et al., (2019). Epidemiology of West Nile Virus in the Eastern Mediterranean region: A systematic review. *PLoS Negl. Trop. Dis.*, 2019, **13**(1): e0007081.

Fall, G.; Di Paolo, N.; Faye, M.; Dia, M.; Freire, C. C. M.; Loucoubar, C.; Zanotto, P. M. A.; Faye, O.; and Sall, A. A. Biological and phylogenetic characteristics of West African lineages of West Nile virus. *PLoS Neg. Trop. Dis.*, 2017, **11**: e0006078.

Farajollahi, A. and Nelder, M. P. Changes in *Aedes albopictus* (Diptera: Culicidae) populations in New Jersey and implications for arbovirus transmission. *J. Med. Entomol.*, 2009, **46**: 1220–1224.

Fratkin, J. D.; Leis, A. A.; Stokic, D. S.; Slavinski, S. A.; and Geiss, R.W. Spinal cord neuropathology in human West Nile virus infection. *Arch. Pathol. Lab. Med.*, 2004, **128**(5): 533–537.

Glass, W. G.; Lim, J. K.; Cholera, R.; Pletnev, A. G.; Gao, J. L.; and Murphy, P. M. Chemokine receptor CCR5 promotes leukocyte trafficking to the brain and survival in West Nile virus infection. *J. Exp. Med.*, 2005, **202**: 1087–1098.

Glass, W. G.; McDermott, D. H.; Lim, J. K.; Lekhong, S.; Yu, S. F.; Frank, W. A.; Pape, J.; Cheshier, R. C.; and Murphy, P. M. CCR5 deficiency increases risk of symptomatic West Nile virus infection. *J. Exp. Med.*, 2006, **203**: 35–40.

Goodell, J. R.; Puig-Basagoiti, F.; Forshey, B. M.; Shi, P. Y.; and Ferguson, D. M. Identification of compounds with anti-West Nile virus activity. *J. Med. Chem.*, 2006, **49**: 2127–2137.

Gu, B.; Mason, P.; Wang, L.; Norton, P.; Bourne, N.; Moriarty, R.; Mehta, A.; Despande, M.; Shah, R.; and Block, T. Antiviral profiles of novel iminocyclitol compounds against bovine viral diarrhea virus, West Nile virus, dengue virus and hepatitis B virus. *Antivir. Chem. Chemother.*, 2007, **18**: 49–59.

Gu, B.; Ouzunov, S.; Wang, L.; Mason, P.; Bourne, N.; Cuconati, A.; and Block, T. M. Discovery of small molecule inhibitors of West Nile virus using a high-throughput sub-genomic replicon screen. *Antiviral Res.*, 2006, **70**: 39–50.

Guarner, J.; Shieh, W. J.; Hunter, S.; Paddock, C. D.; Morken, T.; Campbell, G. L.; Marfin, A. A.; and Zaki, S. R. Clinicopathologic study and laboratory diagnosis of 23 cases with West Nile virus encephalomyelitis. *Hum. Pathol.*, 2004, **35**: 983–990.

Guarner, J.; Shieh, W. J.; Hunter, S.; Paddock, C. D.; Morken, T.; Campbell, G. L.; and Zaki, S. R. Clinicopathologic study and laboratory diagnosis of 23 cases with West Nile virus encephalomyelitis. *Human Pathol.*, 2004, **35**(8): 983–990.

Guirakhoo, F.; Heinz, F. X.; Mandl, C. W.; Holzmann, H.; and Kunz, C. Fusion activity of flaviviruses: Comparison of mature and immature (prM-containing) tick-borne encephalitis virions. *J. Gen. Virol.*, 1991, **72**(6): 1323–1329.

Hamer, G. L. *Culex pipiens* (Diptera: Culicidae): A bridge vector of West Nile virus to humans. *J. Med. Entomol.*, 2008, **45**: 125–128.

Hayes, E. B. and O'Leary, D. R. West Nile virus infection: A pediatric perspective. *Pediatrics*, 2004, **113**(5): 1375–1381.

Hayes, E. B.; Komar, N.; Nasci, R. S.; Montgomery, S. P.; O'Leary, D. R.; and Grant, L. Epidemiology and transmission dynamics of West Nile virus disease. *Campbell*, 2005, **11**(8): 1167–1173.

Heinz, F. X.; Stiasny, K.; Puschner-Auer, G.; Holzmann, H.; Allison, S. L.; Mandl, C. W.; and Kunz, C. Structural changes and functional control of the tick-borne encephalitis virus glycoprotein E by the heterodimeric association with protein prM. *Virology*, 1994, **198**(2): 109–117.

Hinckley, A. F.; O'Leary, D. R.; and Hayes, E. B. Transmission of West Nile virus through human breast milk seems to be rare. *Pediatrics*, 2007, **119**(3): e666–e671.

Holick, J.; Kyle, A.; Ferraro, W.; Delaney, R. R.; and Iwaseczko, M. Discovery of *Aedes albopictus* infected with West Nile virus in southeastern Pennsylvania. *J. Am. Mosq. Control Assoc.*, 2002, **18**: 131.

Hrobowski, Y. M.; Garry, R. F.; and Michael, S. F. Peptide inhibitors of dengue virus and West Nile virus infectivity. *Virol. J.*, 2005, **2**: 49.

https://www.nhp.gov.in/disease/communicable-disease/west-nile-fever.

https://www.who.int/news-room/fact-sheets/detail/west-nile-virus.

Hubálek, Z. and Halouzka, J. West Nile fever—a reemerging mosquito-borne viral disease in Europe. *Emerg. Infect. Dis.*, 1999, **5**(5): 643–650. doi: 10.3201/eid0505.990505.

Hunt, A. R.; Hall, R. A.; Kerst, A. J.; Nasci, R. S.; Savage, H. M.; Panella, N. A.; and Roehrig, J. T. Detection of West Nile virus antigen in mosquitoes and avian tissues by a monoclonal antibody-based capture enzyme immunoassay. *J. Clin. Microbiol.*, 2002, **40**(6): 2023–2030.

Ionescu, R. E.; Cosnier, S.; Herrmann, S.; and Marks, R. S. Amperometric immunosensor for the detection of anti-West Nile virus IgG. *Anal. Chem.*, 2007, **79**(22): 8662–8668.

Iwamoto, M.; Jernigan, D. B.; Guasch, A. Trepka, M. J.; Blackmore, C. G.; Hellinger, W. C.; Pham, S. M.; Zaki, S.; Lanciotti, R. S.; Lance-Parker, S. E.; DiazGranados, C. A.; Winquist, A. G.; Perlino, C. A.; Wiersma, S.; Hillyer, K. L.; Goodman, J. L.; Marfin, A. A.; Chamberland, M. E.; Petersen, L. R.; and West Nile Virus in Transplant Recipients Investigation Team. Transmission of West Nile virus from an organ donor to four transplant recipients. *N. Eng. J. Med.*, 2003, **348**(22): 2196–2203.

Jean, C. M.; Honarmand, S.; Louie, J. K.; and Glaser, C. A. Risk factors for West Nile virus neuro-invasive disease, California, 2005. *Emerg. Infect. Dis.*, 2007, **13**(12): 1918–1920.

Johnston, P. A.; Phillips, J.; Shun, T. Y.; Shinde, S.; Lazo, J. S.; Huryn, D. M.; Myers, M. C.; Ratnikov, B. I.; Smith, J. W.; Su, Y.; Dahl, R.; Cosford, N. D.; Shiryaev, S. A.; and Strongin, A. Y. HTS identifies novel and specific uncompetitive inhibitors of the two-component NS2B-NS3 proteinase of West Nile virus. *Assay Drug Dev. Technol.*, 2007, **5**: 737–750.

Jordan, I.; Briese, T.; Fischer, N.; Lau, J. Y.; and Lipkin, W. I. Ribavirin inhibits West Nile virus replication and cytopathic effect in neural cells. *J. Infect. Dis.*, 2000, **182**: 1214–1217.

Khromykh, A. A. and Westaway, E. G. RNA binding properties of core protein of the flavivirus Kunjin. *Arch. Virol.*, 1996, **141**(3–4): 685–699.

Kilpatrick, M.; Kramer, L. D.; Campbell, S. R.; Alleyne, E. O.; Dobson, A. P.; and Daszak, P. West Nile virus risk assessment and the bridge vector paradigm. *Emerg. Infect. Dis.*, 2005, **11**(3): 425–429.

Kinney, R. M.; Huang, C. Y.; Rose, B. C.; Kroeker, A. D.; Dreher, T. W.; Iversen, P. L.; and Stein, D. A. Inhibition of dengue virus serotypes 1 to 4 in vero cell cultures with morpholino oligomers. *J. Virol.*, 2005, **79**: 5116–5128.

Kleinschmidt-DeMasters, B. K.; Marder, B. A.; Levi, M. E.; Laird, S. P.; McNutt, J. T.; Escott, E. J.; Everson, G. T.; and Tyler, K. L. Naturally acquired West Nile virus encephalomyelitis in transplant recipients: Clinical, laboratory, diagnostic, and neuropathological features. *Arch. Neurol.*, 2004, **61**: 1210–1220.

Kopel, E.; Amitai, Z.; Bin, H.; Shulman, L. M.; Mendelson, E.; and Shefer, R. Surveillance of West Nile virus disease, Tel Aviv district, Israel, 2005 to 2010. *Eurosurveillance*, 2011, **16**(25).

Koraka, P.; Zeller, H.; Niedrig, M.; Osterhaus, A. D.; and Groen, J. Reactivity of serum samples from patients with a flavivirus infection measured by immunofluorescence assay and ELISA. *Microbes Infect.*, 2002, **4**(12): 1209–1215.

Koraka, P.; Zeller, H.; Niedrig, M.; Osterhaus, A. D.; and Groen, J. Reactivity of serum samples from patients with a flavivirus infection measured by immunofluorescence assay and ELISA. *Microbes Infect.*, 2002, **4**(12): 1209–1215.

Kulkarni, A. B.; Mullbacher, A.; and Blanden, R. V. Functional analysis of macrophages, B cells and splenic dendritic cells as antigen-presenting cells in West Nile virus-specific murine T lymphocyte proliferation. *Immunol. Cell Biol.*, 1991, **69**: 71–80.

Kumar, P.; Lee, S. K.; Shankar, P.; and Manjunath, N. A single siRNA suppresses fatal encephalitis induced by two different flaviviruses. *PLoS Med.*, 2006, **3**: e96.

Kyung, M. C.; Thompson, B. S.; Fremont, D. H.; and Diamond, M. S. Antibody recognition of cell surface-associated NS1 triggers Fc-γ receptor-mediated phagocytosis and clearance of West Nile virus-infected cells. *J. Virol.*, 2007, **81**(17): 9551–9555.

Ladbury, G. A. F.; Gavana, M.; Danis, K.; Papa, A.; Papamichail, D.; Mourelatos, S.; Gewehr, S.; Theocharopoulos, G.; Bonovas, S.; Benos, A.; and Panagiotopoulos, T. Population sero-prevalence study after a West Nile virus lineage 2 epidemic, Greece, 2010. *PLoS One*, 2013, **8**(11): e80432.

Lanteri, M. C.; O'Brien, K. M.; Purtha, W. E.; Cameron, M. J.; Lund, J. M.; Owen, R. E.; Heitman, J. W.; Custer, B.; Hirschkorn, D. F.; Tobler, L. H.; Kiely, N., Prince, H. E., Ndhlovu, L. C., Nixon, D. F., Kamel, H. T., Kelvin, D. J., Busch, M. P., Rudensky, A. Y., Diamond, M. S., and Norris, P. J. Tregs control the development of symptomatic West Nile virus infection in humans and mice. *J. Clin. Invest.*, 2009, **119**: 3266–3277.

Laurent Rolle, M.; Boer, E. F.; Lubick, K. J.; Wolfinbarger, J. B.; Carmody, A. B.; Rockx, B.; Liu, W.; Ashour, J.; Shupert, W. L.; Holbrook, M. R.; Barrett, A. D.; Mason, P. W.; Bloom, M. E.; García-Sastre, A.; Khromykh, A. A.; and Best, S. M. The NS5 protein of the virulent West Nile virus NY99 strain is a potent antagonist of type I interferon-mediated JAK-STAT signaling. *J. Virol.*, 2010, **84**(7): 3503–3515.

Le Bon, A.; Thompson, C.; Kamphuis, E.; Durand, V.; Rossmann, C.; Kalinke, U.; and Tough, D. F. Cutting edge: Enhancement of antibody responses through direct stimulation of B and T cells by type I IFN. *J. Immunol.*, 2006, **176**: 2074–2078.

Lim, P. Y.; Behr, M. J.; Chadwick, C. M.; Shi, P. Y.; and Bernard, K. A. Keratinocytes are cell targets of West Nile virus in vivo. *J. Virol.*, 2011, **85**: 5197–5201.

Lin, Y. L.; Huang, Y. L.; Ma, S. H.; Yeh, C. T.; Chiou, S. Y.; Chen, L. K.; and Liao, C. L. Inhibition of Japanese encephalitis virus infection by nitric oxide: Antiviral effect of nitric oxide on RNA virus replication. *J. Virol.*, 1997, **71**: 5227–5235.

Lindsey, N. P.; Erin Staples, J.; Lehman, J. A.; and Fischer, M. Surveillance for human West Nile virus disease, United States, 1999–2008. *Morb. Mortal. Wkly. Rep.*, 2010, **59**(2): 1–17.

Lindsey, N. P.; Hayes, E. B.; Staples, J. E.; and Fischer, M. West Nile virus disease in children, United States, 1999–2007. *Pediatrics*, 2009, **123**(6): e1084–e1089.

Lindsey, N. P.; Staples, J. E.; Lehman, J. A.; and Fischer, M. Medical risk factors for severe West Nile virus disease, United States, 2008–2010. *Am. J. Trop. Med. Hyg.*, 2012, **87**(1): 179–184.

Liu, W. J.; Wang, X. J.; Mokhonov, V. V.; Shi, P.-Y.; Randall, R.; and Khromykh, A.A. Inhibition of interferon signaling by the New York 99 strain and Kunjin subtype of West Nile virus involves blockage of STAT1 and

STAT2 activation by nonstructural proteins. *J. Virol.*, 2005, **79**(3): 1934–1942.

Loeb, M.; Eskandarian, S.; Rupp, M.; Fishman, N.; Gasink, L.; Patterson, J.; et al. Genetic variants and susceptibility to neurological complications following West Nile virus infection. *J. Infect. Dis.*, 2011, **204**(7): 1031–1037.

Ma, D. D.; Rede, T.; Naqvi, N. A.; and Cook, P. D. Synthetic oligonucleotides as therapeutics: The coming of age. *Biotechnol. Annu. Rev.*, 2000, **5**: 155–196.

Markoff, L.; Falgout, B.; and Chang, A. A conserved internal hydrophobic domain mediates the stable membrane integration of the dengue virus capsid protein. *Virology*, 1997, **233**(1): 105–117.

Melian, E. B.; Edmonds, J. H.; Nagasaki, T. K.; Hinzman, E.; Floden, N.; and Khromykh, A. A. West Nile virus NS2A protein facilitates virus-induced apoptosis independently of interferon response. *J. Gen. Virol.*, 2013, **94**(2): 308–313.

Mentoor, J. L. D.; Lubisi, A. B.; Gerdes, T.; Human, S.; Williams, J. H.; and Venter, M. Full-genome sequence of a neuro-invasive West Nile virus lineage 2 strain from a fatal horse infection in South Africa. *Genome Announc.*, 2016, **4**(4): e00740-16.

Morrey, J. D.; Day, C. W.; Julander, J. G.; Blatt, L. M.; Smee, D. F.; and Sidwell, R. W. Effect of interferon-alpha and interferon-inducers on West Nile virus in mouse and hamster animal models. *Antivir. Chem. Chemother.*, 2004, **15**: 101–109.

Morrey, J. D.; Day, C. W.; Julander, J. G.; Blatt, L. M.; Smee, D. F.; and Sidwell, R. W. Effect of interferon-alpha and interferon-inducers on West Nile virus in mouse and hamster animal models. *Antivir. Chem. Chemother.*, 2004, **15**: 101–109.

Mostashari, F.; Bunning, M. L.; Kitsutani, P. T.; Singer, D. A.; Nash, D.; Cooper, M. J.; Katz, N.; Liljebjelke, K. A.; Biggerstaff, B. J.; Fine, A. D.; Layton, M. C.; Mullin, S. M.; Johnson, A. J.; Martin, D. A.; Hayes, E. B.; and Campbell, G. L. Epidemic West Nile encephalitis, New York, 1999: Results of a household-based seroepidemiological survey. *Lancet*, 2001, **358**(9278): 261–264.

Mueller, N. H.; Pattabiraman, N.; Ansarah-Sobrinho, C.; Viswanathan, P.; Pierson, T. C.; and Padmanabhan, R. Identification and biochemical characterization of small-molecule inhibitors of West Nile virus serine protease by a high-throughput screen. *Antimicrob. Agents Chemother.*, 2008, **52**: 3385–3393.

Muñoz-Jordán, J. L. and Fredericksen, B. L. How flaviviruses activate and suppress the interferon response. *Viruses*, 2010, **2**(2): 676–691. doi:10.3390/v2020676.

Muñoz-Jordán, J. L.; Laurent-Rolle, M.; and Ashouretal, J. Inhibition of alpha/beta interferon signaling by the NS4B protein of flaviviruses. *J. Virol.*, 2005, **79**(13): 8004–8013.

Murray, K. O.; Walker, C.; and Gould, E. The virology, epidemiology, and clinical impact of West Nile virus: A decade of advancements in research since its introduction into the Western Hemisphere. *Epidemiol. Infect.*, 2011, **139**: 807–817.

Nash, D.; Mostashari, F.; Fine, A.; Miller, J.; O'Leary, D.; Murray, K.; Huang, A.; Rosenberg, A.; Greenberg, A.; Sherman, M.; Wong, S.; Layton, M.; and 1999 West Nile Outbreak Response Working Group. The outbreak of West Nile virus infection in the New York City area in 1999. *N. Eng. J. Med.*, 2001, **344**(24): 1807–1814.

Nouiery, A. O.; Olivo, P. D.; Slomczynska, U.; Zhou, Y.; Buscher, B.; Geiss, B.; Engle, M.; Roth, R. M.; Chung, K. M.; Samuel, M. A.; and Diamond, M. S. The identification of novel small molecule inhibitors of West Nile virus infection. *J. Virol.*, 2007, **81**(21): 11992–12004.

Pardridge, W. M. Brain metabolism: A perspective from the blood–brain barrier. *Physiol. Rev.*, 1983, **63**: 1481–1535.

Parida, M.; Posadas, G.; Inoue, S.; Hasebe, F.; and Morita, K. Real-time reverse transcription loop-mediated isothermal amplification for rapid detection of West Nile virus. *J. Clin. Microbiol.*, 2004, **42**(1): 257–263.

Patnaik, J. L.; Harmon, H.; and Vogt, R. L. Follow-up of 2003 human West Nile virus infections, Denver, Colorado. *Emerg. Infect. Dis.*, 2006, **12**(7): 1129–1131.

Pealer, L. N.; Marfin, A. A.; Petersen, L. R.; Lanciotti, R. S.; Page, P. L.; Stramer, S. L.; Stobierski, M. G.; Signs, K.; Newman, B.; Kapoor, H.; Goodman, J. L.; Chamberland, M. E.; and West Nile Virus Transmission Investigation Team. Transmission of West Nile virus through blood transfusion in the United States in 2002. *N. Eng. J. Med.*, 2003, **349**(13): 1236–1245.

Pestka, S.; Krause, C. D.; and Walter, M. R. Interferons, interferon-like cytokines, and their receptors. *Immunol. Rev.*, 2004, **202**: 8–32.

Petersen, L. R. and Marfin, A. A. West Nile virus: A primer for the clinician. *Ann. Intern. Med.*, 2002, **137**: 173–179.

Petersen, L. R.; Carson, P. J; Biggerstaff, B. J.; Custer, B.; Borchardt, S. M.; and Busch, M. P. Estimated cumulative incidence of West Nile virus infection in US adults, 1999–2010. *Epidemiol. Infect.*, 2013, **141**(3): 591–595.

Peterson, R. K.; Macedo, P. A.; and Davis, R. S. A human-health risk assessment for West Nile virus and insecticides used in mosquito management. Environ. *Health Perspect.*, 2006, **114**(3): 366–372. doi: 10.1289/ehp.8667.

Pfeffer, M. and Dobler, G. Emergence of zoonotic arboviruses by animal trade and migration. *Parasit. Vectors*, 2010, **3**(1): 35. doi: 10.1186/1756-3305-3-35.

Phipps, L. P.; Gough, R. E.; Ceeraz, V.; Cox, W. J.; and Brown, I. H. Detection of West Nile virus in the tissues of specific pathogen free chickens and serological response to laboratory infection: a comparative study. *Avian Pathol.*, 2007, **36**: 301–305.

Platanias, L. C. Mechanisms of type-I- and type-II-interferon-mediated signalling. *Nat. Rev. Immunol.*, 2005, **5**: 375–386.

Prince, H. E. and Hogrefe, W. R. Assays for detecting West Nile virus antibodies in human serum, plasma, and cerebrospinal fluid. *Clin. Appl. Immunol. Rev.*, 2005, **5**(1): 45–63.

Puig-Basagoiti, F.; Tilgner, M.; Forshey, B. M.; Philpott, S. M.; Espina, N. G.; Wentworth, D. E.; Goebel, S. J.; Masters, P. S.; Falgout, B.; Ren, P.; Ferguson, D. M.; and Shi, P. Y. Triaryl pyrazoline compound inhibits flavivirus RNA replication. *Antimicrob. Agents Chemother.*, 2006, **50**: 1320–1329.

Reisen, W. K.; Fang, Y.; and Martinez, V. M. Vector competence of *Culiseta incidens* and *Culex thriambus* for West Nile virus. *J. Am. Mosq. Control Assoc.*, 2006, **22**: 662–665.

Rice, C. M.; Lenches, E. M.; Eddy, S. R.; Shin, S. J.; Sheets, R. L.; and Strauss, J. H. Nucleotide sequence of yellow fever virus: Implications for flavivirus gene expression and evolution. *Science*, 1985, **229**(4715): 726–733.

Richter, J.; Tryfonos, C.; Tourvas, A.; Floridou, A.; Paphitou, N. I.; Christodoulou, C. Complete genome sequence of West Nile virus (WNV) from the first human case of neuroinvasive WNV infection in Cyprus. *Genome Announc.*, 2017, **5**(43): e01110-17.

Rios, J. J.; Fleming, J. G.; Bryant, U. K.; Carter, C. N.; Huber Jr, J. C.; Long, M. T.; et al. OAS1 polymorphisms are associated with susceptibility to West Nile encephalitis in horses. *PloS One*, 2010, **5**(5): e10537.

Rios, M.; Zhang, M. J.; Grinev, A.; Srinivasan, K.; Daniel, S.; Wood, O.; Hewlett, I. K.; and Dayton, A. I. Monocytes-macrophages are a potential target in human infection with West Nile virus through blood transfusion. *Transfusion*, 2006, **46**: 659–667.

Rizzoli, A.; Jimenez-Clavero, M. A.; Barzon, L.; Cordioli, P.; Figuerola, J.; Koraka, P., Martina, B.; Moreno, A.; Nowotny, N.; Pardigon, N.; Sanders, N.; Ulbert, S.; and Tenorio, A. The challenges of West Nile virus in Europe: Knowledge gaps and research priorities. *Euro Surveill.*, 2015, **20**: 21135.

Sampathkumar, P. West Nile virus: Epidemiology, clinical presentation, diagnosis, and prevention. *Mayo Clin. Proc.*, 2003, **78**: 1137–1144.

Samuel, M. A. and Diamond, M. S. Pathogenesis of West Nile virus infection: A balance between virulence, innate and adaptive immunity, and viral evasion. *J. Virol.*, 2006, **80**(19): 9349–9360.

Samuel, M. A.; Morrey, J. D.; and Diamond, M. S. Caspase 3-dependent cell death of neurons contributes to the pathogenesis of West Nile virus encephalitis. *J. Virol.*, 2007, **81**: 2614–2623.

Sardelis, M. R.; Turell, M. J.; O'Guinn, M. L.; Andre, R. G.; and Roberts, D. R. Vector competence of three North American strains of *Aedes albopictus* for West Nile virus. *J. Am. Mosq. Control Assoc.*, 2002, **18**: 284–289.

Saxena, S. K.; Singh, A.; and Mathur, A. Antiviral effect of nitric oxide during Japanese encephalitis virus infection. *Int. J. Exp. Pathol.*, 2000, **81**: 165–172.

Schuffenecker, I.; Peyrefitte, C. N.; el Harrak, M.; Murri, S.; Leblond, A.; and Zeller, H. G. West Nile virus in Morocco, 2003. *Emerg. Infect. Dis.*, 2005, **11**: 306–309.

Schul, W.; Liu, W.; Xu, H. Y.; Flamand, M.; and Vasudevan, S. G. A dengue fever viremia model in mice shows reduction in viral replication and suppression of the inflammatory response after treatment with antiviral drugs. *J. Infect. Dis.*, 2007, **195**: 665–674.

Sejvar, J. J.; Bode, A. V.; Marfin, A. A.; Campbell, G. L.; Ewing, D.; Mazowiecki, M.; et al. West Nile virus-associated flaccid paralysis. *Emerg. Infect. Dis.*, 2005, **11**(7): 1021–1027. doi:10.3201/eid1107.040991.

Senne, D. A.; Pedersen, J. C.; Hutto, D. L.; Taylor, W. D.; Schmitt, B. J.; and Panigrahy, B. Pathogenicity of West Nile virus in chickens. *Avian Dis.*, 2000, **44**: 642–649.

Shi, P. Y.; Kauffman, E. B.; Ren, P.; Felton, A.; Tai, J. H.; Dupuis, A. P.; and Ebel, G. D. High-throughput detection of West Nile virus RNA. *J. Clin. Microbiol.*, 2001, **39**(4), 1264–1271.

Shieh, W. J.; Guarner, J.; Layton, M.; Fine, A.; Miller, J.; Nash, D.; and Zaki, S. R. The role of pathology in an investigation of an outbreak of West Nile encephalitis in New York, 1999. *Emerg. Infect. Dis.*, 2000, **6**(4): 370.

Shirafuji, H.; Kanehira, K.; Kubo, M.; Shibahara, T.; and Kamio, T. Experimental West Nile virus infection in aigamo ducks, a cross between wild ducks (Anas platyrhynchos) and domestic ducks (Anas platyrhynchos var. domesticus). *Avian Dis.*, 2009, **53**: 239–244.

Shrestha, B. and Diamond, M. S. Role of CD8+ T cells in control of West Nile virus infection. *J. Virol.*, 2004, **78**: 8312–8321.

Shrestha, B.; Samuel, M. A.; and Diamond, M. S. CD8+ T cells require perforin to clear West Nile virus from infected neurons. *J. Virol.*, 2006, **80**: 119–129.

Smithburn, K. C.; Hughes, T. P.; Burke, A. W.; and Paul, J. H. A neurotropic virus isolated from the blood of a native of Uganda. *Am. J. Trop. Med. Hyg.*, 1940, **20**: 471–492. doi: 10.4269/ajtmh.1940.s1-20.471.

Stadler, K.; Allison, S. L.; Schalich, J.; and Heinz, F. X. Proteolytic activation of tick-borne encephalitis virus by furin. *J. Virol.*, 1997, **71**(11): 8475–8481.

Stone, W. B.; Okoniewski, J. C.; Therrien, J. E.; Kramer, L. D.; Kauffman, E. B.; and Eidson, M. VecTest as diagnostic and surveillance tool for West Nile virus in dead birds. Emerg. *Infect. Dis.*, 2004, **10**(12): 2175–2181. doi:10.3201/eid1012.040836.

Swayne, D. E.; Beck, J. R.; and Zaki, S. Pathogenicity of West Nile virus for turkeys. *Avian Dis.*, 2000, **44**(4): 932–937.

Tiawsirisup, S.; Platt, K. B.; Evans, R. B.; and Rowley, W. A. A comparison of West Nile virus transmission by *Ochlerotatus trivittatus* (COQ.), *Culex pipiens* (L.), and *Aedes albopictus* (Skuse). *Vector Borne Zoonotic Dis.*, 2005, **5**: 40–47.

Totani, M.; Yoshii, K.; Kariwa, H.; and Takashima, I. Glycosylation of the envelope protein of West Nile Virus affects its replication in chicks. *Avian Dis.*, 2011, **55**: 561–568.

Turell, M. J.; Dohm, D. J.; Sardelis, M. R.; O'guinn, M. L.; Andreadis, T. G.; and Blow, J. A. An update on the potential of North American mosquitoes (Diptera: Culicidae) to transmit West Nile virus. *J. Med. Entomol.*, 2005, **42**(1): 57–62.

Turell, M. J.; Sardelis, M. R.; O'Guinn, M. L.; and Dohm, D. J. Potential vectors of West Nile virus in North America, In: *Japanese Encephalitis and West Nile Viruses* (Mackenzie, J. S., Barrett, A. D. T.; and Deubel, V.; eds.), New York: Springer-Verlag, 2002: 241–252.

Unlu, I; Mackay, A. J.; Roy, A.; Yates, M. M.; and Foil, L. D. Evidence of vertical transmission of West Nile virus in field-collected mosquitoes. *J. Vector Ecol.*, 2010, **35**: 95–99.

Vanlandingham, D. L. Relative susceptibilities of South Texas mosquitoes to infection with West Nile virus. *Am. J. Trop. Med. Hyg.*, 2007, **77**: 925–928.

Vitek, C. J.; Richards, S. L.; Mores, C. N.; Day, J. F.; and Lord, C. C. Arbovirus transmission by *Culex nigripalpus* in Florida, 2005. *J. Med. Entomol.*, 2008, **45**: 483–493.

Wang, Y.; Lobigs, M.; Lee, E.; and Mullbacher, A. CD8+ T cells mediate recovery and immunopathology in West Nile virus encephalitis. *J. Virol.*, 2003, **77**: 13323–13334.

Watson, J. T.; Pertel, P. E.; Jones, R. C.; Siston, A. M.; Paul, W. S.; Austin, C. C.; and Gerber, S. I. Clinical characteristics and functional outcomes of West Nile fever. *Ann. Intern. Med.*, 2004, **141**: 360–365.

Wedin, C. Complete Genome Sequences and Phylogeny of West Nile Virus Isolates from Southeastern United States, 2003-2012. Graduate Theses and Dissertations, 2013. http://scholarcommons.usf.edu/etd/4958.

Welte, T.; Reagan, K.; Fang, H.; Machain-Williams, C.; Zheng, X.; Mendell, N.; Chang, G. J.; Wu, P.; Blair, C. D.; and Wang, T. Toll-like receptor 7-induced immune response to cutaneous West Nile virus infection. *J. Gen. Virol.*, 2009, **90**: 2660–2668.

Wen, J. L.; Hua, B. C.; Xiang, J. W.; Huang, H.; and Khromykh, A. A. Analysis of adaptive mutations in Kunjin virus replicon RNA reveals a novel role for the flavivirus nonstructural protein NS2A in inhibition of beta interferon promoter driven transcription. *J. Virol.*, 2004, **78**(22): 12225–12235.

Wengler, G and Wengler, G. 1991. The carboxy-terminal part of the NS 3 protein of the West Nile flavivirus can be isolated as a soluble protein after proteolytic cleavage and represents an RNA-stimulated NTPase. *Virology*, 1991, **184**(2): 707–715.

Whitby, K.; Pierson, T. C.; Geiss, B.; Lane, K.; Engle, M.; Zhou, Y.; Doms, R. W.; and Diamond, M. S. Castanospermine, a potent inhibitor of dengue virus infection in vitro and in vivo. *J. Virol.*, 2005, **79**: 8698–8706.

Wilson, J. R.; De Sessions, P. F.; Leon, M. A.; and Scholle, F. West Nile virus nonstructural protein 1 inhibits TLR3 signal transduction. *J. Virol.*, 2008, **82**(17): 8262–8271.

Wu, S. F.; Lee, C. J.; Liao, C. L.; Dwek, R. A.; Zitzmann, N.; and Lin, Y. L. Antiviral effects of an iminosugar derivative on flavivirus infections. *J. Virol.*, 2002, **76**: 3596–3604.

Zeller, H. G. and Schuffenecker, I. West Nile virus: An overview of its spread in Europe and the Mediterranean basin in contrast to its spread in the Americas. *Eur. J. Clin. Microbiol. Infect. Dis.*, 2004, **23**: 147–56.

Zeller, H.; Zientara, S.; Hars, J.; Languille, J.; Mailles, A.; Tolou, H.; Paty, M.-C.; Schaffner, F.; Armengaud, A.; Gaillan, P.; Legras, J.-F.; and Hendrix, P. West Nile outbreak in horses in southern France: September 2004. *Eurosurveillance Wkly.*, 2004, **8**(41).

Zhang, H.; Harpster, M. H.; Park, H. J.; Johnson, P. A.; and Wilson, W. C. Surface-enhanced Raman scattering detection of DNA derived from the West Nile virus genome using magnetic capture of Raman-active gold nanoparticles. *Anal. Chem.*, 2010, **83**(1): 254–260.

Zhang, Y.; Corver, J.; Chipman, P. R.; Zhang, W.; Pletnev, S. V.; Sedlak, D.; Baker, T. S.; Strauss, J. H.; Kuhn, R. J.; and Rossmann, M. G. Structures of immature flavivirus particles. *EMBO J.*, 2003, **22**(11): 2604–2613.

Zhang, Y.; Kaufmann, B.; Chipman, P. R.; Kuhn, R. J.; and Rossmann, M. G. Structure of immature West Nile virus. *J. Virol.*, 2007, **81**(11): 6141–6145.

Zou, S.; Foster, G. A.; Dodd, R. Y.; Petersen, L. R.; and Stramer, S. L. West Nile fever characteristics among viremic persons identified through blood donor screening. *J. Infect. Dis.*, 2010, **202**(9): 1354–1361.

Index

acetylsalicylic acid 177
Actinobacteria 47
acute encephalitis syndrome 239
acute viral hemorrhagic disease 216
aedeagus 10, 14
Aedes 2–5, 21–24, 27–29, 31–34, 36–41, 43–49, 51–54, 81, 82, 89–91, 99, 100, 116, 118–120, 128, 129, 137, 138, 173, 174, 222, 223
Aedes aegypti 12, 16, 22–27, 29, 32–34, 36–38, 165, 173, 180, 216, 224
Aedes africanus 91, 196, 197, 224
Aedes albopictus 22, 29–33, 39, 40, 42, 46, 86, 129, 165, 173, 180
Aedes atropalus 38–40
Aedes cantator 52, 53
Aedes cinereus 53
Aedes furcifer 169, 196
Aedes hensilli 193
Aedes japonicus 22, 41–43, 45
Aedes koreicus 22, 43–46
Aedes nivalis 91
Aedes polynesiensis 49, 50
Aedes pseudoscutellaris 129
Aedes scutellaris 91
Aedes triseriatus 47–49, 242
Aedes vexans 51–53
Aedes vigilax 53
Aedes vittatus 22, 33–37
alphavirus 85, 169, 170, 180
amino acid 106, 108, 113, 195
Anopheles subpictus 241
antibody 124, 130, 131, 133, 135, 136, 175, 198, 201–205, 216, 225
antigen-presenting cell 244
antisense oligomer 247
antiviral drug 177, 226
arbidol 178
arbovirus 83–85, 94, 194, 197, 198, 204
arthralgia 86, 192, 193, 203
arthritis 180
arthropod 81, 84, 85
  blood-sucking 83
  hematophagous 84

biosensor 132–137, 205, 206, 208
blood 10, 11, 13, 15–17, 24, 27, 48, 52, 140, 173, 203, 204, 207, 244, 245, 249
blood–brain barrier 94, 244, 245, 249
blood meal 15–17, 33, 40, 46, 83, 120, 175, 197, 244
blood transfusion 90, 127, 197, 207, 244
Bunyamwera virus 53
Bunyavirus 85, 87, 95

Cache Valley virus 32, 43
carboxy methyl cellulose 130
CCHF *see* Crimean–Congo hemorrhagic fever
cell 118, 120–122, 124, 170, 198–200, 220, 229, 244, 248, 249
  antigen-presenting 244
  endothelial 124, 244
  infected 124, 200
  kidney 126
  mast 124
  nerve 207
  neuron 248

cerebral spinal fluid 204
chemotherapy 86, 91, 93, 126, 201, 208, 230, 247
chikungunya fever 2, 46, 166, 168, 170, 172, 174, 176, 178, 180
chikungunya virus (CHIKV) 31, 32, 37, 46, 50, 85, 86, 95, 165–167, 169, 170, 173–175, 177–180
CHIKV *see* chikungunya virus
Chinese hamster ovary DHFR system 126
conjunctivitis 176, 192, 193
Crimean–Congo hemorrhagic fever (CCHF) 82, 84, 85, 88
Culex albopictus 29
Culex bitaeniorhynchus 241
Culex nigripalpus 242
Culex pipiens 52
Culex quinquefasciatus 242
Culex restuans 52
Culex salinarius 242
Culex stigmatosoma 242
Culex sylvestris 51
Culex tarsalis 242, 248
Culex thriambus 242
Culex vexans 51
Culex vishnui 241, 242
Culex whitmorei 241
cytokine 124, 139, 140
cytoplasm 106, 107, 114, 121, 122, 176, 198, 199

dead-end host 89
dengue fever (DF) 36, 37, 99, 100, 102–104, 106, 108, 110, 112, 114, 116, 118, 120, 122, 124, 126, 136–138
dengue hemorrhagic fever (DHF) 91, 99–102, 105, 116, 122, 124, 140
dengue infection 26, 117, 122, 124, 136
dengue shock syndrome (DSS) 91, 99–102, 122, 123, 140
dengue virus (DENV) 36–38, 90, 91, 95, 99, 101, 103–105, 107–114, 116–124, 126–133, 135–137, 140, 198, 203, 204
DENV *see* dengue virus
DF *see* dengue fever
DHF *see* dengue hemorrhagic fever
disease 21, 24, 27, 28, 82–84, 86–90, 99–102, 122, 127, 138, 140, 177, 180, 202, 203, 207, 216
  allergic 139
  asymptomatic 122
  neuroinvasive 176, 249
  symptomatic 122
drug 127, 177, 180, 202, 216, 230
  anti-depressant 177
  anti-inflammatory 127, 178
  antipyretic 127
  viral-inhibiting 178
DSS *see* dengue shock syndrome

Ebola 82, 230
ELISA *see* enzyme-linked immuno-sorbent assay
encephalitis 53, 84, 86, 89, 94, 180, 215, 226, 230
  arboviral 86
  Eastern equine 32, 40, 43, 48, 52, 85–87, 95
  Japanese 43, 46, 84, 86, 194
  Powassan 86
  Venezuelan equine (VEE) 32, 48, 53, 84–86, 95
  Western equine 32, 48, 52, 85, 86, 95
endocytosis 121
  receptor-mediated 176, 177, 201, 224
enzyme-linked immuno-sorbent assay (ELISA) 131, 178–180, 203, 208, 246

epidemic 34, 36, 94, 100, 101, 105, 166, 169, 192, 193, 216, 218, 222–224, 229, 230, 237
Escherichia coli 126

flaccid paralysis 243
  acute 176
flavivirus 85, 90, 91, 94, 95, 108, 113, 170, 194, 201, 202, 204, 206, 207, 222, 229
FNV *see* French neurotropic vaccine
French neurotropic vaccine (FNV) 217, 226
French viscerotropic virus 226

Gammaproteobacteria 46
GB *see* Golgi body
gene 33, 37, 40, 43, 49, 105, 108, 126, 173, 240, 248
Georgecraigius atropalpus 38
glycoprotein 87, 108, 109, 118, 133, 170, 171, 195, 196
Golgi apparatus 221
Golgi body (GB) 109, 121, 122, 172
gonotrophic cycle 15, 16, 175
Guillain-Barré syndrome 176, 207

headache 87, 89, 120, 122, 176, 180, 227, 230, 243
host 15–17, 46, 121, 122, 172, 175, 195, 196, 220, 221, 224, 228, 240, 243, 244
host cell 118, 121, 122, 170, 177, 196, 221, 224
Hulecoeteomyia japonica japonica 41
Hulecoeteomyia koreica 44
hydroxychloroquine 178
hypovolemic shock 122, 140

immune response 28, 109, 178, 227, 241
  adaptive 244
  elicited 198
  human 196
  innate 200
immune system 124, 172, 195, 196, 230, 248
immunosensor 133
  amperometric 246
  optical fiber 136
infection 89, 90, 93, 124, 131, 135, 136, 179, 180, 198, 199, 202–204, 207, 216, 226, 229, 230, 244, 245
  acute 176
  asymptomatic 175, 243, 249
  chronic phase 178
  documented 174
  experimental 174
  mild 89
  primary 124, 125, 244
  rare 224
  secondary 124, 125, 204, 226
infectious disease 82, 83, 94, 247
insect 13, 15, 81, 83, 84, 100, 101, 118, 228, 245
integrated vector management strategy 137

Jamestown Canyon virus 32, 53
Japanese encephalitis virus (JEV) 32, 40, 43, 92, 95, 248
jaundice 216, 225
JEV *see* Japanese encephalitis virus
juxtamembrane stem 109

kidney 197, 215, 219, 226, 244, 245
kunda 26, 27, 139

La Crosse virus 32, 43, 48, 49, 89
larvae 3, 5–7, 17, 27, 33, 36, 39, 45–48, 173
Lassa fever 230
leucopenia 122, 140, 225

liver 89, 120, 123, 124, 197, 219, 223, 226, 245
lymphadenopathy 88
lymphatic system 120

macrophage 125, 220, 244, 248
Mayaro virus 32
membrane 94, 114, 118, 121, 194, 195, 198, 199, 220, 241
  endosomal 220
  lipid 133, 170, 195, 199
  pinocytic 120
  plasma 118
  vesicle 198
  viral 195
meningitis 90, 94, 240
meningoencephalitis 176, 238, 243
mesonotum 22, 23, 29, 34, 49
mesothorax 10, 11
metathorax 10, 11
monoclonal antibody 136, 202, 206, 229
monoclonal antibody E16 247
monocyte 120, 125, 220
mosquito 1–3, 15–17, 21, 22, 27, 28, 32, 33, 37, 38, 53, 81–84, 89–91, 93, 94, 116–121, 128, 173–175, 196, 197, 216, 217, 222, 224
  adult 11, 13, 14, 22, 29, 33, 38, 41, 44, 47, 49, 51
  American tree-hole 47, 48
  arboreal 174
  Asian bush 41
  Asian rock pool 41, 42
  Asian tiger 29
  brown salt marsh 53
  eastern tree-hole 47
  forest's 29
  inland floodwater 51
  swamp 51
mouse models 178, 201, 248
mycophenolic acid 247

Nairoviruses 88
NASBA *see* nucleic acid sequence-based amplification
nervous system 12, 13, 118, 119, 199, 244, 248
neurotropic disease 227, 249
nucleic acid sequence-based amplification (NASBA) 131, 136
nucleocapsid 85, 87, 90, 94, 113, 121, 122, 168, 170

Ochlerotatus atropalpus 38
Ochlerotatus koreicus 44
optical biosensor 135, 137
  marker-free 136
Oropouche virus 32
outbreak 32, 34, 37, 40, 82, 87, 88, 166, 167, 169, 174, 192, 193, 215, 216, 222, 238
ovipositioning 15, 16, 33

pain 86, 122, 123, 177, 180, 203, 227
  joint 86, 167, 175
  muscular 86, 87
  neuropathic 177
paracetamol 127, 177, 180
parasite 2, 32, 52
pathogen 1, 2, 27, 81–83, 85, 244
pathogenicity 122, 165, 216
patient 86, 91, 93, 94, 122, 123, 127, 133, 139, 140, 202, 203, 225, 226, 230, 243, 244, 247, 249
PCR *see* polymerase chain reaction
Phlebovirus 87, 89
Pichia pastoris 126
piezoelectric biosensor 134, 135
plaque neutralization assay 248
plaque reduction neutralization test (PRNT) 130, 179, 246
polymerase chain reaction (PCR) 131, 179, 193, 204, 229, 246, 249

Potosi virus 32
PRNT *see* plaque reduction neutralization test
proboscis 12, 118, 120
protein 37, 38, 40, 43, 46, 49, 87, 91, 103, 104, 106–111, 113, 170–173, 194–196, 199, 221, 240
  anti-viral 198
  apoptosis protein 1-like 49
  capsid 85, 104, 107, 121, 126, 169, 170, 241
  dengue virus-2 113
  envelope 103, 108, 109, 113, 121, 135, 140, 195, 199, 200, 206, 240
  furin 221
  hydrophilic 110, 172
  hydrophobic 172
  membrane 105, 107, 113, 195, 240
  membrane-linked 103, 140
  nonstructural 94, 103–105, 109–111, 135, 140, 168, 170, 171, 177, 194–196, 199, 220, 240, 241,
  nucleocapsid 103, 106, 140, 240
  odorant-binding 33
  ovarian 118
  pre-membrane (PrM) 105–107, 241
  salivary 48, 120, 248
  structural 103, 106, 109, 140, 168–171, 177, 194, 195, 220, 240
  viral 112, 199, 244
pupa 3, 7, 8, 17, 47, 52

radiotherapy 230
rash 86, 123, 176, 180, 192, 193, 203
  petechial 88
receptor 27, 28, 118, 119, 121, 122, 195, 198, 200, 220, 227
recombinase polymerase amplification 229
replication 28, 118, 119, 124, 165, 171, 176, 196, 198, 199, 238
reservoir host 85, 86, 90
ribavirin 177, 180, 216, 226, 247
ribosome-binding site 170
Rift Valley fever 31, 43, 52, 84, 85, 89, 95
RNA 85, 87, 121, 122, 131, 169–171, 173, 177, 194, 196, 199, 204, 206, 220, 240, 241
Ross river virus 32, 50, 53, 192

Saccharomyces cerevisiae 126
saliva 12, 17, 49, 204, 248
salivary gland 12, 17, 27, 118, 119, 197, 244
sensor 133–136, 206, 246
serotypes of dengue 102, 111, 126, 129, 131, 133–135
serum 192, 202, 206, 229, 238
species 2, 3, 5, 22–24, 28, 29, 31–34, 36–43, 45–54, 82, 83, 194, 196, 242
  bacterial 118
  forest-dwelling 90
  insect 10
  non-refractory 33
  ornithophilous 242
  peri-urban/sylvatic 34
  reactive oxygen 244
  re-introductive 23
  rural 34
SPF chicken 242
Stegomyia albopicta 29
strain 33, 46, 94, 102, 104, 115, 116, 124, 139
  Asibi yellow fever 94, 223, 227
  inactive 126
  Liverpool 27
  multiple 124
  prototype 114, 115
  sylvatic 228

urban 116
wild 227
sylvatic cycle 91, 116, 139, 174, 197, 222
syndrome 207
acute encephalitis 239
cubital tunnel 176
dengue shock 91, 99, 100

therapy 222
antiviral 172
cytokine-mediated 216
intravenous fluid 127
thorax region 10, 22, 29, 33, 38, 41, 45, 47, 51
tissue 89, 118, 119, 121, 225, 244, 246
Togavirus 85, 94
transmission 16, 27, 28, 32, 36, 37, 82–84, 117, 120, 173, 177, 216, 222, 223, 238, 242
horizontal 116
non-vector 196
transovarial 90, 116, 166
transplacental 173, 198, 244
venereal 84
vertical 4, 40, 48, 116, 196, 243

vaccination 202, 217–220, 225–227, 230
vaccine 86, 89, 90, 93, 94, 126, 177, 178, 201, 208, 215, 217–219, 226, 227, 230
attenuated 215
smallpox 226
subunit-based 126
tetravalent 126
virus-based 178
VEE *see* encephalitis, Venezuelan equine
vertebrate host 2, 15, 16, 83, 84, 93, 116
viral genome 104, 131, 171, 194, 196, 199, 200, 206, 220, 229
viral infection 33, 94, 124, 179, 198, 223, 224, 238, 249
viral pathogen 81, 82, 84, 86, 88, 90, 92, 94
viral polyprotein 199, 220, 241
viral protease 106, 110, 121, 172, 240
viremia 83, 93, 125, 174, 178, 197, 198, 242, 247
virion 87, 105, 121, 122, 198, 200
invaginated 121
mature 105, 109, 176, 200, 201
nascent 177
parent 121
single 103
virus-like particle (VLP) 126, 178, 180
virus replication 118, 119, 173, 200, 224
visual analogue scale 177
VLP *see* virus-like particle

water 1, 3–6, 15, 24, 40, 47, 139
domestic 26
dripping 27
lodged 138
sheet 52
West Nile virus (WNV) 31, 40, 43, 48, 53, 90, 92, 94, 95, 206, 237–249
WNV *see* West Nile virus

Yap epidemic 193
yellow fever virus (YFV) 31, 37, 85, 90, 93, 95, 196, 215–230, 248
YFV *see* yellow fever virus

Zika virus (ZIKV) 37, 82, 90, 95, 192–199, 201–208
ZIKV *see* Zika virus